International Conference on Mathematics (ICM 2018)

Recent Advances in Algebra, Numerical Analysis, Applied Analysis and Statistics

ITM Web of Conferences Volume 20 (2018)

Ho Chi Minh City, Vietnam
18 – 20 December 2018

Editors:

R. Bris
Gyu Whan Chang
Chu Duc Khanh
M. Razzaghi
K. Stempak
Thanh Toan Phan

ISBN: 978-1-5108-7313-1

Printed from e-media with permission by:

Curran Associates, Inc.
57 Morehouse Lane
Red Hook, NY 12571

Some format issues inherent in the e-media version may also appear in this print version.

Printed by Curran Associates, Inc. (2018)

For additional information, please contact EDP Sciences – Web of Conferences at the address below.

EDP Sciences – Web of Conferences
17, Avenue du Hoggar
Parc d'Activité de Courtabœuf
BP 112
F-91944 Les Ulis Cedex A
France

Phone: +33 (0) 1 69 18 75 75
Fax: +33 (0) 1 69 28 84 91

contact-edps@webofconferences.org

Additional copies of this publication are available from:

Curran Associates, Inc.
57 Morehouse Lane
Red Hook, NY 12571 USA
Phone: 845-758-0400
Fax: 845-758-2633
Email: curran@proceedings.com
Web: www.proceedings.com

TABLE OF CONTENTS

Message from the Host University

18th December, 2018

Welcome to the International Conference on Mathematics (ICM 2018) at Ton Duc Thang University (TDTU), Ho Chi Minh City, Vietnam. This conference aims to offer a forum for university academics, scholars, and researchers from around the world to share their knowledge and latest results in the advanced topics of Pure and Applied Mathematics. The conference is held every two years at Ton Duc Thang University. This year, the themes of the conference have been selected to cover the areas of Algebra, Numerical Analysis, Applied Analysis and Statistics.

Founded in 1997, TDTU has become a visionary, pioneering, and fast-growing university in Vietnam. Over the past 21 years, TDTU has demonstrated excellence in academic performance. As a comprehensive university, TDTU focuses to improve the quantity and quality of research output by establishing many research centers and cooperating with international scholars. As a result, TDTU is now ranked among the largest and fastest developing universities in Vietnam in all aspects. In 2017, TDTU was ranked second among Vietnamese universities and first among those with regards to scientific research. In 2018, The High Council for Evaluation of Research and Higher Education (HCÉRES) recognized that TDTU complies with French and European standards for universities. In the same year, TDTU was honored to be the first public university in Vietnam to be officially rated 4 out of 5 stars under the QS Stars (UK) standard.

On behalf of TDTU, the host institution of ICM 2018, I would like to express my sincere appreciation to our honorable partners: Incheon National University, Republic of Korea; Mississippi State University, USA; VŠB – Technical University of Ostrava, the Czech Republic; and Wrocław University of Science and Technology, Poland, for their great efforts in co-organizing this conference. I would also like to send my special thanks to conference chairs, committees, reviewers, speakers, authors and participants around the world for their contributions and interest in our event.

I believe that you will have an interesting and fruitful conference in Vietnam. I really look forward to welcoming all of you to our campus and hope that this conference will start a long-term partnership between you and our university.

Prof. Le Vinh Danh, Ph.D.
President
Ton Duc Thang University, Vietnam

Preface

ICM 2018 (International Conference on Mathematics) is held in Ho Chi Minh City, Vietnam and is promoted by the following institutions: Ton Duc Thang University in Vietnam, Incheon National University in Republic of Korea, Mississippi State University in USA, VŠB – Technical University of Ostrava in Czech Republic, and Wrocław University of Science and Technology in Poland.

Hosted by Ton Duc Thang University, ICM 2018 aims to provide a forum for university academics, scholars, and researchers around the world to share their knowledge and latest results in advanced topics of Pure and Applied Mathematics. Starting from this year, such a conference is planned to be held every two years at Ton Duc Thang University. For this first time, the themes of ICM 2018 have been selected to cover the areas of Algebra, Numerical Analysis, Applied Analysis and Statistics.

This book includes 26 papers presented at ICM 2018 from prestigious authors coming from all over the world including Australia, Austria, Czech Republic, Germany, Italy, Malaysia, Poland, Republic of Korea, USA, and Vietnam. These are selected papers which are evaluated by the scientific committee of the conference and reviewed by experts in the relevant fields.

We would like to thank all authors, reviewers, and participants of the conference for their excellent contributions to the success of the conference.

The website of the conference is http://icm2018.tdtu.edu.vn/.

Editors
Radim Briš, Gyu Whan Chang
Chu Duc Khanh, Mohsen Razzaghi
Krzysztof Stempak, Phan Thanh Toan

Organization

Honorary Chairs

Prof. Le Vinh Danh, President of Ton Duc Thang University, Vietnam

Prof. Cezary Madryas, President of Wroclaw University of Science and Technology, Poland

Prof. Václav Snášel, Rector of VŠB – Technical University of Ostrava, Czech Republic

Conference Chairs

Prof. Mohsen Razzaghi, Mississippi State University, USA

Conference Co-chairs

Prof. Radim Briš, VŠB – Technical University of Ostrava, Czech Republic

Prof. Gyu Whan Chang, Incheon National University, Republic of Korea

Dr. Chu Duc Khanh, Ton Duc Thang University, Vietnam

Prof. Krzysztof Stempak, Wroclaw University of Science and Technology, Poland

Local Organizing Committee

Dr. Tran Trong Dao, Ton Duc Thang University, Vietnam

Dr. Phan Dao, Ton Duc Thang University, Vietnam

Dr. Chu Duc Khanh, Ton Duc Thang University, Vietnam

Dr. Le Van Ut, Ton Duc Thang University, Vietnam

Conference Scientific Committee

Ayman Badawi, American University of Sharjah, UAE

Petr Beremlijski, VŠB – Technical University of Ostrava, Czech Republic

Radim Briš, VŠB – Technical University of Ostrava, Czech Republic

Gyu Whan Chang, Incheon National University, Republic of Korea

Conference Secretary

edp sciences

Web of Conferences

Statement of Peer review

In submitting conference proceedings to *Web of Conferences*, the editors of the proceedings certify to the Publisher that

1. They adhere to its **Policy on Publishing Integrity** in order to safeguard good scientific practice in publishing.
2. All articles have been subjected to peer review administered by the proceedings editors.
3. Reviews have been conducted by expert referees, who have been requested to provide unbiased and constructive comments aimed, whenever possible, at improving the work.
4. Proceedings editors have taken all reasonable steps to ensure the quality of the materials they publish and their decision to accept or reject a paper for publication has been based only on the merits of the work and the relevance to the journal.

Title, date and place of the conference:

International Conference on Mathematics: Recent Advances in Algebra, Numerical Analysis, Applied Analysis and Statistics
December 18-20, 2018, Ho Chi Minh City, Vietnam

Proceedings editor(s) and editor's signature:

Dr. Chu Duc Khanh
Phan Thanh Toan
Gyu Whan Chang
RADIM BRIŠ
Mohsen Razzaghi
Prof. dr hab. Krzysztof Stempak

Date: September 6, 2018

ITM Web of Conferences **20**, 01001 (2018) https://doi.org/10.1051/itmconf/20182001001
ICM 2018

Recent results on weakly factorial domains

Gyu Whan Chang[1,*]

[1]Department of Mathematics Education, Incheon National University, Incheon 22012, Republic of Korea

Abstract. In this paper, we will survey recent results on weakly factorial domains base on the results of [11, 13, 14]. Let D be an integral domain, X be an indeterminate over D, $d \in D$, $R = D[X, \frac{d}{X}]$ be a subring of the Laurent polynomial ring $D[X, \frac{1}{X}]$, Γ be a nonzero torsionless commutative cancellative monoid with quotient group G, and $D[\Gamma]$ be the semigroup ring of Γ over D. Among other things, we show that R is a weakly factorial domain if and only if D is a weakly factorial GCD-domain and $d = 0$, d is a unit of D or d is a prime element of D. We also show that if char$(D) = 0$ (resp., char$(D) = p > 0$), then $D[\Gamma]$ is a weakly factorial domain if and only if D is a weakly factorial GCD domain, Γ is a weakly factorial GCD semigroup, and G is of type $(0, 0, 0, \dots)$ (resp., $(0, 0, 0, \dots)$ except p).

Introduction

Let D be an integral domain with quotient field K, X be an indeterminate over D, Γ be a nonzero torsionless commutative cancellative monoid (written additively) with quotient group G, and $D[\Gamma]$ be the semigroup ring of Γ over D. A nonzero nonunit element $p \in D$ is called a *primary element* if $p|ab$ for each $a, b \in D$ implies that $p|a$ or $p|b^n$ for some positive integer n; equivalently, pD is a primary ideal. We say that D is a *unique factorization domain* (UFD) (resp., *weakly factorial domain* (WFD)) if each nonzero nonunit of D can be written as a finite product of prime (resp. primary) elements. Clearly, a prime element is primary, and hence a UFD is a WFD. The concept of a WFD was first introduced by Anderson and Mahaney [3]. Chang introduced the notion of weakly factorial semigroups [11] in order to study when $D[\Gamma]$ is a WFD. A nonzero nonunit element $s \in \Gamma$ is *primary* if, for each $a, b \in \Gamma$, $a + b \in s + \Gamma$ implies that $a \in s + \Gamma$ or $nb \in s + \Gamma$ for some positive integer n. It is clear that $s \in \Gamma$ is primary if and only if $s + \Gamma$ is a primary ideal. We say that Γ is a *weakly factorial semigroup* if each nonunit of Γ is a finite sum of primary elements.

Chang has worked on WFDs [6–8, 11, 13, 14] and its generalization to rings with zero divisors [10, 12]. In this paper, among them, we survey the recent results of [11, 13, 14]. In Section 1, for easy reference, we review some definitions and preliminary results on t-operations, monoinds, semigroup rings, and weakly Krull domains. Let $d \in D$ and $R = D[X, \frac{d}{X}]$; so R is a subring of the Laurent polynomial ring $D[X, \frac{1}{X}]$ containing $D[X]$. In Section 2, we show that R is a weakly factorial domain if and only if D is a weakly factorial GCD-domain and $d = 0$, d is a unit of D or d is a prime element of D. We also show that if D is a weakly factorial GCD-domain, p is a prime element of D, and $n \geq 2$ is an integer,

*e-mail: whan@inu.ac.kr

then $D[X, \frac{p^n}{X}]$ is an almost weakly factorial domain with $Cl(D[X, \frac{p^n}{X}]) = \mathbb{Z}_n$, the cyclic group of order n. In Section 3, we completely characterize when $D[\Gamma]$ is a WFD. We first study the notion of weakly factorial GCD-semigroups. Then, among other things, we prove that if char$(D) = 0$, then $D[\Gamma]$ is a WFD if and only if D is a weakly factorial GCD domain, Γ is a weakly factorial GCD semigroup, and G is of type $(0, 0, 0, \dots)$. We also show that if char$(D) = p > 0$, then $D[\Gamma]$ is a WFD if and only if D is a weakly factorial GCD domain, Γ is a weakly factorial GCD semigroup, and G is of type $(0, 0, 0, \dots)$ except p. However, we omit the proofs of the results; so the reader who is interested in the proofs can refer to [11, 13, 14].

1 Definitions and preliminary results

In this section, we review some definitions and preliminary results on the t-operations, semigroups, semigroup rings, and weakly Krull domains.

1.1 The t-operations

Let D be an integral doman with quotient field K and $F(D)$ be the set of nonzero fractional ideals of D. For $I \in F(D)$, let $I^{-1} = \{x \in K \mid xI \subseteq D\}$; then $I^{-1} \in F(D)$. Hence, we can define the v- and t-operations as follows: $I_v = (I^{-1})^{-1}$, and $I_t = \bigcup\{J_v \mid J \subseteq I$ and $J \in F(D)$ is finitely generated$\}$. An $I \in F(D)$ is called a t-ideal if $I_t = I$. A t-ideal of D is a maximal t-ideal if it is maximal among proper integral t-ideals of D. Let t-Max(D) be the set of maximal t-ideals of D. It is well known that t-Max$(D) \neq \emptyset$ if D is not a field; each ideal in t-Max(D) is a prime ideal; each prime ideal minimal over a t-ideal is a t-ideal (which implies that every height-one prime ideal is a t-ideal); and $D = \bigcap_{P \in t\text{-Max}(D)} D_P$. We say that D is *of finite t-character* if every nonzero nonunit of D is contained in only finitely many maximal t-ideals of D.

An $I \in F(D)$ is said to be t-invertible if $(II^{-1})_t = D$. Let $T(D)$ be the group of t-invertible fractional t-ideals of D under $I * J = (IJ)_t$, and $Prin(D)$ be its subgroup of principal fractional ideals. Then $Cl(D) = T(D)/Prin(D)$, called the t-class group of D, is an abelian group. We say that D is a *Prüfer v-multiplication domain* (PvMD) if each nonzero finitely generated ideal of D is t-invertible. It is known that D is a PvMD if and only if $D[X]$ is a PvMD [26, Corollary 4], if and only if $D[X, \frac{1}{X}]$ is a PvMD [24, Theorem 3.10]. Also, D is a GCD-domain if and only if D is a PvMD with $Cl(D) = \{0\}$ [9, Proposition 2].

1.2 Semigroups

Let Γ be a nonzero torsionless grading monoid, i.e., a nonzero torsionless commutative cancellative monoid (written additively), and $\langle\Gamma\rangle = \{a - b \mid a, b \in \Gamma\}$ be the quotient group of Γ; so $\langle\Gamma\rangle$ is a torsionfree abelian group and Γ is given a total order compatible with the monoid operation [22, page 123]. Let G be a torsionfree abelian group. We say that G is *of type* $(0, 0, 0, \dots)$ if G satisfies the ascending chain condition on its cyclic subgroups (equivalently, for each nonzero element $g \in G$, there exists a largest positive integer n_g such that $n_g x = g$ is solvable in G) [17, Theorem 14.10]. For a prime number p, G is said to be of type $(0, 0, 0, \dots)$ except p if G satisfies the following two conditions; for each nonzero element $g \in G$, (i) the number of prime numbers dividing g is finite and (ii) for each prime number $q \neq p$, q^n does not divide g for some positive integer n. The notion of type $(0, 0, 0, \dots)$ except p was introduced by Matsuda [23, 25] in order to study when $K[G]$, where K is a field with char$(K) = p$, is a generalized Krull domain. Clearly, a torsionfree abelian group of type $(0, 0, 0, \dots)$ is of type $(0, 0, 0, \dots)$ except p for all prime numbers p. As in the case of integral domains, we can define the t-operation on Γ. For more on definitions and basic results (e.g., maximal t-ideals, class groups, weakly factorial semigroups), see [20, Chapter 11], [15], or [11].

1.3 Semigroup rings

For a nonzero torsionless commutative cancellative monoid Γ and an integral domain D, let $D[\Gamma]$ be the semigroup ring of Γ over D. Then $D[\Gamma]$ is an integral domain [17, Theorem 8.1] and each nonzero element $f \in D[\Gamma]$ can be written uniquely as $f = a_1X^{s_1} + \cdots + a_nX^{s_n}$ for $0 \neq a_i \in D$ and $s_i \in \Gamma$ with $s_1 < \cdots < s_n$ [17, Corollary 3.4]. Each nonzero element of the form $aX^s \in D[\Gamma]$ is said to be homogeneous with $\deg(aX^s) = s$. Let H be the set of nonzero homogeneous elements of $D[\Gamma]$. Then $D[\Gamma]_H = K[\langle\Gamma\rangle]$, and hence $D[\Gamma]_H$ is a completely integrally closed GCD-domain [18, Theorem 6.4]. For more on semigroup rings, the reader can refer to [17].

1.4 Weakly Krull domains

Let $X^1(D)$ be the set of height-one prime ideals of an integral domain D. We say that D is a weakly Krull domain if $D = \bigcap_{P\in X^1(D)} D_P$ and this intersection has finite character, i.e., each nonzero nonunit of D is a unit in D_P except finitely many primes in $X^1(D)$.

Lemma 1.1. *D is a weakly Krull domain if and only if $D = \bigcap_{P\in X^1(D)} D_P$ and D is of finite t-character. In this case, t-$\dim(D) = 1$, i.e., t-$Max(D) = X^1(D)$.*

Proof. [4, Lemma 2.1]. □

A nonzero prime ideal Q of $D[X]$ is called an upper to zero in $D[X]$ if $Q \cap D = (0)$, and we say that D is a UMT-domain if each upper to zero in $D[X]$ is a maximal t-ideal of $D[X]$. It is known that D is a PvMD if and only if D is an integrally closed UMT-domain [21, Proposition 3.2].

Proposition 1.2. *The following statements are equivalent for D.*

1. *$D[X]$ is a weakly Krull domain.*
2. *$D[X, X^{-1}]$ is a weakly Krull domain.*
3. *D is a weakly Krull UMT-domain.*

Proof. [2, Propositions 4.7 and 4.11]. □

Clearly, Krull domains are weakly Krull domains. Recall that D is a UFD if and only if D is a Krull domain and $Cl(D) = \{0\}$ [16, Proposition 6.1].

Theorem 1.3. *D is a WFD if and only if D is a weakly Krull domain and $Cl(D) = \{0\}$.*

Proof. [5, Theorem]. □

An *almost weakly factorial domain* (AWFD) is an integral domain D in which for each $0 \neq d \in D$, there is an integer $n = n(d) \geq 1$ such that d^n can be written as a finite product of primary elements of D. It is known that D is an AWFD if and only if D is a weakly Krull domain with $Cl(D)$ torsion [4, Theorem 3.4]; hence

$$\text{UFD} \Rightarrow \text{WFD} \Rightarrow \text{AWFD} \Rightarrow \text{weakly Krull domain.}$$

It is easy to see that if N is a multiplicative subset of a weakly Krull domain (resp., WFD, an AWFD), then D_N satisfies the corresponding property.

2 Weakly factorial generalized Rees rings

Let D be an integral domain, I be a proper ideal of D, and t be an indeterminate over D. Then $R = D[tI, t^{-1}]$ is a subring of $D[t, t^{-1}]$, called the *generalized Rees ring of D with respect to I*. In [27], Whithman proved that if I is finitely generated, then R is a UFD if and only if D is a UFD and t^{-1} is a prime element of R. Let $I = dD$ for $d \in D$ and $t^{-1} = X$; so $R = D[X, \frac{d}{X}]$. In [1], the authors studied several kinds of divisibility properties of R including Krull domains, UFDs, and GCD-domains. In this section we study when $R = D[X, \frac{d}{X}]$ is a WFD.

Proposition 2.1. *Let $d \in D$ be a nonzero element and $R = D[X, \frac{d}{X}]$. Then the following statements are equivalent.*

1. *X is irreducible (resp., prime, primary) in R.*
2. *$\frac{d}{X}$ is irreducible (resp., prime, primary) in R.*
3. *d is a nonunit (resp., prime, primary) in D.*

Proof. For the properties of irreducible and prime, see [1, Proposition 1]. For the property of primary, see [13, Proposition 2]. □

Let $0 \neq a \in D$. It is easy to see that a is primary if and only if $\sqrt{aD}$ is a maximal t-ideal [7, Lemma 2.1]. Hence, by Proposition 2.1, we have

Corollary 2.2. *[13, Corollary 3] Let $d \in D$ be a nonzero nonunit and $R = D[X, \frac{d}{X}]$. If dD is primary in D, then $\sqrt{XR}$ is a maximal t-ideal of R and $(X, \frac{d}{X})_v = R$.*

The next lemma can be easily proved by Proposition 1.2 which is necessary for the proof of Thorem 2.4.

Lemma 2.3. *[13, Proposition 4] Let $d \in D$ be a nonzero nonunit and $R = D[X, \frac{d}{X}]$. Then R is a weakly Krull domain if and only if D is a weakly Krull UMT-domain.*

By using the results of Corollary 2.2 and Lemma 2.3, we can completely characterize when $R = D[X, \frac{d}{X}]$ is a WFD.

Theorem 2.4. *[13, Theorem 6] Let $d \in D$ and $R = D[X, \frac{d}{X}]$. Then the following statements are equivalent.*

1. *R is a WFD.*
2. *R is a weakly factorial GCD-domain.*
3. *D is a weakly factorial GCD-domain and $d = 0$, d is a unit of D, or d is a prime element of D.*

A *ring of Krull type* is a PvMD of finite t-character [19, Theorem 7]. It is known that D is a ring of Krull type if and only if $D[X]$ is a ring of Krull type, or equivalently, $D[X, \frac{1}{X}]$ is a ring of Krull type (cf. [19, Propositions 9 and 12]).

Theorem 2.5. *[13, Theorem 7] Let $d \in D$ and $R = D[X, \frac{d}{X}]$. Then R is a ring of Krull type if and only if D is a ring of Krull type.*

A generalized Krull domain is an integral domain D such that (i) D_P is a valuation domain for all $P \in X^1(D)$, (ii) $D = \bigcap_{P \in X^1(D)} D_P$, and (iii) this intersection has finite character. Clearly, D is a generalized Krull domain if and only if D is a weakly Krull PvMD, if and only if D is a ring of Krull type with t-dim$(D) = 1$. Hence, by Lemma 2.3 andTheorem 2.5, we have

Corollary 2.6. *[13, Corollary 8] Let $d \in D$ and $R = D[X, \frac{d}{X}]$. Then R is a generalized Krull domain if and only if D is a generalized Krull domain.*

It is known that if D is a UFD, p is a prime element of D, and $n \geq 1$ is an integer, then $D[X, \frac{p^n}{X}]$ is a Krull domain with $Cl(D[X, \frac{p^n}{X}]) = \mathbb{Z}_n$ [1, Theorems 8 and 16]. The next result is an AWFD analog.

Corollary 2.7. *[13, Corollary 9] Let D be a weakly factorial GCD-domain, p be a prime element of D, and $n \geq 2$ be an integer. Then $R = D[X, \frac{p^n}{X}]$ is an AWFD with $Cl(R) = \mathbb{Z}_n$.*

Let V be a rank-one nondiscrete valuation domain, y be an indeterminate over V, and $D = V[y]$. Then D is a weakly factorial GCD-domain and y is a prime of D. Thus, by Theorem 2.4 and Corollary 2.7, $R = D[X, \frac{y}{X}]$ is a WFD and $D[X, \frac{y^n}{X}]$ is an AWFD with $Cl(D[X, \frac{y^n}{X}]) = \mathbb{Z}_n$ for all integers $n \geq 2$.

3 Weakly factorial semigroup rings

Let D be an integral domain with quotient field K, Γ be a torsionless grading monoid with quotient group G, $D[\Gamma]$ be the semigroup ring of Γ over D, and H be the set of nonzero homgeneous elements of $D[\Gamma]$.

In order to study when $D[\Gamma]$ is a UFD, we first need the notion of factorial semigroups. As in the domain case, an $\alpha \in \Gamma$ is called a *prime element* if $\alpha + \Gamma$ is a prime ideal of Γ, and we say that Γ is a *factorial semigroup* if each nonunit of Γ can be written as a finite sum of prime elements of Γ.

Theorem 3.1. *$D[\Gamma]$ is a UFD if and only if D is a UFD, Γ is a factorial semigroup, and the quotient group of Γ is of type $(0, 0, 0, \dots)$.*

Proof. [18, Theorem 7.17 and Lemma 7.15]. □

Theorem 3.1 is the motivation of the results in this section, i.e., we completely characterize when $D[\Gamma]$ is a WFD.

3.1 Weakly factorial semigroups

Lemma 3.2. *[11, Lemma 1] Let $\alpha \in \Gamma$ be a nonunit. Then the following statements are equivalent.*

1. *X^α is primary in $D[\Gamma]$.*
2. *$\sqrt{X^\alpha D[\Gamma]}$ is a maximal t-ideal of $D[\Gamma]$.*
3. *$D[\sqrt{\alpha + \Gamma}]$ is a maximal t-ideal of $D[\Gamma]$.*
4. *$\sqrt{\alpha + \Gamma}$ is a maximal t-ideal of Γ.*
5. *α is primary in Γ.*

It is well known that D is a GCD-domain if and only if $aD \cap bD$ is principal, if and only if $(a, b)_v$ is principal for each pair $0 \neq a, b \in D$. Similarly, note that if Γ is a GCD-semigrup, then $gcd(\alpha, \beta) = a \Leftrightarrow (\alpha + \Gamma) \cap (\beta + \Gamma) = \alpha + \beta - a + \Gamma$, $\Leftrightarrow ((\alpha + \Gamma) \cup (\beta + \Gamma))_v = a + \Gamma$ for each pair $\alpha, \beta \in \Gamma$. Thus, Γ is a GCD-semigroup if and only if $\bigcap_{i=1}^k (\alpha_i + \Gamma)$ is principal, if and only if $(\bigcup_{i=1}^k (\alpha_i + \Gamma))_v$ is principal for any $\alpha_1, \dots, \alpha_k \in \Gamma$ (cf. [20, Exercise 1 (page 117) and Theorem 11.5]). The next result is the semigroup analog of [3, Theorem 18] that a WFD D is a GCD-domain if and only if D_P is a valuation domain for each height-one prime ideal P of D.

Proposition 3.3. *[11, Proposition 5] Suppose that Γ is a weakly factorial semigroup. Then Γ is a GCD-semigroup if and only if $\alpha + \Gamma \subseteq \beta + \Gamma$ or $\beta + \Gamma \subseteq \alpha + \Gamma$ for all primary elements α and β of Γ with $\sqrt{\alpha + \Gamma} = \sqrt{\beta + \Gamma}$.*

It is clear that a factorial semigroup is a GCD-semigroup. Also, prime elements are primary, and hence a factorial semigroup is a weakly factorial GCD-semigroup. However, the next example shows that a weakly factorial GCD-semigroup whose quotient group is of type $(0, 0, 0, \dots)$ need not be a factorial semigroup.

Example 3.4. *[11, Example 12]* Let $G = \{m + n\sqrt{2} \mid m, n \in \mathbb{Z}\}$ be the additive group and $\Gamma = \{\alpha \in G \mid \alpha \geq 0\}$ be a subsemigroup of G.

1. Γ is a torsionless grading monoid with quotient group G.
2. G is of type $(0, 0, 0, \dots)$.
3. Γ is a weakly factorial GCD-semigroup.
4. Γ is not a factorial semigroup.

Proof. (1), (2) and (3). These are easy exercises. (4) This can be proved by the fact that $\sqrt{2}$ is irrational. □

3.2 When G is of type $(0, 0, 0, \dots)$

Let $D[X]$ be the polynomial ring over D and $\mathbb{N}_0$ be the semigroup of nonnegative integers. Then (i) $D[X] = D[\mathbb{N}_0]$ and (ii) $D[X]$ is a WFD if and only if D is a weakly factorial GCD-domain (Theorem 2.4). Note that $\mathbb{N}_0$ is a torsionless grading monoid whose quotient group $\mathbb{Z}$ is of type $(0, 0, 0, \dots)$; hence the next result is a natural generalization of the result (ii) to $D[\Gamma]$ with G of type $(0, 0, 0, \dots)$.

Theorem 3.5. *[11, Theorem 9] Assume that G is of type $(0, 0, 0, \dots)$. Then $D[\Gamma]$ is a WFD if and only if D is a weakly factorial GCD-domain and Γ is a weakly factorial GCD-semigroup.*

Since a factorial semigroup is a weakly factorial GCD semigroup, by Theorem 3.5, we have

Corollary 3.6. *[11, Corollary 10] Assume that G is of type $(0, 0, 0, \dots)$ and Γ is a factorial semigroup. Then $D[\Gamma]$ is a WFD if and only if D is a weakly factorial GCD-domain.*

Set $H_\alpha = \mathbb{N}_0$ for each α, and let $\Gamma = \sum_\alpha H_\alpha$. Then $K[\Gamma] = K[\{X_\alpha\}]$ is a factorial domain, and hence Γ is a factorial semigroup whose quotient group is of type $(0, 0, 0, \dots)$. Thus, by Corollary 3.6, we have

Corollary 3.7. *[11, Corollary 11] Let $\{X_\alpha\}$ be a nonempty set of indeterminates over D. Then $D[\{X_\alpha\}]$ is a WFD if and only if D is a weakly factorial GCD-domain.*

Example 3.8. Let $\mathbb{Z}$ be the ring of integers and $\Gamma = \{m + n\sqrt{2} \mid m, n \in \mathbb{Z}$ and $m + n\sqrt{2} \geq 0\}$. Then $\mathbb{Z}[\Gamma]$ is a WFD but not a UFD by Theorems 3.1, 3.5 and Example 3.4.

The next example shows that Theorem 3.5 is not true when G is not of type $(0, 0, 0, \dots)$.

Example 3.9. *[11, Example 13]* Let $\mathbb{R}$ be the field of real numbers and Γ be the additive semigroup of nonnegative rational numbers.

1. Γ is a torsion-free grading monoid.

2. Γ is a weakly factorial GCD-semigroup.

3. $\mathbb{R}[\Gamma]$ has the (Krull) dimension one, and hence t-dim($\mathbb{R}[\Gamma]$) = 1.

4. $\mathbb{R}[\Gamma]$ is not a WFD.

Proof. (1) Clear. (2) This is an easy exercise. (3) Let $\mathbb{N}_0$ be the semigroup of nonnegative integers. Then $\mathbb{R}[\Gamma]$ is integral over $\mathbb{R}[\mathbb{N}_0]$ and $\dim(\mathbb{R}[\mathbb{N}_0]) = 1$. Thus, $\dim(\mathbb{R}[\Gamma]) = t$-$\dim(\mathbb{R}[\Gamma]) = 1$. (4) It can be shown that $1 - X \in \mathbb{R}[\Gamma]$ is contained in infinitely many height-one prime ideals of $\mathbb{R}[\Gamma]$. □

3.3 char(D) = 0

In this section, we study the WFD property of $D[\Gamma]$ under the assumption that char(D) = 0. It is known that if char(D) = 0, then $D[G]$ is a ring of Krull type if and only if D is a ring of Krull type and G is of type $(0, 0, 0, \ldots)$ [25, Theorem 3].

Lemma 3.10. *[14, Theorem 3.1] Let G be a torsionfree abelian group, $D[G]$ be the group ring of G over D, and assume that char(D) = 0. Then $D[G]$ is of finite t-character if and only if D is of finite t-character and G is of type $(0, 0, 0, \ldots)$.*

Corollary 3.11. *[14, Corollary 3.2] Let K be a field, G be a torsionfree abelian group, and assume that char(K) = 0. Then the following statements are equivalent.*

1. *$K[G]$ is a WFD.*

2. *$K[G]$ is a weakly Krull domain.*

3. *$K[G]$ is of finite t-character.*

4. *G is of type $(0, 0, 0 \ldots)$.*

5. *$K[G]$ is a UFD.*

6. *$K[G]$ is a Krull domain.*

7. *$K[G]$ is a generalized Krull domain.*

Proof. (1) ⇒ (2) ⇒ (3) Clear. (3) ⇒ (4) Lemma 3.10. (4) ⇒ (5) Theorem 3.1. (5) ⇒ (1) Clear. (5) ⇒ (6) ⇒ (7) ⇒ (3) Clear. □

Theorem 3.12. *[14, Theorem 3.4] Assume that char(D) = 0. Then $D[\Gamma]$ is a WFD if and only if D is a weakly factorial GCD-domain, Γ is a weakly factorial GCD-semigroup, and G is of type $(0, 0, 0, ...)$.*

Proof. If $D[\Gamma]$ is a WFD, then $D[\Gamma]_H$ is a WFD. Note that $D[\Gamma]_H = K[G]$; hence G is of type $(0, 0, 0 \ldots)$ by Corollary 3.11. Thus, the result follows directly from Theorem 3.5. □

3.4 char(D) = $p > 0$

In this section, we study when $D[\Gamma]$ is a WFD under the assumption that char(D) = $p > 0$. Recall that if char(D) = $p > 0$, then $D[G]$ is a ring of Krull type if and only if D is a ring of Krull type and G is of type $(0, 0, 0, \ldots)$ except p [25, Theorem 5].

Lemma 3.13. *[14, Theorem 4.2] Let G be a torsionfree abelian group, $D[G]$ be the group ring of G over D, and assume that char(D) = $p > 0$. Then $D[G]$ is of finite t-character if and only if D is of finite t-character and G is of type $(0, 0, 0, \ldots)$ except p.*

Corollary 3.14. *[14, Corollary 4.3] Let K be a field, G be a torsionfree abelian group, and assume that char(K) = $p > 0$. Then the following statements are equivalent.*

1. *$K[G]$ is a WFD.*
2. *$K[G]$ is a weakly Krull domain.*
3. *$K[G]$ is of finite t-character.*
4. *G is of type $(0, 0, 0 \ldots)$ except p.*
5. *$K[G]$ is a generalized Krull domain.*

We next give a complete characterization of a WFD $D[\Gamma]$ with char(D) = $p > 0$.

Theorem 3.15. *[14, Theorem 4.5] Assume that char(D) = $p > 0$. Then $D[\Gamma]$ is a WFD if and only if D is a weakly factorial GCD-domain, Γ is a weakly factorial GCD-semigroup, and G is of type $(0, 0, 0, \ldots)$ except p.*

Example 3.16. *[14, Example 4.6]* Let K be a field, p be a prime number, $G = \bigcup_{n=1}^{\infty}(1/p^n)\mathbb{Z}$, and $K[G]$ be the group ring of G over K.

1. G is of type $(0, 0, 0 \ldots)$ except p but not of type $(0, 0, 0 \ldots)$.
2. If char(K) = $p > 0$, then $K[G]$ is a weakly factorial domain.
3. If char(K) $\neq p$ (e.g., char(K) = 0), then $K[G]$ is not a WFD.

Hence, $\mathbb{R}[G]$ is not a WFD, while $\mathbb{Z}_p[G]$ is a WFD, where $\mathbb{Z}_p$ is a field of p elements.

Proof. For (1), see [25, Proposition 12]. For (2) and (3), see (1) and Theorems 3.5 and 3.15. □

References

[1] D.D. Anderson and D.F. Anderson, *The ring $R[X, r/X]$*, in: Zero-Dimensional Commutative Rings, in: Lecture Notes in Pure and Appl. Math., vol. 171, (Marcel Dekker, New York, 1995), pp. 95-113.

[2] D.D. Anderson, E. Houston, and M. Zafrullah, *t-linked extensions, the t-class group, and Nagata's theorem*, J. Pure Appl. Algebra **86**, 109-124 (1993).

[3] D.D. Anderson and L.A. Mahaney, *On primary factorizations*, J. Pure Appl. Algebra **54**, 141-154 (1988).

[4] D.D. Anderson, J. Mott, and M. Zafrullah, *Finite character representations for integral domains*, Bollettino U.M.I. **6-B**(7), 613-630 (1992).

[5] D.D. Anderson and M. Zafrullah, *Weakly factorial domains and groups of divisiblity*, Proc. Amer. Math. Soc. **109**, 907-913 (1990).

[6] D.F. Anderson and G.W. Chang, *Homogeneous splitiing sets of a graded integral domain*, J. Algebra **288**, 527-544 (2005).

[7] D.F. Anderson, G.W. Chang, and J. Park, *Generalized weakly factorial domains*, Houston J. Math. **29**, 1-13 (2003).

[8] D.F. Anderson, G.W. Chang, J. Park, *Weakly Krull and related domains of the form $D+M$, $A+XB[X]$, and $A+X^2B[X]$*, Rocky Mountain J. Math. **36**, 1-22 (2006).

[9] A. Bouvier, *Le groupe des classes d'un anneau intégré*, 107ème Congrès National des Sociétés Savantes, Brest, fasc. **IV**, 85-92 (1982).

[10] G.W. Chang, *Weakly factorial rings with zero divisors*, in: Ideal Theoretic Methods in Commutative Algebra, in: Lecture Notes in Pure and Appl. Math., vol. 220, (Marcle Dekker, New York, 2001), pp. 119-131.

[11] G.W. Chang, *Semigroup rings as wekly factoral domains*, Comm. Algebra **37**, 3278-3287 (2009).

[12] G.W. Chang, *Some examples of weakly factorial rings*, Korean J. Math. **21**, 319-323 (2013).

[13] G.W. Chang, *Weakly factorial property of a generalized Rees ring $D[X, \frac{d}{X}]$*, Rocky Mountain J. Math., to appear.

[14] G.W. Chang and D.Y. Oh, *Semigroup rings as wekly factoral domains, II*, Preprint.

[15] S. El Baghdadi, L. Izelgue, and S. Kabbaj, *On the class group of a graded domain*, J. Pure Appl. Algebra **171**, 171-184 (2002).

[16] R. Fossum, *The Divisor Class Group of a Krull Domain*, (Springer, New York, 1973).

[17] R. Gilmer, *Commutative Semigroup Rings*, (The Univ. of Chicago Press, Chicago, 1984).

[18] R. Gilmer and T. Parker, *Divisiblility properties in semigroup rings*, Michigan Math. J. **21**, 65-86 (1974).

[19] M. Griffin, *Some results on v-multiplication rings*, Canad. Math. J. **19**, 710-722 (1967).

[20] F. Halter-Koch, *Ideal System: An introduction to Multiplicative Ideal Theory*, (Marcel Dekker, New York, 1998).

[21] E. Houston and M. Zafrullah, *On t-invertibility II*, Comm. Algebra **17**(8), 1955-1969 (1989).

[22] D.G. Northcott, *Lessons on Rings, Modules, and Multiplicities*, (Cambridge Univ. Press, Cambridge, 1968).

[23] R. Matsuda, *Torsion-free abelian semigroup rings III*, Bull. Fac. Sci. Ibaraki Univ. **9**, 1-49 (1977).

[24] R. Matsuda, *Torsion-free abelian semigroup rings IV*, Bull. Fac. Sci. Ibaraki Univ. Ser. A **10**, 1-27 (1978).

[25] R. Matsuda, *Krull properties of semigroup rings*, Bull. Fac. Sci. Ibaraki Univ. **14**, 1-12 (1982).

[26] H. Uda, *A characterization of Prüfer v-multiplication domains in terms of polynomial grade*, Hiroshima Math. J. **16**, 115-120 (1986).

[27] D. Whitman, *A note on unique factorization in Rees rings*, Math. Japon. **17**, 13-14 (1972).

Some strange behaviors of the power series ring $R[[X]]$

Phan Thanh Toan[1,*]

[1]Fractional Calculus, Optimization and Algebra Research Group, Faculty of Mathematics and Statistics, Ton Duc Thang University, Ho Chi Minh City, Vietnam

Abstract. Let R be a commutative ring with identity. Let $R[X]$ and $R[[X]]$ be the polynomial ring and the power series ring respectively over R. Being the completion of $R[X]$ (under the X-adic topology), $R[[X]]$ does not always share the same property with $R[X]$. In this paper, we present some known strange behaviors of $R[[X]]$ compared to those of $R[X]$.

1 Introduction

In this paper, a ring means a commutative ring with identity. Let R be a ring. Let $R[X]$ and $R[[X]]$ be the polynomial ring and the power series ring respectively over R. Being the completion of $R[X]$ (under the X-adic topology), $R[[X]]$ does not always share the same property with $R[X]$. In this paper, we present some strange behaviors of $R[[X]]$ compared to those of $R[X]$. Aspects of $R[[X]]$ that will be considered include Krull dimension, transcendence degree, and Noetherian property.

Let R be a ring. If there exists a chain $P_0 \subset P_1 \subset \cdots \subset P_n$ of $n+1$ prime ideals of R, but no such chain of $n+2$ prime ideals, then we say that the Krull dimension of R is n (or R is n-dimensional) and write Krull-dim $R = n$. Otherwise, we say that the Krull dimension of R is infinite (or R is infinite-dimensional) and write Krull-dim $R = \infty$. For a cardinal number α, we say that $\dim R = \alpha$ if R has a chain of prime ideals with length α but no longer chains. (The length of a chain $\mathcal{P}$ of prime ideals of R is defined by $|\mathcal{P}| - 1$, where $|\mathcal{P}|$ denotes the cardinality of $\mathcal{P}$. For two chains $\mathcal{P}$ and Q of prime ideals, we say that $\mathcal{P}$ is longer than Q if $|\mathcal{P}| > |Q|$). We also say that $\dim R \geq \alpha$ if there is a chain of prime ideals of R with length $\geq \alpha$ and that $\dim R \leq \alpha$ if every chain of prime ideals of R has length $\leq \alpha$. Hence, $\dim R = \alpha$ if and only if $\dim R \geq \alpha$ and $\dim R \leq \alpha$. We note that if Krull-dim$R = n < \infty$, then $\dim R =$ Krull-dim R. Furthermore, $\dim R \geq \aleph_0$ implies Krull-dim$R = \infty$ but not vice versa.

The Krull dimension of the polynomial ring $R[X]$ is fairly well-known for a finite-dimensional ring R. For example, $\dim R[X_1, \ldots, X_n] = \dim R + n$ if R is a Noetherian ring or a Prüfer domain [16, 17]. In general, $\dim R + 1 \leq \dim R[X] \leq 2\dim R + 1$ [16]. For the power series ring $R[[X]]$, it is shown by J.T. Arnold that $\dim R[[X]] \geq \aleph_0$ if R is a non-SFT ring. He defines an ideal I of a ring R to be an SFT ideal if there exist a finitely generated ideal $J \subseteq I$ and $k \in \mathbb{N}$ such that $a^k \in J$ for each $a \in I$ and he calls a ring R an SFT ring if every ideal of R is an SFT ideal. Typical examples of non-SFT domains are finite-dimensional nondiscrete valuation domains and non-Noetherian almost Dedekind domains. Kang and Park [9] showed that $\dim V[[X]] \geq |\mathbb{R}|$ and hence $\dim V[[X]] \geq \aleph_1$ for a 1-dimensional nondiscrete

*e-mail: phanthanhtoan@tdtu.edu.vn

valuation domain V. Kang and Toan [12] proved that $\dim V[[X]] \geq 2^{\aleph_1}$ for a 1-dimensional nondiscrete valuation domain V. Loper and Lucas [14, 15] also achieved the result that $\dim D[[X]] \geq 2^{\aleph_1}$ for a non-Noetherian almost Dedekind domain or a 1-dimensional nondiscrete valuation domain D. Recently, Toan and Kang [11] generalize the above results by showing that $\dim R[[X]] \geq 2^{\aleph_1}$ if R is a non-SFT domains. In Section 2, we are going to introduce their technique in proving this result.

Let D be an integral domain with quotient field K. It is easy to see that the polynomial rings $D[X]$ and $K[X]$ have the same quotient field. However, as mentioned in [3], it is rare that $D[[X]]$ and $K[[X]]$ have the same quotient field. Except for the trivial case when D is a field, the only example showing that $D[[X]]$ and $K[[X]]$ have the same quotient field is given by Gilmer in [7]. For two integral domains $D_1 \subseteq D_2$, denote by $tr.d.(D_2/D_1)$ the transcendence degree of the quotient field of D_2 over that of D_1. Hence, for a cardinal number α, $tr.d.(D_2/D_1) \geq \alpha$ if there exists a subset of the quotient field of D_2 with cardinality at least α that is algebraically independent over the quotient field of D_1. Sheldon showed in [18] that if D contains a nonzero element a such that $\cap_{i=1}^{\infty} a^i D = (0)$, then $tr.d.(D[[x/a]]/D[[X]]) \geq \aleph_0$ and hence $tr.d.(K[[X]]/D[[X]]) \geq \aleph_0$ since $D[[x/a]] \subseteq D[1/a][[X]] \subseteq K[[X]]$. In [2], Arnold and Boyd made a great improvement of this result by showing that if $K[[X]]$ and $D[[X]]$ have different quotient fields, then $tr.d.(K[[X]]/D[[X]]) \geq \aleph_0$. We present here the technique in [8], where the authors showed that whenever $K[[X]]$ and $D[[X]]$ have different quotient fields, $tr.d.(K[[X]]/D[[X]]) \geq \aleph_1$. Note that in general the bound $\aleph_1$ is the greatest lower bound that one can obtain since under the continuum hypothesis the cardinality of the quotient field of $K[[X]]$ is exactly $\aleph_1$ provided that K is countable.

Let α be an infinite cardinal number (e.g., $\alpha = \aleph_0, \aleph_1, \ldots$). An ideal I of a ring R is called an α-generated ideal if I can be generated by a set with cardinality $\leq \alpha$. R is called an α-generated ring if every ideal of R is an α-generated ideal. By definition, an $\aleph_0$-generated ring is a ring whose ideals are countably generated. Trivial examples of $\aleph_0$-generated rings are those that have only countably many elements (so that each ideal has itself as a countable generating set). Every Noetherian ring is obviously an $\aleph_0$-generated ring. However, the converse does not hold. Polynomial rings $R[X_1, X_2, \ldots, X_n, \ldots]$ in countably infinite indeterminates over countable rings R, the ring O of algebraic integers, the ring $\mathrm{Int}(\mathbb{Z})$ of integer-valued polynomials on $\mathbb{Z}$, and 1-dimensional nondiscrete valuation domains are good examples of $\aleph_0$-generated rings that are not Noetherian rings.

Even though the class of $\aleph_0$-generated rings is strictly larger than the class of Noetherian rings, it is shown in [10] that when restricted to power series rings, they are actually the same. In other words, the concepts "$\aleph_0$-generated ring" and "Noetherian ring" are the same for the power series ring $R[[X]]$. This shows a strange behavior of the power series ring $R[[X]]$ compared to that of the polynomial ring $R[X]$. Indeed, for any infinite cardinal number α, R is an α-generated ring if and only if $R[X]$ is an α-generated ring, which is an analogue of Hilbert Basis Theorem stating that R is a Noetherian ring if and only if $R[X]$ is a Noetherian ring.

2 Krull dimension of $R[[X]]$

2.1 Krull dimension of $R[X]$

The aim of this section is to show that $\dim R[[X]] \geq 2^{\aleph_1}$ if R is a non-SFT domain, which contrasts with the fact that the Krull dimension of the polynomial ring $R[X]$ is always finite provided that R is. The following result is from [16, Theorems 2 and 9].

Theorem 2.1 *If R is a finite-dimensional ring, then*

$$\dim R + 1 \leq \dim R[X] \leq 2 \dim R + 1.$$

Furthermore, if R is a Noetherian ring, then $\dim R[X] = \dim R + 1$.

When R is a Prüfer domain, one has the following.

Theorem 2.2 ([17, Theorem 4]) *If R is a finite-dimensional Prüfer domain, then*

$$\dim R[X_1, \ldots, X_n] = \dim R + n.$$

2.2 Construction of an η_1-set $\mathcal{A}$

We first construct an η_1-set $\mathcal{A}$. This η_1-set will be the index set of an infinite chain of prime ideals in $D[\![X]\!]$ when D is a non-SFT domain.

- Let $\mathbb{N} = \{1, 2, 3, \ldots\}$ be the set of positive integers and let $\mathcal{U}$ be the set of all subsets U of $\mathbb{N}$ such that $U = \{n, n+1, \ldots\}$ for some $n \in \mathbb{N}$.
- For two sequences of positive integers $s = \{s_n\}$, $t = \{t_n\}$, define $s \gg t$ (or $t \ll s$) if for each positive integer k, there is a set $U \in \mathcal{U}$ (depending on k) such that $s_n > kt_n$ for each $n \in U$, i.e., $s_n > kt_n$ for all large n.
- Let $\mathcal{S}$ be the collection of all $\mathcal{A}$ such that $\mathcal{A}$ has the following properties.
  - $\mathcal{A}$ is a nonempty collection of strictly increasing sequences $s = \{s_n\}$ of positive integers.
  - If $s \in \mathcal{A}$, then $s \gg b$, where $b = \{b_n\}$ is the sequence defined by $b_n := n$ for all n.
  - If $s, t \in \mathcal{A}$ and $s \neq t$, then $s \gg t$ or $t \gg s$.

If u is the sequence defined by $u_n := b_n^2$ for each n, then it is easy to see that $u \gg b$. It follows that the set $\mathcal{S}$ is nonempty. Order $\mathcal{S}$ by set-theoretic inclusion. By Zorn's Lemma, there exists a maximal element in $\mathcal{S}$.

Definition 2.3 Let $(\mathcal{Y}, \leqslant)$ be a totally ordered set and B, C be subsets of $\mathcal{Y}$. We say that $B \ll C$ if $b \ll c$ for each $b \in B$ and each $c \in C$. A totally ordered set $(\mathcal{Y}, \leqslant)$ is an η_1-set if for any two countable subsets B, C of $\mathcal{Y}$ such that $B \ll C$, there exists an element $a \in \mathcal{Y}$ such that $B \ll a \ll C$, i.e., $b \ll a \ll c$ for each $b \in B$ and each $c \in C$.

Let $\mathcal{A}$ be a maximal element in $\mathcal{S}$. This choice of $\mathcal{A}$ will be fixed through the rest of this section. For $s, t \in \mathcal{A}$, we define $s \underline{\ll} t$ if and only if $s = t$ or $s \ll t$. Then $(\mathcal{A}, \underline{\ll})$ becomes a totally order set.

Theorem 2.4 ([11, Theorem 2.3]) *The set $(\mathcal{A}, \underline{\ll})$ is an η_1-set.*

2.3 Chains of prime ideals in $D[[X]]$

Suppose that D is a non-SFT domain. Then there is an ideal I of D which is not SFT. As in [1, p. 300], there exists a sequence $a_0, a_1, \ldots, a_n, \ldots$ of $I \subseteq D$ such that

$$a_m^m \notin (a_0, a_1, \ldots, a_{m-1})$$

for each $m \geq 1$. Fix this sequence $a_0, a_1, \ldots, a_n, \ldots$. For an element $s = \{s_n\} \in \mathcal{A}$, let

$$f_s := a_0 + a_1 X^{s_1} + \cdots + a_n X^{s_n} + \cdots$$

in $D[[X]]$. For elements $v_1, v_2, \ldots, v_n$ (not necessarily distinct) in $\mathcal{A}$, we let $I_{v_1, v_2, \ldots, v_n}$ be the ideal of $D[[X]]$ generated by all power series f_s such that $s \gg v_i$ for each $i = 1, 2, \ldots, n$,

$$I_{v_1, v_2, \ldots, v_n} := \langle \{f_s \mid s \in \mathcal{A} \text{ and } s \gg \{v_1, v_2, \ldots, v_n\}\} \rangle.$$

Define

$$\frac{I_{v_1,v_2,\ldots,v_n}}{f_{v_1}f_{v_2}\cdots f_{v_n}} := \left\{ \frac{f}{f_{v_1}f_{v_2}\cdots f_{v_n}} \middle| f \in I_{v_1,v_2,\ldots,v_n} \right\}$$

and

$$S := \bigcup_{\substack{n\in\mathbb{N} \\ v_1,v_2,\ldots,v_n\in\mathcal{A}}} \frac{I_{v_1,v_2,\ldots,v_n}}{f_{v_1}f_{v_2}\cdots f_{v_n}}.$$

Then S is a subset of the quotient field of $D[[X]]$. For elements $v_1,\ldots,v_n$ (not necessarily distinct) in $\mathcal{A}$, let $B = \{v_1,\ldots,v_n\}$ be a multiset. We also let $||B|| := n$. Define

$$I_B := I_{v_1,\ldots,v_n} \quad \text{and} \quad f_B := \prod_{v\in B} f_v = f_{v_1}\cdots f_{v_n}.$$

Then the set S can be rewritten as

$$S = \bigcup \frac{I_B}{f_B},$$

where the union is taken over all multisets B consisting of elements in $\mathcal{A}$. Let $D[[X]][S]$ be the ring generated by S over $D[[X]]$. Denote by $SD[[X]][S]$ the ideal of $D[[X]][S]$ generated by S. Then this ideal is not the unit ideal ([11, Lemma 3.4]).

Theorem 2.5 ([6, Theorem 19.6]) *Let P be a prime ideal of an integral domain R. There exists a valuation overring W of R with maximal ideal Q such that $Q \cap R = P$.*

Let $R := D[[X]][S]$. Since SR is not the unit ideal of R, one can choose a prime ideal P of R such that P contains SR. By Theorem 2.5, there exists a valuation overring W of R with maximal ideal Q such that $Q \cap R = P$. For each $\alpha \in \mathcal{A}$, let $I_\alpha W$ be the ideal of W generated by I_α and let

$$P_\alpha := \sqrt{I_\alpha W} \cap D[[X]].$$

Since W is a valuation domain, $\sqrt{I_\alpha W}$ is a prime ideal of W. Hence, P_α is a prime ideal of $D[[X]]$. For $\alpha,\beta \in \mathcal{A}$, it can be shown that $\alpha \ll \beta$ if and only if $P_\alpha \supset P_\beta$. Hence, the set $\mathcal{P} := \{P_\alpha\}_{\alpha\in\mathcal{A}}$ is a chain of prime ideals in $D[[X]]$. Furthermore, it is an η_1-set since $\mathcal{A}$ is.

A totally ordered set A is *Dedekind-complete* provided that every nonempty subset of A that has an upper bound has a supremum. Finally we can now present the main result of [11].

Theorem 2.6 *If D is a non-SFT domain, then* $\dim D[[X]] \geq 2^{\aleph_1}$.

Proof. Let

$$\mathcal{P}^* := \{\cup_{\alpha\in\mathcal{D}} P_\alpha \mid \emptyset \neq \mathcal{D} \subseteq \mathcal{A}\}.$$

Then $\mathcal{P}^*$ is a chain of prime ideals in $D[[X]]$ and $\mathcal{P}^*$ is Dedekind-complete. $\mathcal{P}^*$ contains an η_1-set, namely, $\mathcal{P}$. Therefore, by [5, Corollary 13.24], $\mathcal{P}^*$ has cardinality at least $2^{\aleph_1}$.

3 Transcendence degree in power series rings

3.1 Transcendence degree in polynomial rings

Theorem 3.1 *Let D be an integral domain with quotient field K. Then $D[X]$ and $K[X]$ have the same quotient field. Hence,*

$$tr.d.(K[X]/D[X]) = 1.$$

Proof. This is obvious.

3.2 Construction of an upper fathomless set $\mathcal{B}$

In this subsection, we construct an uncountable set $\mathcal{B}$. In the next subsection, we will construct a set $\{\alpha_f\}_{f\in\mathcal{B}}$ of power series in $K[[x]]$ that is algebraically independent over the quotient field of $D[[x]]$ when $K[[x]]$ and $D[[x]]$ have different quotient fields.

Let $\mathbb{N} = \{1, 2, \ldots\}$ be the set of positive integers. Set $\mathcal{S}$ be the collection of all functions $f : \mathbb{N} \to \mathbb{N}$ satisfying the following three conditions (see [18]).

- $f(i) > i$ for all $i \in \mathbb{N}$.
- $f(i + 1) > f(i) + 1$ for all $i \in \mathbb{N}$.
- For each $k \geq 1$, there exists an $I \geq 1$ (depending on k) such that $f(i + 1) > k \cdot f(i)$ for all $i \geq I$.

The function $f(i) = 2^{2^i}$ satisfies these three conditions. Hence, the set $\mathcal{S}$ is nonempty. For a function f in $\mathcal{S}$, denote

$$\mathrm{im}(f|_{\geq I}) = \{f(i) \mid i \geq I\}.$$

Hence, $\mathrm{im}(f|_{\geq 1}) = \mathrm{im}(f)$ is the image of f.

Definition 3.2 For two functions f and g in $\mathcal{S}$, we define $g \gg f$ (we also write $f \ll g$) if there exists an integer $I \geq 1$ such that the following hold.

- $\mathrm{im}(g|_{\geq I}) \subseteq \mathrm{im}(f)$ and $g(i) > f(i)$ for all $i \geq I$.
- If $s_1 < s_2$ are both in $\mathrm{im}(g|_{\geq I})$, then they are not adjacent in $\mathrm{im}(f)$, i.e., there is no i such that $s_1 = f(i)$ and $s_2 = f(i + 1)$.

Remark 3.3 For $f, g, h \in \mathcal{S}$, if $f \ll g$ and $g \ll h$, then $f \ll h$.

Let $C(\mathcal{S})$ denote the collection of all nonempty subsets $\mathcal{B}$ of $\mathcal{S}$ satisfying: for any two different functions f and g in $\mathcal{B}$, either $f \gg g$ or $g \gg f$. We order $C(\mathcal{S})$ by set-theoretic inclusion. By Zorn's Lemma, there exists a maximal element in $C(\mathcal{S})$. Let $\mathcal{B}$ be a maximal element in $C(\mathcal{S})$. This choice of $\mathcal{B}$ will be fixed through the rest of the section. For $f, g \in \mathcal{B}$, we define $f \underline{\ll} g$ if and only if $f = g$ or $f \ll g$. Then $(\mathcal{B}, \underline{\ll})$ becomes a totally ordered set.

Definition 3.4 A totally ordered set $(\mathcal{Y}, \underline{\ll})$ is called an *upper fathomless set* if for every nonempty countable subset C of $\mathcal{Y}$, there exists an element $y \in \mathcal{Y}$ such that $y \gg c$ for all $c \in C$.

By definition, every upper fathomless set is an uncountable set. The following is [8, Theorem 7].

Theorem 3.5 *The set $(\mathcal{B}, \underline{\ll})$ is an upper fathomless set and hence is uncountable.*

3.3 Transcendence degree in power series rings

Suppose that D is an integral domain with quotient field K such that the quotient field of $K[[X]]$ properly contains the quotient field of $D[[X]]$. Then by [7, Theorem 1], there exists a sequence $\{a_i\}_{i=1}^{\infty}$ of nonzero elements of D such that $\cap_{i=1}^{\infty}(a_i) = (0)$. Replacing $\{(a_i)\}_{i=1}^{\infty}$ by a (suitable) strictly descending subsequence if necessary, we may assume that the sequence $\{(a_i)\}_{i=1}^{\infty}$ is strictly descending (see [7, Remark 1]), i.e.,

$$(a_1) \supset (a_2) \supset \cdots \supset (a_i) \supset \cdots .$$

Therefore, if $i_1 \geq i_2$ then $(a_{i_1}) \subseteq (a_{i_2})$ and hence a_{i_1} is a multiple of a_{i_2}. Let $\mathcal{B}$ be the upper fathomless set constructed in Subsection 3.2. For each $f \in \mathcal{B}$, define a power series $\boldsymbol{a}_f$ in $K[[X]]$ by

$$\boldsymbol{a}_f = \sum_{i=1}^{\infty} (X/a_i)^{f(i)}.$$

We will give sketch of proof of the following main result of the section. For a power series $\boldsymbol{p}$ in $K[[X]]$, the support of $\boldsymbol{p}$, denoted by $\mathrm{supp}(\boldsymbol{p})$, is the set of all nonnegative integer s for which the coefficient of X^s in $\boldsymbol{p}$ is nonzero.

Theorem 3.6 ([8, Theorem 12]) *Suppose that D is an integral domain with quotient field K such that $D[[X]]$ and $K[[X]]$ have different quotient fields. Then the quotient field of $K[[X]]$ has uncountable transcendence degree over the quotient field of $D[[X]]$, i.e, $t.r.(K[[X]]/D[[X]]) \geq \aleph_1$.*

Sketch of proof. We show that the set $\{\boldsymbol{a}_f\}_{f\in\mathcal{B}}$ is algebraically independent over the quotient field of $D[[X]]$. Suppose on the contrary that there are some $f_1, f_2, \ldots, f_J \in \mathcal{B}$ such that $\boldsymbol{a}_{f_1}, \boldsymbol{a}_{f_2}, \ldots, \boldsymbol{a}_{f_J}$ are algebraically dependent over the quotient field of $D[[X]]$. We may assume that

$$f_1 \ll f_2 \ll \cdots \ll f_J.$$

Let $P(t_1, \ldots, t_J)$ be a nonzero polynomial with coefficients in the quotient field of $D[[X]]$ such that

$$P(\boldsymbol{a}_{f_1}, \ldots, \boldsymbol{a}_{f_J}) = 0.$$

Multiplying both sides of this equation by a suitable element in $D[[X]]$, we may assume that the coefficients of $P(t_1, \ldots, t_J)$ are in $D[[X]]$. Let M be the set of all power series of the form $\boldsymbol{a}_{f_1}^{e(1)} \ldots \boldsymbol{a}_{f_J}^{e(J)}$ corresponding to those monomials $t_1^{e(1)} \ldots t_J^{e(J)}$ which occur in $P(t_1, \ldots, t_J)$ with nonzero coefficient. Then the relation $P(\boldsymbol{a}_{f_1}, \ldots, \boldsymbol{a}_{f_J}) = 0$ is actually a linear combination of the elements of M with coefficients in $D[[X]]$. Order the elements of M lexicographically, using the exponents of $\boldsymbol{a}_{f_1}$, then $\boldsymbol{a}_{f_2}$, and so forth. Write $M = \{\boldsymbol{m}_0, \boldsymbol{m}_1, \ldots, \boldsymbol{m}_R\}$, where $\boldsymbol{m}_0$ is the largest term in this ordering. Then $P(\boldsymbol{a}_{f_1}, \ldots, \boldsymbol{a}_{f_J}) = 0$ has the form

$$\boldsymbol{d}_0\boldsymbol{m}_0 + \boldsymbol{d}_1\boldsymbol{m}_1 + \cdots + \boldsymbol{d}_R\boldsymbol{m}_R = 0$$

for some $\boldsymbol{d}_r \in D[[X]]$ $(r = 0, 1, \ldots, R)$, $\boldsymbol{d}_0 \neq 0$. Denote by k the index of the first nonzero coefficient $d_{0,k}$ of $\boldsymbol{d}_0$. The proof is finished by showing a contradiction that $0 \neq d_{0,k} \in \cap_{i=1}^{\infty}(a_i)$, which is technically complicated and is omitted.

4 $\aleph_0$-generated power series rings

The purpose of this section is to show that the power series ring $R[\![X]\!]$ is an $\aleph_0$-generated ring if and only if $R[\![X]\!]$ is a Noetherian ring (and hence) if and only if R is a Noetherian ring. In order to prove this result, we only need to show that if R is a non-Noetherian ring, then the power series ring $R[\![X]\!]$ is not an $\aleph_0$-generated ring since if R is a Noetherian ring, then $R[\![X]\!]$ is also a Noetherian ring (for example, see [13, Theorem 71]) and hence an $\aleph_0$-generated ring. Suppose that R is a non-Noetherian ring. Our task is to construct an ideal J of $R[\![X]\!]$ that cannot be generated by any countable subset of J. Using the η_1-set $(\mathcal{A}, \leqslant)$ in Section 2, we construct generators for the J of $R[\![X]\!]$ as follows.

Let R be a non-Noetherian ring. Then there exists a sequence $a_0, a_1, \ldots, a_m, \ldots$ of elements in R such that

$$a_m \notin (a_0, a_1, \ldots, a_{m-1})$$

for each $m \geq 1$. For each integer $m \geq 0$, we let $I_m := (a_0, a_1, \dots, a_m)$. Then $a_m \notin I_{m-1}$ for each $m \geq 1$. For each sequence $s = \{s_n\} \in \mathcal{A}$, we define

$$f_s := a_0 + a_1 X^{s_1} + a_2 X^{s_2} + \cdots + a_n X^{s_n} + \cdots \in R[\![X]\!].$$

We let J be the ideal of $R[\![X]\!]$ generated by all f_s with $s \in \mathcal{A}$. The following is the main result of [10].

Theorem 4.1 ([10, Theorem 13]) *For a ring R, the following are equivalent.*

(1) $R[\![X]\!]$ *is an* $\aleph_0$*-generated ring.*

(2) $R[\![X]\!]$ *is a Noetherian ring.*

(3) R *is a Noetherian ring.*

Proof. We only need to prove that (1) implies (3). Suppose that R is not a non-Noetherian ring. We show that $R[\![X]\!]$ is not an $\aleph_0$-generated ring. It suffices to show that the ideal J constructed above is not a countably generated ideal. Suppose on the contrary that J is countably generated. Then there exists a countable subset $\mathcal{B}$ of $\mathcal{A}$ such that J is generated by $\{f_s \mid s \in \mathcal{B}\}$. Since $\mathcal{A}$ is a fathomless set, there exists a sequence $v \in \mathcal{A}$ such that $v \ll \mathcal{B}$. Since $f_v \in J$, f_v is a finite sum of elements of the form $h(s)f_s$,

$$f_v = \sum_s h(s) f_s, \tag{1}$$

where $h(s) \in R[\![X]\!]$ and $s \in \mathcal{B}$. Since $v \ll \mathcal{B}$, by taking a finite intersection of members of $\mathcal{U}$, we can find a set $U \in \mathcal{U}$ such that $v_m < s_m$ for each $m \in U$ and for each s appearing in the finite sum (1). Choose any number $m \in U$. Since $v_m < s_m$, the coefficient of f_s at X^j belongs to I_{m-1} for all $j \leq v_m$. It follows that the coefficient of $h(s)f_s$ at X^{v_m} belongs to I_{m-1}. This holds for every s appearing in the finite sum (1). Therefore, the coefficient of $\sum_s h(s)f_s$ at X^{v_m} belongs to I_{m-1}. This is a contradiction since the coefficient of $f_v = \sum_s h(s)f_s$ at X^{v_m} is a_m and $a_m \notin I_{m-1}$.

Corollary 4.2 *If R is an* $\aleph_0$*-generated ring, then the power series ring* $R[\![X]\!]$ *is not necessarily an* $\aleph_0$*-generated ring.*

Proof. Let R be any $\aleph_0$-generated ring which is not a Noetherian ring (for example, let $R = \mathbb{Q}[X_1, X_2, \dots, X_n, \dots]$, the polynomial ring in countably infinite indeterminates over the field of rational numbers $\mathbb{Q}$). Then $R[\![X]\!]$ is not an $\aleph_0$-generated ring by Theorem 4.1.

The above corollary contrasts with the following famous result for polynomial rings.

Theorem 4.3 *(Hilbert Basis Theorem.) If R is a Noetherian ring, then so is the polynomial ring* $R[X]$.

In comparing with Theorem 4.1, on has the following result, whose its proof follows the standard one of Hilbert Basis Theorem.

Theorem 4.4 ([10, Theorem 22]) *For any infinite cardinal number* α*, a ring R is an* α*-generated ring if and only if the polynomial ring* $R[X]$ *is an* α*-generated ring.*

5 Conclusion

Being the completion of the polynomial ring $R[X]$, the power series ring $R[[X]]$ however has very strange behaviors. The paper presents the following three examples.

- If $\dim R$ is finite, then so is $\dim R[X]$. However, $\dim R[[X]]$ can be infinite, in fact $\geq 2^{\aleph_1}$, even if $\dim R$ is finite.
- If D is an integral domain with quotient field K, then $D[X]$ and $K[X]$ share the same quotient field. However, the quotient field of $K[[X]]$ is much larger than that of $D[[X]]$. In fact, the quotient field of $K[[X]]$ is often has uncountable transcendence degree over that of $D[[X]]$.
- The concepts "Noetherian ring" and "$\aleph_0$-generated ring" are the same for the power series ring $R[[X]]$ while they are different for the polynomial ring $R[X]$. In fact, the polynomial ring $R[X]$ is an α-generated ring if and only if R is an α-generated ring, which is an analogue of Hilbert Basis Theorem stating that the polynomial ring $R[X]$ is a Noetherian ring if and only if R is a Noetherian ring.

Techniques involved in the proofs of the above results are quite complicated. By exploring these, we hope that similar techniques can be further developed to have a better understand of the power series ring $R[[X]]$.

References

[1] J.T. Arnold, Krull dimension in power series rings, Trans. Amer. Math. Soc. **177**, 299–304 (1973).

[2] J.T. Arnold and D.W. Boyd, Transcendence degree in power series rings, J. Algebra **57**, no. 1, 180–195 (1979).

[3] J.W. Brewer, *Power series over commutative rings, Lecture Notes in Pure Appl. Math. 64,* (Marcel Dekker, Inc., New York, 1981).

[4] R. Gilbert, A note on the quotient field of the domain $D[[X]]$, Proc. Amer. Math. Soc. **18**, 1138–1140 (1967).

[5] L. Gillman and M. Jerison, *Rings of continuous functions,* (The University Series in Higher Mathematics, D. Van Nostrand Co., Inc., Princeton, N.J.-Toronto-London-New York, 1960).

[6] R. Gilmer, *Multiplicative ideal theory,* (Marcel Dekker, New York, 1972).

[7] R. Gilmer, On commutative rings of finite rank, Duke Math. J. **39**, 381–383 (1972).

[8] L.T.N. Giau and P.T. Toan, Transcendental Degree in Power Series Rings, J. Algebra **501**, 51–67 (2018).

[9] B.G. Kang and M.H. Park, Krull-dimension of the power series ring over a nondiscrete valuation domain is uncountable, J. Algebra **378**, 12–21 (2013).

[10] B.G. Kang and P.T. Toan, A remark on the Noetherian property of power series rings Pacific J. Math. **283**, 353–363 (2016).

[11] B.G. Kang and P.T. Toan, Krull dimension of power series rings over non-SFT domains, J. Algebra **499**, 516–537 (2018).

[12] B.G. Kang and P.T. Toan, Krull dimension of a power series ring over a valuation domain, preprint.

[13] I. Kaplansky, *Commutative rings, Revised edition,* (The University of Chicago Press, Chicago, Ill.-London, 1974).

[14] K.A. Loper and T.G. Lucas, Constructing chains of primes in power series rings, J. Algebra **334**, 175–194 (2011).

[15] K.A. Loper and T.G. Lucas, Constructing chains of primes in power series rings II, J. Algebra Appl. **12**, no. 1, 1250123, 30 pp. (2013).

[16] A. Seidenberg, A note on the dimension theory of rings, Pacific J. Math. **3**, 505–512 (1953).

[17] A. Seidenberg, On the dimension theory of rings II, Pacific J. Math. **4**, 603–614 (1954).

[18] P.B. Sheldon, How changing $D[[X]]$ changes its quotient field, Trans. Amer. Math. Soc. **159**, 223–244 (1971).

Polynomial functions on subsets of non-commutative rings — a link between ringsets and null-ideal sets

Sophie Frisch[1,*]

[1]Institut für Analysis und Zahlentheorie, Technische Universität Graz, Kopernikusgasse 24, 8010 Graz, Austria

Abstract. Regarding polynomial functions on a subset S of a non-commutative ring R, that is, functions induced by polynomials in $R[x]$ (whose variable commutes with the coefficients), we show connections between, on one hand, sets S such that the integer-valued polynomials on S form a ring, and, on the other hand, sets S such that the set of polynomials in $R[x]$ that are zero on S is an ideal of $R[x]$.

2010 Mathematics Subject Classification: 13F20, 16D25, 16P10, 16S99

Keywords and phrases: polynomial mappings, polynomial functions, integer-valued polynomials, null polynomials, null ideals, matrix algebras, triangular matrices, finite rings, non-commutative rings, ringsets

1 Introduction

In the theory of polynomial mappings on commutative rings, there are two notable subtopics, namely, polynomial functions on finite rings, and rings of integer-valued polynomials. Here, we are concerned with generalizations of these two topics to polynomial mappings on non-commutative rings, as proposed by Loper and Werner [8], and developed further by Werner [11–14], Peruginelli [9, 10], and the present author [3–5], among others. More particularly, we will investigate connections between null ideals of polynomials on finite non-commutative rings and integer-valued polynomials on non-commutative rings.

When we talk about polynomial functions on a non-commutative ring R, we mean functions induced by elements of the usual polynomial ring $R[x]$ whose indeterminate x commutes with the elements of R. Non-commutative rings R for which polynomial functions have been studied include rings of quaternions [7, 11, 14], and matrix algebras [4, 5, 12].

To begin, we introduce the two objects we want to relate, null ideals and rings of integer-valued polynomials, in their original, commutative setting:

When considering polynomial functions on a finite commutative ring R, the first thing one likes to know is the so called null ideal $\mathrm{N}(R)$ of $R[x]$ consisting of all null-polynomials, that is, polynomials such that the function induced on R by substitution of the variable is zero. The null ideal is important, because the residue classes of $R[x]$ mod $\mathrm{N}(R)$ correspond to the different polynomial mappings on R and hence the index $[R[x]\colon \mathrm{N}(R)]$ indicates the number of different polynomial mappings on R.

*e-mail: frisch@math.tugraz.at

Regarding integer-valued polynomials, they are defined, for a domain D with quotient field K, as those polynomials f in $K[x]$ such that the polynomial function defined by f on K takes every element of D to an element of D [1].

We now generalize polynomial functions to non-commutative rings.

Let R be a (possibly non-commutative) ring and

$$f = \sum_k c_k x^k = \sum_k x^k c_k \in R[x].$$

Then, f induces two polynomial functions on R, namely, the right polynomial function $f_r \colon R \to R$ and the left polynomial function $f_l \colon R \to R$, where

$$f_r(s) = \sum_k c_k s^k \quad \text{and} \quad f_l(s) = \sum_k s^k c_k.$$

There are other generalizations of polynomial functions to polynomials with coefficients in non-commutative rings, using polynomials whose indeterminate does not commute with the coefficients. We are not concerned with this kind of generalized polynomials here. Our topic are left and right polynomial functions defined, as above, on a non-commutative ring by polynomials in the usual polynomial ring $R[x]$ whose indeterminate x commutes with the elements of R.

Regarding these left and right polynomial functions on a non-commutative ring R, we notice that they do not, in general, admit a substitution homomorphism.

It may happen, for some s in R, and $f, g \in R[x]$ that

$$f_r(s)g_r(s) \neq (fg)_r(s) \quad \text{and also} \quad f_l(s)g_l(s) \neq (fg)_l(s).$$

In order to generalize null-ideals to polynomials on finite non-commutative rings, we consider the sets of right and left null-polynomials, respectively, on R. We note that, in the absence of a substitution homomorphism, neither set is necessarily an ideal of $R[x]$.

Definition 1.1. Let R a ring and $f \in R[x]$. The polynomial f is called a right null-polynomial on R in case $f_r(s) = 0$ for all $s \in R$, and a left null-polynomial on R in case $f_l(s) = 0$ for all $s \in R$.

We denote the sets of right and left null polynomials on R, respectively, by

$$\mathrm{N}^r(R) = \{f \in R[x] \mid \forall s \in S\ f_r(s) = 0\} \quad \text{and} \quad \mathrm{N}^l(R) = \{f \in R[x] \mid \forall s \in S\ f_l(s) = 0\}$$

It is immediately clear that

Fact 1.2 (Werner [13]). *For every ring R,*

1. $\mathrm{N}^r(R)$ *is a left ideal of* $R[x]$,
2. $\mathrm{N}^l(R)$ *is a right ideal of* $R[x]$.

Proof. Indeed, if $f, g \in R[x]$, $f = \sum_k a_k x^k$, $g = \sum_k b_k x^k$, and $s \in R$, then

$$(gf)_r(s) = \sum_{k,l} b_k a_l s^{k+l} = \sum_{k,l} b_k a_l s^l s^k = \sum_k b_k \left(\sum_l a_l s^l\right) s^k = \sum_k b_k \left(f_r(s)\right) s^k \qquad (1)$$

The last expression is zero whenever $f_r(s) = 0$, which makes $\mathrm{N}^r(R)$ a left ideal of $R[x]$. Similarly, by interchanging left and right, we see that $\mathrm{N}^l(R)$ is always a right ideal of $R[x]$. □

Whether $\mathrm{N}^r(R)$ is also a right ideal of $R[x]$, and thus an ideal, for any finite ring R, is an open question, and similarly the question whether $\mathrm{N}^l(R)$ is a left ideal and therefore an ideal. There are no known counterexamples.

Werner has found many sufficient conditions on R for $\mathrm{N}^r(R)$ to be a right ideal [13], but none of them are necessary. If we take R as the ring of upper triangular matrices over a commutative ring T, we can, by judicious choice of T, find examples of rings violating all of Werner's necessary conditions, which nevertheless satisfy that $\mathrm{N}^r(R)$ is a right ideal and $\mathrm{N}^l(R)$ is a left ideal of $R[x]$ [5]. Such examples can also be found among rings of integer-valued polynomials on quaternions [14].

Now, when we generalize integer-valued polynomials to polynomials with coefficients in a non-commutative ring, the usual setup (as introduced in [4]) is

Definition 1.3. Let D be a domain with quotient field K and A a finitely generated, torsion-free D-algebra. Let $B = A \otimes_D K$. To avoid certain pathologies, we stipulate that $A \cap K = D$ when A and K are canonically embedded in B.

Then, the set of right integer-valued polynomials (with coefficients in B) on A is

$$\mathrm{Int}^{\mathrm{r}}{}_B(A) = \{f \in B[x] \mid \forall a \in A\ f_r(a) \in A\}$$

and the set of left integer-valued polynomials (with coefficients in B) on A is

$$\mathrm{Int}^{\mathrm{l}}{}_B(A) = \{f \in B[x] \mid \forall a \in A\ f_l(a) \in A\}.$$

We remark that it is not a priori clear that $\mathrm{Int}^{\mathrm{r}}{}_B(A)$ and $\mathrm{Int}^{\mathrm{l}}{}_B(A)$ are rings, because, in the absence of a substitution homomorphism, closure under multiplication is not a given. Yet, there are no known counterexamples. Whether $\mathrm{Int}^{\mathrm{r}}{}_B(A)$ and $\mathrm{Int}^{\mathrm{l}}{}_B(A)$ are rings for any D-algebra A as in Definition 1.3 remains an open question.

In some cases it is possible to describe $\mathrm{Int}^{\mathrm{r}}{}_B(A)$ and $\mathrm{Int}^{\mathrm{l}}{}_B(A)$ via their relation to the commutative ring $\mathrm{Int}_K(A) = \mathrm{Int}^{\mathrm{r}}{}_B(A) \cap K[x] = \mathrm{Int}^{\mathrm{l}}{}_B(A) \cap K[x]$, for instance, when $A = \mathrm{M}_n(D)$ is the ring of $n \times n$ matrices over D. Here, $\mathrm{Int}^{\mathrm{l}}{}_{\mathrm{M}_n(K)}(\mathrm{M}_n(D))$ coincides with $\mathrm{Int}^{\mathrm{r}}{}_{\mathrm{M}_n(K)}(\mathrm{M}_n(D))$ (shown to be a ring by Werner [12]), and is canonically isomorphic to $\mathrm{M}_n(\mathrm{Int}_K(\mathrm{M}_n(D)))$ [4]. The algebras for which $\mathrm{Int}_B(A) \simeq \mathrm{Int}_K(A) \otimes_D A$ thus holds have been characterized by Peruginelli and Werner [10].

For $\mathrm{T}_n(D)$ the ring of $n \times n$ upper triangular matrices with entries in D, $\mathrm{Int}^{\mathrm{r}}{}_{\mathrm{T}_n(K)}(\mathrm{T}_n(D))$ is isomorphic to the algebra of matrices whose entries in position (j, k) are in $\mathrm{Int}_K(\mathrm{T}_{n-k+1}(D))$, and $\mathrm{Int}^{\mathrm{l}}{}_{\mathrm{T}_n(K)}(\mathrm{T}_n(D))$ is isomorphic to the algebra of matrices whose entries in position (j, k) are from $\mathrm{Int}_K(\mathrm{T}_j(D))$ [5]. The commutative rings of integer-valued polynomials on upper (or lower) triangular matrices (with coefficients in K), $\mathrm{Int}_K(\mathrm{T}_n(D))$, are of interest in their own right [2].

Again, Werner has given different sufficient conditions on A for $\mathrm{Int}^{\mathrm{r}}{}_B(A)$ to be a ring, but we know that these conditions are not necessary. Taking A as the ring of upper triangular matrices over a judiciously chosen domain D we can find examples where $\mathrm{Int}^{\mathrm{r}}{}_B(A)$ and $\mathrm{Int}^{\mathrm{l}}{}_B(A)$ are rings, but all known sufficient conditions are violated [5]. Also, such examples can be found among rings of integer-valued polynomials over quaternion algebras [14].

2 A connection between ringsets and null-ideal sets

We do not know whether $\mathrm{Int}^{\mathrm{r}}{}_B(A)$ is always closed under multiplication; nor do we know whether $\mathrm{N}^r(R)$ is always an ideal of $R[x]$. As a way out of this quandary, we widen the scope of our investigation. Following Werner [14], we consider integer-valued polynomials on subsets of A.

Here, in addition to integer-valued polynomials on subsets, we will also consider null-polynomials on subsets, and demonstrate a connection between the two.

In what follows, we will often confine ourselves to right polynomial functions, with the understanding that everything also holds, mutatis mutandis, for left polynomial functions. In the context of right polynomial functions, $f(c)$ means $f_r(c)$.

Definition 2.1. Let A be a D-algebra and everything as in Definition 1.3 and $S \subseteq A$. The set of right integer-valued polynomials on S is

$$\mathrm{Int}^r_B(S, A) = \{f \in B[x] \mid \forall s \in S \ f_r(s) \in A\}.$$

S is called a *right ringset* if $\mathrm{Int}_B(S, A)$ is closed under multiplication, and hence a ring.

Likewise, the set of left integer-valued polynomials on S is

$$\mathrm{Int}^l_B(S, A) = \{f \in B[x] \mid \forall s \in S \ f_l(s) \in A\}.$$

S is called a *left ringset* if $\mathrm{Int}_B(S, A)$ is closed under multiplication, and hence a ring.

It is easy to give examples, both of ringsets and of sets that are not ringsets: For any D-algebra satisfying that $\bigcap_{d \in D\setminus\{0\}} dA = (0)$, which is, for instance, the case if A is a free D-module and D is a Noetherian or Krull domain, Werner [14] showed that a singleton $\{s\} \subseteq A$ is a ringset if and only if s is in the center of A. To see "only if," suppose that s does not commute with some $t \in A$. Let $d \in D \setminus \{0\}$ such that $ts - st \notin dA$, and let $f(x) = d^{-1}(x - s)$. Then, both f and t are in $\mathrm{Int}_B(\{s\}, A)$, but ft is not. Indeed, $(ft)(x) = d^{-1}(tx - st)$ and $(ft)(s) = d^{-1}(ts - st) \notin A$.

Note that an arbitrary union of ringsets is always a ringset, by the fact that an intersection of rings is a ring.

Definition 2.2. Let R be a ring and S a subset of R. We denote by $N^r_R(S)$ the set of polynomials $f \in R[x]$ such that for all $s \in S$, $f_r(s) = 0$. We abbreviate $N^r_R(R)$ by $\mathrm{N}^r(R)$.

Likewise, we denote by $N^l_R(S)$ the set of polynomials $f \in R[x]$ such that for all $s \in S$, $f_l(s) = 0$ and abbreviate $N^l_R(R)$ by $\mathrm{N}^l(R)$.

Remark 2.3. Note that $N^r_R(S)$ is always a left ideal of $R[x]$. This is demonstrated, just like the fact that $\mathrm{N}^r(R)$ is a left ideal, by equation 1.

The question is: for which sets S is $N^r_R(S)$ a right ideal?

Likewise, $N^l_R(S)$ is always a right ideal of $R[x]$, and the question is: for which sets S is $N^l_R(S)$ a left ideal?

Definition 2.4. We say that S (as a subset of R) is a right null-ideal set if $N^r_R(S)$ is a right ideal of $R[x]$, and hence an ideal of $R[x]$.

We say that S (as a subset of R) is a left null-ideal set if $N^l_R(S)$ is a left ideal of $R[x]$, and hence an ideal of $R[x]$.

We will now give a criterion for ringsets in terms of null-ideal sets. For this purpose, we introduce null-polynomials modulo an ideal. We will later rephrase everything using null polynomials in the strict sense.

Definition 2.5. Let R be a ring and S a subset of R and I an ideal of R.

A polynomial $f \in R[x]$ is called a right null polynomial modulo I on S if $f_r(s) \in I$ for every $s \in S$.

A polynomial $f \in R[x]$ is called a left null-polynomial modulo I on S if $f_l(s) \in I$ for every $s \in S$.

We denote by $N^r_{(R \bmod I)}(S)$ the set of right null-polynomials mod I on S and by $N^l_{(R \bmod I)}(S)$ the set of left null-polynomials mod I on S.

Note that $N^r_{(R \mod I)}(S)$ is always a left ideal of $R[x]$ – again, this can be seen by equation 1 – and that $N^l_{(R \mod I)}(S)$ is always a right ideal of $R[x]$.

Definition 2.6. We call a subset S of R a right null-ideal set modulo I if $N^r_{(R \mod I)}(S)$ is a right ideal, and hence an ideal, of $R[x]$.

We call a subset S of R a left null-ideal set modulo I if $N^l_{(R \mod I)}(S)$ is a left ideal, and hence an ideal, of $R[x]$.

For basic facts about division with remainder in rings of polynomials over non-commutative rings, we refer to Hungerford [6]. In particular, recall that a polynomial f has a right factor $(x - s)$ if and only if $f_r(s) = 0$, and that the remainder of f under polynomial division by $(x - s)$ from the right is $f_r(s)$.

In the special case of $D = \mathbb{Z}$ and $S = A$ the following has been shown, by a different argument, by Werner [13, Thm. 2.4]

Theorem 2.7. *Let A be a D-algebra and everything as in Definition 1.3. Let $S \subseteq A$.*

Then, S is a right ringset if and only if S is a right null-ideal-set modulo dA for all non-zero $d \in D$.

Similarly, S is a left ringset if and only if S is a left null-ideal-set modulo dA for all non-zero $d \in D$.

Proof. We show the statement for right polynomial mappings. We write $f(c)$ for $f_r(c)$ in this context. (The left case is similar.)

Suppose S is a right null-ideal-set modulo dA for all non-zero $d \in D$. Let $F, G \in \mathrm{Int}^r{}_B(S, A)$. To show: $(FG) \in \mathrm{Int}^r{}_B(S, A)$.

We write F and G as $F = f/d$, $G = g/c$, such that $f, g \in A[x]$, $c, d \in D \setminus \{0\}$. For every $s \in S$, $f(s) \in dA$, and $g(s) \in cA$. Note, in particular, that f is a right null-polynomial modulo dA on S. Having represented FG as $FG = (fg)/(dc)$, we need to show, for an arbitrary $s \in S$, that $(fg)(s) \in dcA$.

By division with remainder in $A[x]$ by $(x - s)$ from the right, we get

$$g(x) = q(x)(x - s) + g(s)$$

for some $q \in A[x]$. We know that $g(s) = ca = ac$ for some $a \in A$. For this $a \in A$,

$$(fg)(x) = f(x)q(x)(x - s) + f(x)ac.$$

We set $h(x) = f(x)a$.

f being a right null-polynomial modulo dA on S implies, by the fact that S is a right null-ideal-set modulo dA, that $h = fa$ is also a right null-polynomial modulo dA on S, and that, therefore, $h(s) \in dA$. Finally, we see that $(fg)(s) = h(s)c$ is in dcA, as required.

For the reverse implication, suppose that S is not a right null-ideal-set modulo dA for some fixed $d \in D \setminus \{0\}$. To show: S is not a right ringset.

Let f, g in $A[x]$ such that f is a right null-polynomial modulo dA on S, but (fg) is not. Now consider $F = d^{-1}f \in B[x]$ and $G = g \in A[x]$. Both F and G are in $\mathrm{Int}^r{}_B(S, A)$, but their product $FG = d^{-1}fg$ is not. □

We now rephrase the link between ring sets and null ideal sets using null polynomials in the strict sense. Let $S \subseteq R$ and I an ideal of R. We denote by $S + I$ the set of residue classes of elements of S in A/I, that is, $S + I := \{s + I \mid s \in S\}$.

Let $f \in R[x]$, and $\bar{f}$ image of f in $(R/I)[x]$ under canonical projection. Then f is a right null polynomial modulo I on S if and only if $\bar{f} \in N_{(A/I)}(S + I)$. In other words,

$$N^r_{(R \mod I)}(S) = \pi^{-1}(N_{(A/I)}(S + I)),$$

where $\pi\colon R[x] \to (R/I)[x]$ is the canonical projection.

This shows that S is a right null-ideal set modulo I if and only if $S + I = \{s + I \mid s \in S\}$, as a subset of R/I, is a right null ideal set (and similarly with right replaced by left).

Theorem 2.8 (Version of Thm 2.7). *Let A be a D-algebra and everything as in Definition 1.3. Let $S \subseteq A$. Then, S is a right ringset if and only if, for all non-zero $d \in D$, $S + dA$ as a subset of A/dA is a right null-ideal-set.*

3 A common framework for ringsets and null-ideal sets

We state most facts of this section only for right polynomial functions, with the understanding that everything also holds when left is interchanged with right throughout. From now on, $f(c)$ abbreviates $f_r(c)$, the result of substituting c for x in f to the right of the coefficients.

Among the known sufficient conditions on A and R, respectively, for $\mathrm{Int}^r_B(A)$ to be a ring, and for $\mathrm{N}^r(R)$ to be an ideal of $R[x]$, there are identical properties that have been shown, independently, to be sufficient conditions for both questions.

We will now sketch a common generalization for $\mathrm{Int}^r_B(A)$ and $\mathrm{N}^r(R)$ that allows a unified treatment of both objects.

Definition 3.1. Let R be a ring, T a subring of R, and I an ideal of T. We denote by $\mathrm{pol}^r_R(T, I)$ the set of polynomials in $R[x]$ that map every element of T to an element of I, under right substitution.

$$\mathrm{pol}^r_R(T, I) = \{f \in R[x] \mid \forall t \in T\ f_r(t) \in I\}$$

More generally, let S be a subset of T. Then we define

$$\mathrm{pol}^r_R(S, I) = \{f \in R[x] \mid \forall s \in S\ f_r(s) \in I\}$$

Note that the subring T of R is still subtly present in the definition of $\mathrm{pol}^r_R(S, I)$, since I is assumed to be an ideal of T.

In the special case where $R = T$ and $I = (0)$, we get $\mathrm{pol}^r_R(R, (0)) = \mathrm{N}^r(R)$, the set of right null-polynomials on R.

When $R = T$, $S \subseteq T$ and $I = (0)$, we get $\mathrm{pol}^r_R(S, (0)) = \mathrm{N}^r_R(S)$, the set of right null-polynomials on a subset S of R.

When A is a D-algebra and $B = A \otimes_D K$, as in Definition 1.3, and we set $R = B$ and $T = I = A$, we have $\mathrm{pol}^r_B(A, A) = \mathrm{Int}^r{}_B(A)$, the set of right integer-valued polynomials on A.

Likewise, to recover integer-valued polynomials on subsets, we set $R = B$ and $T = I = A$, and let S be a subset of A. Then, $\mathrm{pol}^r_B(S, A) = \mathrm{Int}^r{}_B(S, A)$ is the set of right integer-valued polynomials on S.

We now give an example of how integer-valued polynomials and null polynomials can be treated together in a more general setting.

Theorem 3.2. *Let $C \subseteq R[x]$ and $S \subseteq T$. Then, for $\mathrm{pol}^r_R(S, I)$ to be closed under right multiplication by elements of C, it is sufficient that it is closed under right multiplication by the images $c_r(s)$ with $c \in C$ and $s \in S$.*

Proof. Assume $\mathrm{pol}^r_R(S, I)$ is closed under right multiplication by elements of the form $c(s)$ with $c \in C$ and $s \in S$. Let $f \in \mathrm{pol}^r_R(S, I)$ and $c \in C$. For an arbitrary $s \in S$, we have to show that $(fc)(s) \in I$.

By division with remainder in $R[x]$ of c by $(x - s)$ from the right, we get

$$c(x) = q(x)(x - s) + c(s)$$

for some $q \in R[x]$. Now

$$(fc)(x) = f(x)q(x)(x-s) + f(x)c(s).$$

Let $h(x) = f(x)c(s)$. Then $(fc)(s) = h(s)$. By assumption, $h \in \mathrm{pol}^r_R(S,I)$, and, therefore, $(fc)(s) \in I$. □

Corollary 3.3 (Werner [14, Prop. 6.13]). *$\mathrm{Int}^{\mathrm{r}}{}_B(S,A)$ is a ring if and only if $\mathrm{Int}^{\mathrm{r}}{}_B(S,A)$ is closed under right multiplication by elements of A.*

Corollary 3.4 (Werner [13, Lemma 2.3]). *$\mathrm{N}^r_R(S)$ is an ideal if and only if it is closed under right multiplication by elements of R.*

By the above corollaries, the two questions,

1. whether $\mathrm{Int}^{\mathrm{r}}{}_B(A)$ is a ring, and
2. whether $\mathrm{N}^r_R(R)$ is an ideal of $R[x]$,

can now be subsumed under a single question

3. is $\mathrm{pol}^r_R(T,I)$ a right T-module (with the restricion of the multiplication of $R[x]$ as scalar multiplication)?

We illustrate the principle of treating null-ideals and rings of integer-valued polynomials in one common setting by another one of Werner's sufficient conditions.

Theorem 3.5. *Let R be a ring, T a subring of R and I an ideal of T. If T is generated by units as an algebra over its center, then*

1. *$\mathrm{pol}^r_R(T,I)$ is a right T-module.*
2. *More generally, for every subset S of T that is closed under conjugation by units of T, $\mathrm{pol}^r_R(S,I)$ is a right T-module.*

Proof. Let $f = \sum_k c_k x^k \in \mathrm{pol}^r_R(T,I)$, and u a unit of T. Then $fu \in \mathrm{pol}^r_R(S,I)$, because, for any $t \in S$, t can be written as $t = u^{-1}\tau u$ with $\tau = utu^{-1} \in S$, and then

$$(fu)(t) = \sum_k c_k u t^k = \sum_k c_k u u^{-1} \tau^k u = f(\tau)u,$$

where $f(\tau)u \in I$, because $f(\tau) \in I$ and I is an ideal of T.

Therefore, $\mathrm{pol}^r_R(S,I)$ is closed under multiplication from the right by units of T. Also, $\mathrm{pol}^r_R(S,I)$ is certainly closed under multiplication from the right by elements in the center of T (thanks to the fact that S is a subset of T), and closed under addition and subtraction. Since every element of T is a finite sum of products of central elements and units of T, we may conclude that $\mathrm{pol}^r_R(T,I)$ is closed under multiplication from the right by elements of T. □

Corollary 3.6 (Werner [14, Prop. 6.13]). *If A is generated by units as an algebra over its center and $S \subseteq A$ is closed under conjugation by units of A, then $\mathrm{Int}^{\mathrm{r}}{}_B(S,A)$ is a ring, i.e., S is a right ringset.*

Corollary 3.7. *If R is generated by units as an algebra over its center and $S \subseteq R$ is closed under conjugation by units of R, then $\mathrm{N}^r_R(S)$ is an ideal of $R[x]$, i.e., S is a right null-ideal set.*

Acknowledgement. S. Frisch is supported by the Austrian Science Fund (FWF): P 27816-N26.

References

[1] P.-J. Cahen and J.-L. Chabert, *Integer-valued polynomials*, vol. 48 of Mathematical Surveys and Monographs, Amer. Math. Soc., 1997.

[2] S. Evrard, Y. Fares, and K. Johnson, *Integer valued polynomials on lower triangular integer matrices*, Monatsh. Math. 170, 147–160 (2013).

[3] S. Frisch, *Integer-valued polynomials on algebras – a survey*, Actes des rencontres du CIRM (electronic) 2, 27–32 (2010).

[4] S. Frisch, *Integer-valued polynomials on algebras*, J. Algebra 373 (2013), 414–425, see also the corrigendum 412, p282 (2014).

[5] S. Frisch, *Polynomial functions on upper triangular matrix algebras*, Monatsh. Math. 184, 201–215 (2017).

[6] T. W. Hungerford, *Algebra*, vol. 73 of Graduate Texts in Mathematics, Springer-Verlag, New York-Berlin, 1980. Reprint of the 1974 original.

[7] K. Johnson and M. Pavlovski, *Integer-valued polynomials on the Hurwitz ring of integral quaternions*, Comm. Algebra 40, 4171–4176 (2012).

[8] K. A. Loper and N. J. Werner, *Generalized rings of integer-valued polynomials*, J. Number Theory 132, 2481–2490 (2012).

[9] G. Peruginelli and N. J. Werner, *Non-triviality conditions for integer-valued polynomial rings on algebras*, Monatsh. Math. 183, 177–189 (2017).

[10] G. Peruginelli and N. J. Werner, *Decomposition of integer-valued polynomial algebras*, J. Pure Appl. Algebra 222, 2562–2579 (2018).

[11] N. J. Werner, *Integer-valued polynomials over quaternion rings*, J. Algebra 324, 1754–1769 (2010).

[12] N. J. Werner, *Integer-valued polynomials over matrix rings*, Comm. Algebra 40, 4717 – 4726 (2012).

[13] N. J. Werner, *Polynomials that kill each element of a finite ring*, J. Algebra Appl. 13, 1350111, 12 (2014).

[14] N. J. Werner, *Integer-valued polynomials on algebras: a survey of recent results and open questions*, in Rings, polynomials, and modules, Springer, Cham, 353–375, 2017.

Irreducible polynomials in Int($\mathbb{Z}$)

Austin Antoniou[1,*], *Sarah* Nakato[2,**], and *Roswitha* Rissner[3,***]

[1]Ohio State University, 231 W. 18th Ave., MW 505, Columbus, OH 43210, USA
[2]Graz University of Technology, Kopernikusgasse 24, 8010 Graz, Austria
[3]Alpen-Adria-Universität Klagenfurt, Universitätsstraße 65-67, 9020 Klagenfurt, Austria

Abstract. In order to fully understand the factorization behavior of the ring $\text{Int}(\mathbb{Z}) = \{f \in \mathbb{Q}[x] \mid f(\mathbb{Z}) \subseteq \mathbb{Z}\}$ of integer-valued polynomials on $\mathbb{Z}$, it is crucial to identify the irreducible elements. Peruginelli [8] gives an algorithmic criterion to recognize whether an integer-valued polynomial $\frac{g}{d}$ is irreducible in the case where d is a square-free integer and $g \in \mathbb{Z}[x]$ has fixed divisor d. For integer-valued polynomials with arbitrary composite denominators, so far there is no algorithmic criterion known to recognize whether they are irreducible. We describe a computational method which allows us to recognize all irreducible polynomials in Int($\mathbb{Z}$). We present some known facts, preliminary new results and open questions.

1 Introduction

The ring $\text{Int}(\mathbb{Z}) = \{f \in \mathbb{Q}[x] \mid f(\mathbb{Z}) \subseteq \mathbb{Z}\}$ of integer-valued polynomials on $\mathbb{Z}$ is well-known to be a non-unique factorization domain. The investigation of the factorization in this ring has been subject to research over the last three decades, see for example [1], [2], [3], [4], [5] and [8].

Peruginelli [8] described a computational method the determine whether an image-primitive polynomial $\frac{g}{d} \in \text{Int}(\mathbb{Z})$ is irreducible in the case where the d is square-free. His method uses so-called p-coverings and is based on the primary decomposition of the ideal $\{g \in \mathbb{Z}[x] \mid g(\mathbb{Z}) \subseteq p\mathbb{Z}\}$ which he described in [7].

So far, there is no similar computational method to verify the irreducibility of an image-primitive polynomial with a general composite denominator.

In this conference paper we develop such a method, which we refer to as "table method". This is work in progress and we want to emphasize that this method is not very efficient in terms of run-time or memory. It is merely a first step towards recognizing irreducible polynomials in a computational way. A second goal that we pursue in our research is to use the table method to construct irreducible polynomials with prescribed fixed divisors. An overall summary of our current open research questions can be found in Section 4 at the end of this short paper.

As for the structure of this paper, we give a few basic facts on the fixed divisor of a polynomial in Section 1 and then proceed with factorizations in Int($\mathbb{Z}$) in Section 2. Next,

*e-mail: antoniou.6@buckeyemail.osu.edu
**e-mail: snakato@tugraz.at
***e-mail: roswitha.rissner@aau.at

in Section 3 we explain the table criterion and state the main result together with a few immediate corollaries. Finally, we conclude this short note in Section 4 with open questions that we want to answer in a full version.

Preliminaries

Let $g \in \mathbb{Z}[x]$ be a polynomial. It is easily seen that for $d \in \mathbb{Z}$, $\frac{g}{d}$ is integer-valued on $\mathbb{Z}$ if and only if d divides all values $g(a)$ with $a \in \mathbb{Z}$. This motivates the definition of the fixed divisor.

Definition 1.1. For $f \in \mathrm{Int}(\mathbb{Z})$, we define the *fixed divisor* $\mathrm{d}(f)$ to be

$$\mathrm{d}(f) = (f(a) \mid a \in \mathbb{Z}).$$

We say f is *image-primitive* if $\mathrm{d}(f) = \mathbb{Z}$.

Remark 1.2. It is well-known that there exist primitive polynomials in $\mathbb{Z}[x]$ which are irreducible in both $\mathbb{Z}[x]$ and $\mathbb{Q}[x]$ but factor non-trivially in $\mathrm{Int}(\mathbb{Z})$, for example,

$$x^3 - x + 3 = 3 \cdot \frac{x^3 - x + 3}{3}.$$

Note that $x^3 - x + 3$ is primitive, but not image-primitive since $\mathrm{d}(x^3 - x + 3) = 3\mathbb{Z}$. More generally, Chapman and McClain [3, Thm. 2.4] have shown that a primitive polynomial $g \in \mathbb{Z}[x]$ is irreducible in $\mathrm{Int}(\mathbb{Z})$ if and only if g is image-primitive and irreducible in $\mathbb{Z}[x]$.

Remark 1.3. It is easily seen that $\mathrm{d}(fg) \subseteq \mathrm{d}(f)\,\mathrm{d}(g)$ for polynomials $f, g \in \mathrm{Int}(\mathbb{Z})$. In general, equality does not hold. For example, $\mathrm{d}(x) = \mathrm{d}(x-1) = \mathbb{Z}$ but $\mathrm{d}(x(x-1)) = 2\mathbb{Z}$.

Remark 1.4. Assume that $g \in \mathbb{Z}[x]$ is a primitive polynomial with fixed divisor $\mathrm{d}(g) = d\mathbb{Z}$ for an integer d. If p is a prime number with $p \mid d$, then $g(a) \equiv 0 \mod p$ for all $a \in \mathbb{Z}$ or equivalently, the residue class of g modulo p has p different roots in the field $\mathbb{Z}/p\mathbb{Z}$.

In particular, this implies that the prime decomposition of the fixed divisor of a primitive polynomial $g \in \mathbb{Z}[x]$ can only contain prime numbers p with $p \leq \deg(g)$.

This observation follows from a result of Pólya [9], cf. [10, Thm. 3.1].

2 Factorizations in Int($\mathbb{Z}$) and the σ notation

Notation 2.1. Let $\mathbb{P} \subseteq \mathbb{Z}$ denote the set of (positive) prime elements and for each prime number $p \in \mathbb{P}$, we write v_p for the p-adic valuation.

Every polynomial $f \in \mathrm{Int}(\mathbb{Z})$ can be written as

$$f = a\frac{\prod_{i\in I} g_i}{\prod_{p\in\mathbb{P}} p^{k_p}} \tag{1}$$

where $a \in \mathbb{Z}$, $G = (g_i)_{i\in I}$ is a finite family of primitive, irreducible polynomials $g_i \in \mathbb{Z}[x]$ and $k_p \in \mathbb{N}_0$ such that $k_p = 0$ for all but finitely many $p \in \mathbb{P}$.

Remark 2.2. It follows from Remark 1.2 that irreducible polynomials in $\mathrm{Int}(\mathbb{Z})$ are necessarily image-primitive, that is, $a = \pm 1$ in Equation (1). In the context of factorizations, the sign of a is not important. From now on, we assume $a = 1$ whenever we consider image-primitive polynomials in $\mathrm{Int}(\mathbb{Z})$.

Notation 2.3. Let $G = (g_i)_{i\in I}$ be a finite family of polynomials $g_i \in \mathbb{Q}[x]$ and $J \subseteq I$ a subset. For a prime $p \in \mathbb{P}$ and $a \in \mathbb{Z}$, we set

$$\sigma_G^p(J) = \min_{a\in\mathbb{Z}} \sum_{j\in J} \nu_p(g_j(a)).$$

Lemma 2.4. *Let $G = (g_i)_{i\in I}$ be a finite family of polynomials $g_i \in \mathbb{Z}[x]$, $p \in \mathbb{P}$ a prime number and $J \subseteq I$. Then the following assertions hold*

(i) $\sigma_G^p(J) = \nu_p\left(\mathrm{d}\left(\prod_{j\in J} g_j\right)\right)$ *and*

(ii) $\sigma_G^p(I) \geq \sigma_G^p(J) + \sigma_G^p(I \setminus J)$.

Proof. (*i*): For $k \in \mathbb{N}_0$, $k = \nu_p\left(\mathrm{d}\left(\prod_{j\in J} g_j\right)\right)$ if and only if for all $a \in \mathbb{Z}$, $\nu_p\left(\prod_{j\in J} g_j(a)\right) \geq k$ and there exists $b \in \mathbb{Z}$ such that $\nu_p\left(\prod_{j\in J} g_j(b)\right) < k + 1$. It follows immediately that

$$\nu_p\left(\mathrm{d}\left(\prod_{j\in J} g_j\right)\right) = \min_{a\in\mathbb{Z}}\left(\nu_p\left(\prod_{j\in J} g_j(a)\right)\right) = \min_{a\in\mathbb{Z}} \sum_{j\in J} \nu_p(g_j(a)) = \sigma_G^p(J).$$

(*ii*): Since $\sum_{i\in I} \nu_p(g_i(a)) = \sum_{j\in J} \nu_p(g_j(a)) + \sum_{j\in I\setminus J} \nu_p(g_j(a))$ for all $a \in \mathbb{Z}$, the claim follows. □

Lemma 2.4 states that $\sigma_G^p(I)$ is a short-cut notation for the p-adic valuation of the fixed divisor of the polynomial $\prod_{i\in I} g_i$. This notation turns out to be useful. The next lemma states an easy but important result.

Lemma 2.5. *Let $G = (g_i)_{i\in I}$ be a finite family of primitive, irreducible polynomials $g_i \in \mathbb{Z}[x]$ and let $f = \frac{\prod_{i\in I} g_i}{\prod_{p\in\mathbb{P}} p^{k_p}} \in \mathbb{Q}[x]$.*

Then

(i) $f \in \mathrm{Int}(\mathbb{Z})$ *if and only if* $\sigma_G^p(I) \geq k_p$ *for all* $p \in \mathbb{P}$ *and*

(ii) $f \in \mathrm{Int}(\mathbb{Z})$ *is image-primitive if and only if* $\sigma_G^p(I) = k_p$ *for all* $p \in \mathbb{P}$.

Proof. This follows immediately from Lemma 2.4 and Definition 1.1. □

Remark 2.6. Let $G = (g_i)_{i\in I}$ be a finite family of primitive, irreducible polynomials $g_i \in \mathbb{Z}[x]$ such that $f = \frac{\prod_{i\in I} g_i}{\prod_{p\in\mathbb{P}} p^{k_p}} \in \mathrm{Int}(\mathbb{Z})$ is image-primitive.

Assume that f is reducible in $\mathrm{Int}(\mathbb{Z})$, that is, $f = h\tilde{h}$ factors into non-units $h, \tilde{h} \in \mathrm{Int}(\mathbb{Z})$. Since both h and $\tilde{h}$ divide f in $\mathbb{Q}[x]$ which is a unique factorization domain, it follows that h and $\tilde{h}$ are of the form

$$h = c\prod_{j\in J} g_j \quad \text{and} \quad \tilde{h} = \tilde{c}\prod_{j\in I\setminus J} g_j$$

for a subset $J \subseteq I$ and some $c, \tilde{c} \in \mathbb{Q}$ with $c\tilde{c} = \prod_{p\in\mathbb{P}} p^{-k_p}$.

Furthermore, f is image-primitive, which implies that neither h nor $\tilde{h}$ is a constant and hence $\emptyset \neq J \subsetneq I$.

Moreover, according to Lemmas 2.4 and 2.5 it follows that

$$k_p = \sigma_G^p(I) \geq \sigma_G^p(J) \geq -\nu_p(c)$$

and similarly

$$k_p \geq -\nu_p(\tilde{c})$$

for all prime numbers $p \in \mathbb{P}$.

On the other hand, $\nu_p(c) + \nu_p(\tilde{c}) = -k_p$ holds for all primes $p \in \mathbb{P}$ which implies that

$$-k_p \leq \nu_p(c), \nu_p(\tilde{c}) \leq 0.$$

We can conclude that every factor of f in $\mathrm{Int}(\mathbb{Z})$ which is a non-unit is of the form

$$\frac{\prod_{j \in J} g_j}{\prod_{p \in \mathbb{P}} p^{m_p}}$$

where $\emptyset \neq J \subseteq I$ and $0 \leq m_p \leq k_p$ for all prime numbers $p \in \mathbb{P}$.

3 Table method

In this section, we develop a computational method to recognize whether an image-primivite polynomial $f \in \mathrm{Int}(\mathbb{Z})$ is irreducible. Moreover, if f is reducible, this technique can be used to find all factorizations of f in $\mathrm{Int}(\mathbb{Z})$.

Following Remark 2.2, we can assume that f is of the form

$$f = \frac{\prod_{i \in I} g_i}{\prod_{p \in \mathbb{P}} p^{k_p}}.$$

Remark 3.1. We can determine $\sigma_G^p(J)$ by computing the values $\sum_{j \in J} \nu_p(g_j(a))$ for only finitely many $a \in \mathbb{Z}$ because every polynomial $g \in \mathbb{Z}[x]$ respects congruences, that is, if $a \equiv b \mod p^k$, then $g(a) \equiv g(b) \mod p^k$. In particular, this implies that if $\sigma_G^p(J) = n$, then there exists a $0 \leq a \leq p^{n+1} - 1$ with $\sum_{j \in J} \nu_p(g_j(a)) = n$.

A strategy that we pursue in this work is to create a table as in Table 1. The last entry of

Table 1. Template for table method

	g_1	g_2	$\dots$	Σ
0	$\nu_p(g_1(0))$	$\nu_p(g_2(0))$	$\dots$	$\sum_{j \in J} \nu_p(g_j(0))$
1	$\nu_p(g_1(1))$	$\nu_p(g_2(1))$	$\dots$	$\sum_{j \in J} \nu_p(g_j(1))$
$\vdots$		$\vdots$		$\vdots$
a	$\nu_p(g_1(a))$	$\nu_p(g_2(a))$	$\dots$	$\sum_{j \in J} \nu_p(g_j(a))$
$\vdots$		$\vdots$		$\vdots$

every row is greater than or equal to $\sigma_G^p(J)$. Say, we encounter the value n at the right end of one row. Then we only need to compute the table rows for $0 \leq a \leq p^n - 1$. Note that it is possible that this value is only reached in a row with $a \geq p^n$ in which case we can stop the computation, see Example 3.2.

Note that the starting value $a = 0$ is chosen randomly.

Example 3.2. We use Lemma 2.5 to determine whether $\frac{x(x-1)(x-2)}{9}$ is in $\mathrm{Int}(\mathbb{Z})$. Let $G = (g_1, g_2, g_3)$ with $g_1 = x$, $g_2 = x - 1$ and $g_3 = x - 2$. We compute $\sigma_G^3(\{1, 2, 3\})$ using the table method explained in Remark 3.1, see Table 2. Once we reach the fourth row we can stop further computations and conclude that $\sigma_G^3(\{1, 2, 3\}) = 1$. Hence $\frac{x(x-1)(x-2)}{9}$ is not integer-valued on $\mathbb{Z}$.

Example 3.3. From Table 2 and Lemma 2.5 we can conclude that $\frac{x(x-1)(x-2)}{3}$ is in $\mathrm{Int}(\mathbb{Z})$. It is also easily checked (for example using an additional table), that $\mathrm{d}(x(x-1)(x-2)) = 6\mathbb{Z}$ and hence $\frac{x(x-1)(x-2)}{6}$ is an image-primitive polynomial in $\mathrm{Int}(\mathbb{Z})$.

Table 2. Table method for Example 3.2

	g_1	g_2	g_3	$\sum$
0	∞	0	0	∞
1	0	∞	0	∞
2	0	0	∞	∞
3	1	0	0	1

Notation 3.4. For a finite family $G = (g_i)_{i\in I}$ of polynomials $g_i \in \mathbb{Z}[x]$ and $J \subseteq I$, we set

$$\sigma_G(J) = \sum_{p\in\mathbb{P}} \sigma_G^p(J).$$

Keep in mind that the sum above is finite, since $\sigma_G^p(J) = 0$ for all prime numbers $p \in \mathbb{P}$ with $p > \sum_{j\in J} \deg(g_j)$, see Remark 1.4.

Theorem 1. *Let $G = (g_i)_{i\in I}$ be a finite family of primitive, irreducible polynomials $g_i \in \mathbb{Z}[x]$ such that $f = \frac{\prod_{i\in I} g_i}{\prod_{p\in\mathbb{P}} p^{k_p}} \in \mathrm{Int}(\mathbb{Z})$ is image-primitive.*

Then the following assertions are equivalent:

(i) f is irreducible in $\mathrm{Int}(\mathbb{Z})$.

(ii) For all subsets $\emptyset \neq J \subsetneq I$ there exists a prime number $p \in \mathbb{P}$ such that

$$\sigma_G^p(J) + \sigma_G^p(I \setminus J) < k_p.$$

(iii) For all subsets $\emptyset \neq J \subsetneq I$,

$$\sigma_G(J) + \sigma_G(I \setminus J) < \sum_{p\in\mathbb{P}} k_p.$$

Proof. *(ii)* $\Leftrightarrow$ *(iii)*: It follows from Lemmas 2.4 and 2.5 that for all $p \in \mathbb{P}$ and all $J \subseteq I$,

$$\sigma_G^p(J) + \sigma_G^p(I \setminus J) \leq \sigma_G^p(I) = k_p$$

holds. The equivalence between assertions *(ii)* and *(iii)* follows.

(iii) $\Rightarrow$ *(i)*: Assume that f is reducible in $\mathrm{Int}(\mathbb{Z})$, that is,

$$f = \frac{\prod_{j\in J} g_j}{\prod_{p\in\mathbb{P}} p^{m_p}} \cdot \frac{\prod_{j\in I\setminus J} g_j}{\prod_{p\in\mathbb{P}} p^{k_p - m_p}}$$

for some subset $\emptyset \neq J \subsetneq I$ and natural numbers $m_p \leq k_p$ (cf. Remark 2.6). By Lemma 2.5 it follows that $\sigma_G^p(J) \geq m_p$ and $\sigma_G^p(I \setminus J) \geq k_p - m_p$ for all primes $p \in \mathbb{P}$. Hence

$$\sigma_G(J) + \sigma_G(I \setminus J) = \sum_{p\in\mathbb{P}} \sigma_G^p(J) + \sigma_G^p(I \setminus J) \geq \sum_{p\in\mathbb{P}} k_p.$$

(i) $\Rightarrow$ *(iii)*: Assume $\emptyset \neq J \subsetneq I$ is a subset such that $\sigma_G(J) + \sigma_G(I \setminus J) \geq \sum_{p\in\mathbb{P}} k_p$. For $p \in \mathbb{P}$, we set $m_p = \sigma_G^p(J)$ and $\tilde{m}_p = \sigma_G^p(I \setminus J)$. By Lemma 2.5 it follows that

$$h = \frac{\prod_{j\in J} g_j}{\prod_p p^{m_p}} \quad \text{and} \quad \tilde{h} = \frac{\prod_{j\in J} g_j}{\prod_p p^{\tilde{m}_p}}$$

are elements of Int($\mathbb{Z}$).

Next, $m_p + \tilde{m}_p \le \sigma_G^p(I) = k_p$ for all primes $p \in \mathbb{P}$ according to Lemmas 2.4 and 2.5. On the other hand,

$$\sum_{p\in\mathbb{P}} m_p + \tilde{m}_p = \sigma_G(J) + \sigma_G(I \setminus J) \ge \sum_{p\in\mathbb{P}} k_p$$

holds, and therefore it follows that $m_p + \tilde{m}_p = k_p$ for all $p \in \mathbb{P}$.

Hence

$$f = h \cdot \tilde{h}$$

is a proper factorization of f in Int($\mathbb{Z}$), that is, f is reducible in Int($\mathbb{Z}$). □

Theorem 1 yields some immediate corollaries. The next one is merely a restatement of the theorem in case only one prime factor appears in the denominator.

Corollary 3.5. *Let $G = (g_i)_{i\in I}$ be a finite family of primitive, irreducible polynomials $g_i \in \mathbb{Z}[x]$, $p \in \mathbb{P}$ a prime number and $k \in \mathbb{N}$ such that $f = \frac{\prod_{i\in I} g_i}{p^k} \in$ Int($\mathbb{Z}$) is image-primitive.*

Then f is irreducible in Int($\mathbb{Z}$) if and only if for all subsets $\emptyset \ne J \subsetneq I$

$$\sigma_G^p(J) + \sigma_G^p(I \setminus J) < k.$$

The next corollary summarizes how to find all factorizations of a polynomial $f \in$ Int($\mathbb{Z}$) into not necessarily irreducible factors.

Corollary 3.6. *Let $G = (g_i)_{i\in I}$ be a finite family of primitive, irreducible polynomials $g_i \in \mathbb{Z}[x]$ such that $f = \frac{\prod_{i\in I} g_i}{\prod_{p\in\mathbb{P}} p^{k_p}} \in$ Int($\mathbb{Z}$) is image-primitive.*

Then every factorization of f (into not necessarily irreducibles) is of the form

$$f = \left(\frac{\prod_{j\in J_1} g_j}{\prod_{p\in\mathbb{P}} p^{m_{1,p}}}\right) \cdots \left(\frac{\prod_{j\in J_t} g_j}{\prod_{p\in\mathbb{P}} p^{m_{t,p}}}\right)$$

where $I = J_1 \uplus \cdots \uplus J_t$ is a partition of I such that for each prime number $p \in \mathbb{P}$ the following conditions hold

(i) $\sigma_G^p(J_i) = m_{i,p}$ for all $1 \le i \le t$,

(ii) $m_{1,p} + \cdots + m_{t,p} = k_p$ and

Remark 3.7. We can use Corollary 3.6 also to find the factorizations into irreducibles of $f \in$ Int($\mathbb{Z}$). For this purpose, we need to find all partitions of I which satisfy the conditions but cannot be "refined" any further.

Example 3.8. We find all factorizations of

$$f = \frac{(x^3 + 2x + 1)(x^2 + 2x + 1)(x^3 - 3x^2 + 8x + 20)(x^3 + 5)}{4}$$

and consider the family of polynomials $G = (g_1, g_2, g_3, g_4)$ with $g_1 = x^3 + 2x + 1$, $g_2 = x^2 + 2x + 1$, $g_3 = x^3 - 3x^2 + 8x + 20$ and $g_4 = x^3 + 5$.

The degree of the product is 11, which means that possible prime factors of the fixed divisor are 2, 3, 5, 7 and 11. By plugging in 0, we can exclude 3 and 7 and by plugging in 1, we can exclude 5 and 11.

We create a table for $p = 2$, see Table 3.

It follows that the $\sigma_G(\{1, 2, 3, 4\}) = \sigma_G^2(\{1, 2, 3, 4\}) = 2$. Therefore, f is image-primitive.

Table 3. Tables for Example 3.8

	g_1	g_2	g_3	g_4	$\sum \nu_2$
0	0	0	2	0	2
1	2	2	1	1	6
2	0	0	5	0	5
3	1	4	2	5	12

There are 3 partitions of $\{1,2,3,4\}$ which satisfy the conditions of Corollary 3.6. These are

$$\begin{aligned}\{1,2,3,4\} &= \{1,3\} \uplus \{2\} \uplus \{4\} \\ &= \{2,3\} \uplus \{1\} \uplus \{4\} \\ &= \{3,4\} \uplus \{1\} \uplus \{2\}\end{aligned}$$

It follows that

$$\begin{aligned} f &= \frac{(x^3+2x+1)(x^3-3x^2+8x+20)}{4}(x^2+2x+1)(x^3+5) \\ &= \frac{(x^2+2x+1)(x^3-3x^2+8x+20)}{4}(x^3+2x+1)(x^3+5) \\ &= \frac{(x^3-3x^2+8x+20)(x^3+5)}{4}(x^3+2x+1)(x^2+2x+1) \end{aligned}$$

are all factorizations of f into irreducibles in $\mathrm{Int}(\mathbb{Z})$.

Corollary 3.9. *Let $G = (g_i)_{i\in I}$ be a finite family of primitive, irreducible polynomials $g_i \in \mathbb{Z}[x]$ such that $f = \frac{\prod_{i\in I} g_i}{\prod_{p\in\mathbb{P}} p^{k_p}} \in \mathrm{Int}(\mathbb{Z})$ is image-primitive.*

If there exists a prime number $p \in \mathrm{Int}(\mathbb{Z})$ such that $\sigma_G^p(J) + \sigma_G^p(I \setminus J) < k_p$ for all subsets $\emptyset \neq J \subsetneq I$, then f is irreducible.

The following example shows that the implication in Corollary 3.9 cannot be reversed in general. A similar observation has already been made by Peruginelli [8] in the case of square-free denominators.

Example 3.10. We consider the polynomial

$$f = \frac{(x^2+4)(x^2-2x+5)(x^3-5x^2+6x+5)}{20}.$$

We set $g_1 = x^2+4$, $g_2 = x^2-2x+5$ and $g_3 = x^3-5x^2+6x+5$, $G = (g_1, g_2, g_3)$ and compute a table with respect to ν_2 and ν_5 using the table method which we explained in Remark 3.1, see Table 4. Possible prime factors of the fixed divisors of the numerator are 2, 3, 5 and 7. By plugging in 0, one can see that 3 and 7 cannot occur.

We can read off these tables that $\sigma_G^2(\{1,2\}) + \sigma_G^2(\{3\}) = 2 = k_2$ and $\sigma_G^5(\{2\}) + \sigma_G^2(\{1,3\}) = 1 = k_5$. Hence the lefthand side of the implication in Corollary 3.9 does not hold.

However, f is irreducible according to Theorem 1, since

$$\begin{aligned} \sigma_G(\{1\}) + \sigma_G(\{2,3\}) &= 0 < 3 = k_2 + k_5 \\ \sigma_G(\{2\}) + \sigma_G(\{1,3\}) &= 1 < 3 \\ \sigma_G(\{3\}) + \sigma_G(\{1,2\}) &= 2 < 3 \end{aligned}$$

Table 4. Tables for Example 3.10

	g_1	g_2	g_3	$\sum v_2$
0	2	0	0	2
1	0	2	0	2
2	3	0	0	3
3	0	3	0	3

	g_1	g_2	g_3	$\sum v_5$
0	0	1	1	2
1	1	0	0	1
2	0	1	1	2
3	0	0	1	1
4	1	0	0	1

Since for all non-empty subsets J of an index set of size 3, either J or its complement is a singleton, there are no more subsets to be checked.

4 Open questions

Absolutely irreducible polynomials in Int($\mathbb{Z}$)

We are interested in a certain subclass of irreducible polynomials in Int($\mathbb{Z}$).

Definition 4.1. A polynomial $f \in \text{Int}(\mathbb{Z})$ is *absolutely irreducible* if for all $n \in \mathbb{N}$, f^n has exactly one factorization into irreducibles, that is,

$$f^n = \underbrace{f \cdots f}_{n \text{ times}}$$

Remark 4.2. It is easily seen that $\text{d}(g^n) = \text{d}(g)^n$. Hence, if $g \in \mathbb{Z}[x]$ is primitive and irreducible with fixed divisor $d\mathbb{Z}$, then $f = \frac{g}{d}$ is absolutely irreducible.

Polynomials of the form described in Remark 4.2 are not the only types of absolutely irreducible polynomial as the following example shows.

Example 4.3. The polynomial $f = \frac{x(x-1)}{2}$ is absolutely irreducible: According to Remark 2.6, any factorization of f^n into two non-units in Int($\mathbb{Z}$) is of the form

$$\left(\frac{x^k(x-1)^\ell}{2^t}\right) \cdot \left(\frac{x^{n-k}(x-1)^{n-\ell}}{2^{n-t}}\right)$$

However, $\text{d}(x) = \text{d}(x^m) = \mathbb{Z}$ and $\text{d}(x-1) = \text{d}((x-1)^m) = \mathbb{Z}$ for all m. Therefore, it follows that $t = \ell = k$ which implies that f is absolutely irreducible.

The following example shows that not every irreducible polynomials is also absolutely irreducible.

Example 4.4. Let $f = \frac{x(x^2+2x+5)}{2}$. Using Remark 3.1, we create tables for the prime numbers 2, see Table 5. The prime number 3 can be excluded by plugging in 1. From this table we

Table 5. Tables for Example 4.4

	x	x^2+2x+5	$\sum v_2$
0	∞	0	∞
1	0	3	3
2	1	0	1

can see, that f is irreducible in Int($\mathbb{Z}$). However, it is not absolutely irreducible since

$$f^2 = f \cdot f = \frac{x^2(x^2+2x+5)}{4} \cdot (x^2+2x+5)$$

Question: Can we use the table method to recognize absolutely irreducible polynomials? For example, assume f is an irreducible polynomial in $\mathrm{Int}(\mathbb{Z})$ such that there are two polynomials in the numerator $g_i \neq g_j$ whose table columns look alike. Then, it is easily seen that they can be interchanged in every factorization of f^n with $n > 1$ and hence such a polynomial cannot be absolutely irreducible.

If not the table method, is there some other way to recognize absolutely irreducible polynomials?

A Irreducible polynomials in $\mathrm{Int}(\mathbb{Z}_{(p)})$

The *length* of factorization of, here a polynomial $f \in \mathrm{Int}(\mathbb{Z}_{(p)})$, into irreducibles is the number of irreducible factors. The collection of all possible lengths of an element is called its *set of lengths*. When we speak of multisets of lengths, then we simply take multiplicities of lengths into account. Note that we want to count multiplicities only for essentially different factorizations, that is, if the factors of the two factorizations are not pairwise associated. For details on factorizations, we refer to Geroldinger and Halter-Koch's textbook [6].

Question. Which multisets of lengths can occur in $\mathrm{Int}(\mathbb{Z}_{(p)})$?

It is known that every finite (multi-)set of natural numbers occurs as (multi-)set of a polynomial in $\mathrm{Int}(\mathbb{Z})$, see Frisch [4] and Frisch, Nakato and Rissner [5]. The construction that was used in these papers cannot immediately be adapted to the local (or semilocal case) because it strongly relies on the existence of infinitely many prime numbers.

In order to overcome the difficulties, we are trying to answer the following question.

Question. How to construct irreducible polynomials in $\mathrm{Int}(\mathbb{Z}_{(p)})$ with a prescribed fixed divisor?

Another question that comes up in that context is whether $\mathrm{Int}(\mathbb{Z}_{(p)})$ is *transfer Krull*, that is, there exists a transfer homomorphism to a block monoid. For definitions and further details, we refer to [6].

It has been shown in [4] that no such homomorphism can exist for $\mathrm{Int}(\mathbb{Z})$. More generally, it is known that $\mathrm{Int}(D)$ is not transfer Krull if D is a Dedekind domain with infinitely many maximal ideals, all of which have finite index, see [5].

Question. Does there exist a transfer homomorphism from the multiplicative monoid of $\mathrm{Int}(\mathbb{Z}_{(p)})$ to a block monoid?

Acknowledgement

Sarah Nakato is supported by the Austrian Science Fund (FWF): P 27816 and P 30934. Roswitha Rissner is supported by the Austrian Science Fund (FWF): P 28466.

References

[1] David F. Anderson and Paul-Jean Cahen and Scott T. Chapman and William W. Smith: *Some factorization properties of the ring of integer-valued polynomials*, in Zero-dimensional commutative rings, Lecture Notes in Pure and Appl. Math.,171:125–142,1995.

[2] Paul-Jean Cahen and Jean-Luc Chabert: *Elasticity for integral-valued polynomials*, J. Pure Appl. Algebra, 103:303–311, 1995.

[3] Scott T. Chapman and Barbara A. McClain: *Irreducible polynomials and full elasticity in rings of integer-valued polynomials*, J. Algebra., 293:595–610, 2005.

[4] Sophie Frisch: *A construction of integer-valued polynomials with prescribed sets of lengths of factorizations*, Monatsh. Math., 171, no. 3-4;341–350, 2013.

[5] Sophie Frisch and Sarah Nakato and Roswitha Rissner: *Sets of lengths of factorizations of integer-valued polynomials on Dedekind domains with finite residue fields*, Submitted.

[6] Alfred Geroldinger and Franz Halter-Koch: *Non-Unique Factorizations: Algebraic, Combinatorial and Analytic Theory*, vol. 278 of Pure and Applied Mathematics. Chapman & Hall/CRC, 2006.

[7] Giulio Peruginelli: *Primary decomposition of the ideal of polynomials whose fixed divisor is divisible by a prime power*, J. Algebra, 298:227–242, 2014.

[8] Giulio Peruginelli: *Factorization of integer-valued polynomials with square-free denominator*, Comm. Alg, 43:197–211, 2015.

[9] Georg Pólya: *Über ganzwertige Polynome in algebraischen Zahlkörpern*, J. reine angew. Math. (Crelle), 149:97–116, 1919.

[10] Władysław Narkiewicz: *Polynomial mappings*, Lecture Notes in Mathematics, Springer Berlin, 1995.

ITM Web of Conferences **20**, 01005 (2018) https://doi.org/10.1051/itmconf/20182001005
ICM 2018

Differential Resultants

Scott McCallum[1,*] and *Franz* Winkler[2,**]

[1]Dept. of Computing, Macquarie University, NSW 2109, Australia
[2]RISC, J.Kepler University, Linz, Austria

Abstract. We review classical concepts of resultants of algebraic polynomials, and we adapt some of these concepts to objects in differential algebra, such as linear differential operators and differential polynomials.

1 Introduction

There are interesting notions of a differential polynomial resultant in the literature. Chardin [3] presented an elegant treatment of resultants and subresultants of (noncommutative) ordinary differential operators. Carra'-Ferro (see for example [1, 2]) published several works on differential resultants of various kinds, with firm algebraic foundations, but the relations to Zwillinger's [18] suggested notion of a differential resultant of a systems of two coupled algebraic ordinary differential equations (AODEs) and also to Chardin's theory might not be immediately clear from glancing through these works. Rueda and Sendra [14] define a linear complete differential resultant and investigate its application to the implicitization problem of systems of differential polynomial parametric equations. Zhang et al. [17] construct a matrix whose determinant contains the resultant of two generic ordinary differential polynomials as a factor.

Here we investigate the relation between the theories of Chardin and Carra'-Ferro. In fact we will see that these approaches are intimately related. It would appear that the common source for the essential basic notion of differential resultant can be traced to work of Ritt [13] in the 1930s. After reviewing relevant background material on elimination theory and differential algebra, in Sections 2 and 3, we will present in Section 4 the concepts of resultant for linear differential operators, for linear homogeneous differential polynomials, and finally for arbitrary differential polynomials. This could be viewed as a simpler and more streamlined account of Carra'-Ferro's theory. The problem posed by Zwillinger seems to be treatable by such methods; but this needs to be worked out in more detail.

Background material and more proof details may be found in [12].

2 Basics of elimination theory

In this section we briefly review basic facts from polynomial elimination theory and differential algebra.

*e-mail: scott.mccallum@mq.edu.au
**e-mail: franz.winkler@risc.jku.at

Let $\mathcal{R}$ be an integral domain (commutative ring with identity element 1, and no zero divisors), with quotient field $\mathcal{K}$. By $\overline{\mathcal{K}}$ we denote the algebraic closure of $\mathcal{K}$.

First we review the basic theory of the Sylvester resultant for algebraic polyomials, with an emphasis on the necessary requirements for the underlying coefficient domain. Let

$$f(x) = \sum_{i=0}^{m} a_i x^i, \quad g(x) = \sum_{j=0}^{n} b_j x^j$$

be polynomials of positive degrees m and n, respectively, in $\mathcal{R}[x]$. If f and g have a common factor $d(x)$ of positive degree, then they have a common root in the algebraic closure $\overline{\mathcal{K}}$ of $\mathcal{K}$; so the system of equations

$$f(x) = g(x) = 0 \tag{1}$$

has a solution in $\overline{\mathcal{K}}$.
On the other hand, if $\alpha \in \overline{\mathcal{K}}$ is a common root of f and g, then $\text{norm}_{\mathcal{K}(\alpha):\mathcal{K}}(x-\alpha)$ is a common divisor of f and g in $\mathcal{K}[x]$. So, by Gauss' Lemma (for which we need $\mathcal{R}$ to be a unique factorization domain) on primitive polynomials there is a similar (only differing by a factor in $\mathcal{K}$) common factor of f and g in $\mathcal{R}[x]$. We summarize these observations as follows:

Proposition 1. *Let $\mathcal{R}$ be a unique factorization domain (UFD) with quotient field $\mathcal{K}$. For polynomials $f(x), g(x) \in \mathcal{R}[x]$ the following are equivalent:*

(i)*f and g have a common solution in $\overline{\mathcal{K}}$, the algebraic closure of $\mathcal{K}$,*

(ii)*f and g have a common factor of positive degree in $\mathcal{R}[x]$.*

So now we want to determine a necessary condition for f and g to have a common divisor of positive degree in $\mathcal{R}[x]$. Suppose that f and g indeed have a common divisor $d(x)$ of positive degree in $\mathcal{R}[x]$; i.e.,

$$f(x) = d(x)\overline{f}(x), \quad g(x) = d(x)\overline{g}(x). \tag{2}$$

Then for $p(x) := \overline{g}(x)$, $q(x) := -\overline{f}(x)$ we have

$$p(x)f(x) + q(x)g(x) = 0. \tag{3}$$

So there are non-zero polynomials p and q with $\deg p < \deg g, \deg q < \deg f$, satisfying equation (3). This leads to the linear system

$$\begin{pmatrix} p_{n-1} & \cdots & p_0 & q_{m-1} & \cdots & q_0 \end{pmatrix} \cdot \begin{pmatrix} A \\ \cdots \\ B \end{pmatrix} = 0\,, \tag{4}$$

where the $n \times (m+n)$ matrix A contains the coefficients of $f(x)$ $(a_m a_{m-1} \dots a_0 0 \dots 0)$ in the first row, which are shifted to the right by $(i-1)$ positions in the i-th row $(0 \dots 0 a_m a_{m-1} \dots a_0 0 \dots 0)$; the $m \times (m+n)$ matrix B is constructed analogously with the coefficients of $g(x)$. Under the given conditions, the system (4) has a non-trivial solution. The matrix of this system (4) is called the *Sylvester matrix* of f and g. Thus, the determinant of the Sylvester matrix of f and g is 0. The *resultant* of f and g, $\text{res}(f, g)$, is this determinant, and it is clear that the resultant is a polynomial expression of the coefficients of f and g, and therefore an element of the integral domain $\mathcal{R}$. This does not require $\mathcal{R}$ to be a UFD.

We summarize this in the following proposition.

Proposition 2. *Let $f, g \in \mathcal{R}[x]$, for $\mathcal{R}$ an integral domain.*
$\mathrm{res}(f, g) = 0$ is a necessary condition for f and g to have a common factor of positive degree; and therefore a common solution in $\overline{\mathcal{K}}$.

If we identify a polynomial of degree d with the vector of its coefficients of length $d + 1$, we may also express this in terms of the linear map

$$\begin{array}{llcl} S: & \mathcal{K}^{m+n} & \longrightarrow & \mathcal{K}^{m+n} \\ & (p_{n-1}, \ldots, p_0, q_{m-1}, \ldots, q_0) & \mapsto & \text{coefficients of } pf + qg \end{array}$$

The existence of a non-trivial linear combination (3) is equivalent to S having a non-trivial kernel, and therefore to S having determinant 0.

A proof of the following may be found, for instance, in [4].

Proposition 3. *The resultant is a constant in the ideal generated by f and g in $\mathcal{R}[x]$; i.e. we can write*

$$\mathrm{res}(f, g) = u(x)f(x) + v(x)g(x), \tag{5}$$

with $u, v \in \mathcal{R}[x]$. Moreover, these cofactors satisfy the degree bounds $\deg(u) < \deg(g)$, $\deg(v) < \deg(f)$.

Another variation on defining the Sylvester resultant of two polynomials with coefficients in a UFD (unique factorization domain) $\mathcal{D}$ is to start instead with two *homogeneous* polynomials

$$F(x, y) = \sum_{i=0}^{m} a_i x^i y^{m-i} \quad , \quad G(x, y) = \sum_{j=0} b_j x^j y^{n-j} .$$

Let us similarly regard the coefficients a_i and b_j as indeterminates. Then the resultant of F and G is defined as $\mathrm{res}(F, G) = \mathrm{res}(f, g)$, where $f(x) = F(x, 1)$ and $g(x) = G(x, 1)$. Then the relation between the vanishing of the resultant and the existence of a common zero may be expressed as follows.

Proposition 4. *After assigning values to the coefficients from a UFD $\mathcal{D}$, $\mathrm{res}(F, G) = 0$ is a necessary and sufficient condition for $F(x, y)$ and $G(x, y)$ to have a common factor of positive degree over $\mathcal{D}$, hence for a common zero to exist over an extension of the quotient field of $\mathcal{D}$.*

The concept of resultant can be extended from 2 polynomials in 1 variable (or 2 homogeneous polynomials in 2 variables) to n homogeneous polynomials in n variables. This multipolynomial resultant is known as *Macaulay's resultant*. For n generic homogeneous polynomials $F_1, \ldots F_n$ in the n variables $x_1, \ldots, x_n$, of positive (total) degrees d_i, there exists a multipolynomial resultant R, which is a polynomial in the indeterminate coefficients of the F_i, with the following property. If the coefficients of the F_i are assigned values from a field K, then the vanishing of R is necessary and sufficient for a nontrivial common zero of the F_i to exist in some extension of K. Here we will only give a very brief survey, referring the reader to more comprehensive sources such as [5, 6, 10, 15, 16] for the full story. We will primarily follow the treatment of this topic in [16].

Given r homogeneous polynomials $F_1, \ldots, F_r$ in $x_1, \ldots, x_n$, with indeterminate coefficients comprising a set A, an integral polynomial T in these indeterminates (that is, $T \in \mathbb{Z}[A]$) is called an *inertia form* for $F_1, \ldots, F_r$ if $x_i^\tau T \in (F_1, \ldots, F_r)$, for suitable i and τ.

Van der Waerden observes that the inertia forms comprise an ideal $\mathcal{I}$ of $\mathbb{Z}[A]$, and he shows further that $\mathcal{I}$ is a *prime* ideal of this ring. It follows from these observations that we may take the ideal $\mathcal{I}$ of inertia forms to be a resultant system for the given $F_1, \ldots, F_r$ in the sense that for special values of the coefficients in $\mathcal{K}$, the vanishing of all elements of the resultant system is necessary and sufficient for there to exist a nontrivial solution to the system $F_1 = 0, \ldots, F_r = 0$ in some extension of $\mathcal{K}$.

Now consider the case in which we have n homogeneous polynomials in the same number n of variables. Let $F_1, \ldots, F_n$ be n generic homogeneous forms in $x_1, \ldots, x_n$ of positive total degrees $d_1, \ldots, d_n$. That is, every possible coefficient of each F_i is a distinct indeterminate, and the set of all such indeterminate coefficients is denoted by A. Let $\mathcal{I}$ denote the ideal of inertia forms for $F_1, \ldots, F_n$. Proofs of the following two propositions my be found in [12].

Proposition 5. *$\mathcal{I}$ is a nonzero principal ideal of $\mathbb{Z}[A]$: $\mathcal{I} = (R)$, for some $R \neq 0$.*

R is uniquely determined up to sign. We call R the *(generic multipolynomial) resultant* of $F_1, \ldots, F_n$.

Proposition 6. *The vanishing of $\mathcal{R}$ for particular $F_1, \ldots, F_n$ with coefficients in a field $\mathcal{K}$ is necessary and sufficient for the existence of a nontrivial zero of the system $F_1 = 0, \ldots, F_n = 0$ in some extension of $\mathcal{K}$.*

The above considerations also lead to the notion of a resultant of n non-nonhomogeneous polynomials in $n-1$ variables. For a given non-homogeneous $f(x_1, \ldots, x_{n-1})$ over $\mathcal{K}$ of total degree d, we may write $f = H_d + H_{d-1} + \cdots + H_0$, where the H_j are homogeneous of degree j. Then H_d is known as the *leading form* of f. Recall that the *homogenization* $F(x_1, \ldots, x_n)$ of f is defined by $F = H_d + H_{d-1}x_n + \cdots + H_0 x_n^d$.

Let $f_1, \ldots, f_n$ be particular non-homogeneous polynomials in $x_1, \ldots, x_{n-1}$ over $\mathcal{K}$ of positive total degrees d_i, and with leading forms H_{i,d_i}. We put

$$\operatorname{res}(f_1, \ldots, f_n) = \operatorname{res}(F_1, \ldots, F_n) ,$$

where F_i is the homogenization of f_i. Then we have (see proof in [11]):

Proposition 7. *The vanishing of $\operatorname{res}(f_1, \ldots, f_n)$ is necessary and sufficient for either the forms H_{i,d_i} to have a common nontrivial zero over an extension of $\mathcal{K}$, or the polynomials f_i to have a common zero over an extension of $\mathcal{K}$.*

Observe that the common zeros of the f_i correspond to the affine solutions of the system, whereas the nontrivial common zeros of the leading forms correspond to the projective solutions on the hyperplane at infinity.

3 Basics of differential algebra

Now let us review some basic differential algebra. For this it is enough to assume $\mathcal{R}$ to be a commutative ring with 1. A *derivation* on $\mathcal{R}$ is a mapping $\partial : \mathcal{R} \to \mathcal{R}$ such that $\partial(a + b) = \partial(a) + \partial(b)$ and $\partial(ab) = \partial(a)b + a\partial(b)$ for all $a, b \in \mathcal{R}$. That $\partial(0) = 0$ and $\partial(1) = 0$ follow readily from these axioms. We sometimes denote the derivative of a $\partial(a)$ by a'. Such a ring (or integral domain or field) $\mathcal{R}$ together with a derivation on $\mathcal{R}$ is called a *differential ring (or integral domain or field, respectively)*. In such a ring $\mathcal{R}$ elements r such that $r' = 0$ are known as *constants* and the set C of constants comprises a subring of $\mathcal{R}$. If $\mathcal{R}$ is a field, C is a subfield of $\mathcal{R}$. An ideal I of such a ring $\mathcal{R}$ is known as a *differential ideal* if $r \in I$ implies $r' \in I$. If $r_1, \ldots, r_n \in R$ we denote by $[r_1, \ldots, r_n]$ the differential ideal generated by $r_1, \ldots, r_n$, that is, the ideal generated by the r_i and all their derivatives.

Example 1. **(a)** *The familiar rings such as $\mathbb{Z}$, $\mathbb{Q}$, $\mathbb{R}$ and $\mathbb{C}$ are differential rings if we set $\partial(a) = 0$ for all elements a.*
(b) *Let $\mathcal{K}$ be a field and t an indeterminate over $\mathcal{K}$. Then $\mathcal{K}[t]$, equipped with the derivation $\partial = d/dt$, is a differential integral domain and its quotient field $\mathcal{K}(t)$ is a differential field,*

again with standard differentiation as its derivation. $\mathcal{K}$ *is the ring (field) of constants of* $\mathcal{K}[t]$ ($\mathcal{K}(t)$).

Let $(\mathcal{R}, \partial)$ be a differential ring. Let $x = x^{(0)}, x^{(1)}, x^{(2)}, \ldots$ be distinct indeterminates over $\mathcal{R}$. Put $\partial(x^{(i)}) = x^{(i+1)}$ for all $i \geq 0$. Then ∂ can be extended to a derivation on the polynomial ring $\mathcal{R}\{x\} := \mathcal{R}[x^{(0)}, x^{(1)}, \ldots]$ in a natural way, and we denote this extension also by ∂. The ring $\mathcal{R}\{x\}$ together with this extended ∂ is a differential ring, called the ring of *differential polynomials* in the differential indeterminate x over $\mathcal{R}$. An element $f(x) = \sum_{i=0}^{m} a_i x^{(i)}$ of $\mathcal{R}\{x\}$ with $a_m \neq 0$ has *order* m and *leading coefficient* a_m.

It may be helpful to think of elements of $\mathcal{R}$ and of $x, x^{(1)}, \ldots$ as functions of an indeterminate t, and to regard ∂ as differentiation with respect to t.) If $(\mathcal{K}, \partial)$ is a differential field then $\mathcal{K}\{x\}$ is a differential integral domain, and its derivation extends uniquely to the quotient field. We write $\mathcal{K}\langle x\rangle$ for this quotient field; its elements are *differential rational functions* of x over $\mathcal{K}$. We similarly denote by $\mathcal{K}\langle\eta\rangle$ the differential extension of $\mathcal{K}$ by η, where η lies in some differential field containing $\mathcal{K}$.

We consider the ring of linear differential operators $\mathcal{R}[\partial]$, where the application of $A = \sum_{i=0}^{m} a_i \partial^i$ to $r \in \mathcal{R}$ is defined as

$$A(r) = \sum_{i=0}^{m} a_i r^{(i)} .$$

Here $r^{(i)}$ denotes the i-fold application of ∂ (that is, $'$) to r. If $a_m \neq 0$, the *order* of A is m and a_m is the *leading coefficient* of A. Now the application of A can naturally be extended to $\mathcal{K}$, and to any extension of $\mathcal{K}$. If $A(\eta) = 0$, with η in $\mathcal{R}$, $\mathcal{K}$ or any extension of $\mathcal{K}$, we call η a *root* of the linear differential operator A.

The ring $\mathcal{R}[\partial]$ is non-commutative, satisfying the relation

$$\partial r = r\partial + r' .$$

From a linear homogeneous ODE $p(x) = 0$, with $p(x) \in \mathcal{R}\{x\}$,

$$p(x) = p_0(t)x + p_1(t)x' + \cdots + p_n(t)x^{(n)} = 0 ,$$

we can extract a linear differential operator

$$O(p) = A = \sum_{i=0}^{n} p_i \partial^i ,$$

such that the given ODE can be written as $A(x) = 0$, in which x is regarded as an unknown element of $\mathcal{R}$, $\mathcal{K}$ or some extension of $\mathcal{K}$. Such a linear homogeneous ODE always has the trivial solution $x = 0$; so a linear differential operator always has the trivial root 0. The ring $\mathcal{K}[\partial]$ is left-Euclidean (see [3]), so every left-ideal ${}_{\mathcal{K}}I$ of the form ${}_{\mathcal{K}}I = (A, B)$ is principal, and is generated by the right-gcd of A and B. As remarked in [3] with reference to [7], under suitable assumptions on $\mathcal{K}$, any linear differential operator of positive order has a root in some extension of $\mathcal{K}$. We state this result precisely.

Theorem 1 (Ritt-Kolchin). *Assume that the differential field* $\mathcal{K}$ *has characteristic 0 and that its field* C *of constants is algebraically closed. Then, for any linear differential operator* A *over* $\mathcal{K}$ *of positive order* n*, there exist* n *roots* $\eta_1, \ldots, \eta_n$ *in a suitable extension of* $\mathcal{K}$*, such that the* η_i *are linearly independent over* C*. Moreover, the field* $\mathcal{K}\langle\eta_1, \ldots, \eta_n\rangle$ $(= \mathcal{K}\langle\eta_1\rangle \ldots \langle\eta_n\rangle)$ *contains no constant not in* C.

This result is stated and proved in [9] using results from [8] and [13]. The field $\mathcal{K}\langle\eta_1,\ldots,\eta_n\rangle$ associated with A is known as a *Picard-Vessiot extension* of $\mathcal{K}$ (for A). Henceforth assume the hypotheses of the Ritt-Kolchin Theorem.

It follows from the Ritt-Kolchin Theorem that if the operators $A, B \in \mathcal{K}[\partial]$ have a common factor F of positive order on the right, i.e.,

$$A = \overline{A}\cdot F, \quad \text{and } B = \overline{B}\cdot F, \tag{6}$$

then they have a non-trivial common root in a suitable extension of $\mathcal{K}$. On the other hand, if A and B have a non-trivial common root η in a suitable extension of $\mathcal{K}$, we see from the properties of a left-Euclidean ring, that they have a common right factor of positive order in $\mathcal{K}[\partial]$. We summarize this in the following proposition.

Proposition 8. *Assume that $\mathcal{K}$ has characteristic 0 and that its field of constants is algebraically closed. Let A, B be differential operators of positive orders in $\mathcal{K}[\partial]$. Then the following are equivalent:*

(i) *A and B have a common non-trivial root in an extension of $\mathcal{K}$,*

(ii) *A and B have a common factor of positive order on the right in $\mathcal{K}[\partial]$.*

The existence of a non-trivial factor (6) is equivalent to the existence of a non-trivial order-bounded linear combination

$$CA + DB = 0\,, \tag{7}$$

with order(C) < order(B) and order(D) < order(A), and $(C, D) \neq (0, 0)$.
For given $A, B \in \mathcal{K}[\partial]$, with m = order(A), n = order(B), consider the linear map

$$\begin{array}{llll} S: & \mathcal{K}^{m+n} & \longrightarrow & \mathcal{K}^{m+n} \\ & (c_{n-1},\ldots,c_0,d_{m-1},\ldots,d_0) & \mapsto & \text{coefficients of } CA+DB \end{array} \tag{8}$$

Obviously the existence of a non-trivial linear combination (7) is equivalent to S having a non-trivial kernel, and therefore to S having determinant 0. Indeed we have the following result.

Proposition 9. $\det(S) = 0$ *if and only if A and B have a common factor (on the right) in $\mathcal{K}[\partial]$ of positive order.*

4 The differential Sylvester resultant

4.1 Resultant of two linear differential operators

So let us see which linear conditions on the coefficients of A and B we get by requiring that (7) has a non-trivial solution of bounded order, i.e.,

$$\text{order}(C) < \text{order}(B) \quad \text{and} \quad \text{order}(D) < \text{order}(A).$$

Example 2. We consider a differential operator A of order 2 and another operator B of order 3. We try to find an operator C of order 2 and an operator D of order 1 s.t. $C\cdot A + D\cdot B = 0$. We make an ansatz and compare coefficients of order 4 down to order 0.

$$(c_2\partial^2 + c_1\partial + c_0)(a_2\partial^2 + a_1\partial + a_0) \;+\; (d_1\partial + d_0)(b_3\partial^3 + b_2\partial^2 + b_1\partial + b_0)$$

order 4:

$$c_2\partial^2 a_2\partial^2 + d_1\partial b_3\partial^3 = 0$$

$$\underline{a_2c_2\partial^4} + 2a_2'c_2\partial^3 + a_2''c_2\partial^2 + \underline{b_3d_1\partial^4} + b_3'd_1\partial^3 = 0$$

order 3:

$$(2a_2'c_2\partial^3 + a_2''c_2\partial^2 \text{ from above}) + c_2\partial^2 a_1\partial + c_1\partial a_2\partial^2 + (b_3'd_1\partial^3 \text{ from above}) + d_1\partial b_2\partial^2 + d_0b_3\partial^3 = 0$$

$$\underline{2a_2'c_2\partial^3} + \underline{a_1c_2\partial^3} + \underline{a_2c_1\partial^3} + a_2''c_2\partial^2 + 2a_1'c_2\partial^2 + a_1''c_2\partial + \underline{b_3'd_1\partial^3} + \underline{b_2d_1\partial^3} + \underline{b_3d_0\partial^3} + b_2'd_1\partial^2 = 0$$

order 2:

$$(a_2''c_2\partial^2 + 2a_1'c_2\partial^2 + a_1''c_2\partial + a_2'c_1\partial^2 \text{ from above}) + c_2\partial^2 a_0 + c_1\partial a_1\partial + c_0a_2\partial^2$$
$$+(b_2'd_1\partial^2 \text{ from above}) + d_1\partial b_1\partial + d_0b_2\partial^2 = 0$$

$$\underline{a_2''c_2\partial^2} + \underline{2a_1'c_2\partial^2} + a_1''c_2\partial + \underline{a_2'c_1\partial^2} + \underline{a_0c_2\partial^2} + 2a_0'c_2\partial + a_0''c_2 + \underline{a_1c_1\partial^2} + a_1'c_1\partial + \underline{a_2c_0\partial^2}$$
$$+\underline{b_2'd_1\partial^2} + \underline{b_1d_1\partial^2} + b_1'd_1\partial + \underline{b_2d_0\partial^2} = 0$$

order 1:

$$(a_1''c_2\partial + 2a_0'c_2\partial + a_0''c_2 + a_1'c_1\partial \text{ from above}) + c_1\partial a_0 + c_0a_1\partial + (b_1'd_1\partial \text{ from above}) + d_1\partial b_0 + d_0b_1\partial = 0$$

$$\underline{a_1''c_2\partial} + \underline{2a_0'c_2\partial} + \underline{a_1'c_1\partial} + \underline{a_0c_1\partial} + a_0'c_1 + \underline{a_1c_0\partial} + a_0''c_2 + \underline{b_1'd_1\partial} + \underline{b_0d_1\partial} + b_0'd_1 + \underline{b_1d_0\partial} = 0$$

order 0:

$$(\underline{a_0'c_1} + \underline{a_0''c_2} \text{ from above}) + \underline{a_0c_0} + (\underline{b_0'd_1} \text{ from above}) + \underline{b_0d_0} = 0.$$

So, finally

$$\begin{pmatrix} c_2 & c_1 & c_0 & d_1 & d_0 \end{pmatrix} \cdot \begin{pmatrix} a_2 & a_1 + 2a_2' & a_0 + 2a_1' + a_2'' & 2a_0' + a_1'' & a_0'' \\ 0 & a_2 & a_1 + a_2' & a_0 + a_1' & a_0' \\ 0 & 0 & a_2 & a_1 & a_0 \\ b_3 & b_2 + b_3' & b_1 + b_2' & b_0 + b_1' & b_0' \\ 0 & b_3 & b_2 & b_1 & b_0 \end{pmatrix} = \begin{pmatrix} 0 & \cdots & 0 \end{pmatrix}.$$

Observe, that the rows of this matrix consist of the coefficients of

$$\partial^2 A,\ \partial A,\ A, \partial B,\ B\,.$$

Theorem 2. *The linear map S in (8) corresponding to (7) is given by the matrix whose rows are $\partial^{n-1}A, \ldots, \partial A,\ A, \partial^{m-1}B, \ldots, \partial B,\ B$.*

Definition 1. *Let A, B be linear differential operators in $\mathcal{R}[\partial]$ of* $\text{order}(A) = m, \text{order}(B) = n$, *with $m, n > 0$.*
By $\partial\text{syl}(A, B)$ we denote the (differential) Sylvester matrix: *i.e., the $(m+n) \times (m+n)$-matrix whose rows contain the coefficients of*

$$\partial^{n-1}A, \ldots, \partial A,\ A, \partial^{m-1}B, \ldots, \partial B,\ B\,.$$

The (differential Sylvester) resultant *of A and B, $\partial\text{res}(A, B)$, is the determinant of $\partial\text{syl}(A, B)$.*

From Propositions 8 and 9 the following analogue of Proposition 2 is immediate.
Theorem 3. *Assume that $\mathcal{K}$ has characteristic 0 and that its field of constants is algebraically closed. Let A, B be linear differential operators over $\mathcal{R}$ of positive orders. Then the condition $\partial\text{res}(A, B) = 0$ is both necessary and sufficient for there to exist a common non-trivial root of A and B in an extension of $\mathcal{K}$.*

We close this subsection by stating an analogue of Proposition 3.
Theorem 4. *Let $A, B \in \mathcal{R}[\partial]$. The resultant of A and B is contained in (A, B), the ideal generated by A and B in $\mathcal{R}[\partial]$. Moreover, $\partial\text{res}(A, B)$ can be written as a linear combination $\partial\text{res}(A, B) = CA + DB$, with* $\text{order}(C) < \text{order}(B)$, *and* $\text{order}(D) < \text{order}(A)$.

4.2 Resultant of two linear homogeneous differential polynomials

The results for differential resultants which we have derived for linear differential operators can also be stated in terms of linear homogeneous differential polynomials. Such a treatment facilitates the generalization to the non-linear algebraic differential case.

Let $(\mathcal{R}, \partial)$ be a differential domain with quotient field $\mathcal{K}$. Then elements of $\mathcal{R}\{x\}$ can be interpreted as algebraic ordinary differential equations (AODEs). For instance, the differential polynomial

$$3\,x\,x^{(1)} + 2t\,x^{(2)} \in \mathbb{C}(t)\{x\}$$

corresponds to the AODE

$$3x(t)x'(t) + 2tx''(t) = 0\,.$$

The next proposition says that linear differential operators correspond to linear homogeneous differential polynomials $\mathcal{R}_{LH}\{x\}$ in a natural way. $\mathcal{R}[\partial]$ and $\mathcal{R}_{LH}\{x\}$ are isomporphic as left $\mathcal{R}$-modules and $\mathcal{K}[\partial]$ and $\mathcal{K}_{LH}\{x\}$ are isomporphic as left vector spaces over $\mathcal{K}$.

Definition 2. *Let $f(x)$ and $g(x)$ be elements of $\mathcal{R}_{LH}\{x\}$ of positive orders m and n, respectively. Then the* (differential) Sylvester matrix *of $f(x)$ and $g(x)$, denoted by $\partial\mathrm{syl}(f, g)$, is $\partial\mathrm{syl}(A, B)$, where $A = O(f)$ and $B = O(g)$. The* (differential Sylvester) resultant *of $f(x)$ and $g(x)$, denoted by $\partial\mathrm{res}(f, g)$, is $\partial\mathrm{res}(A, B)$.*

We may observe that the $m + n$ rows of $\partial\mathrm{syl}(f, g)$ contain the coefficients of

$$f^{(n-1)}(x), \ldots, f^{(1)}(x), f(x), g^{(m-1)}(x), \ldots, g^{(1)}(x), g(x).$$

The following analogue and slight reformulation of Theorem 3 is immediate.

Theorem 5. *Assume that $\mathcal{K}$ has characteristic 0 and that its field of constants is algebraically closed. Let $f(x), g(x)$ be linear homogeneous differential polynomials of positive orders over $\mathcal{R}$. Then the condition $\partial\mathrm{res}(f, g) = 0$ is both necessary and sufficient for there to exist a common non-trivial solution of $f(x) = 0$ and $g(x) = 0$ in an extension of $\mathcal{K}$.*

We have also an analogue and slight reformulation of Theorem 4:

Theorem 6. *Let $f(x), g(x) \in \mathcal{R}_{LH}\{x\}$. Then $x\partial\mathrm{res}(f, g)$ is contained in the differential ideal $[f, g]$.*

4.3 Resultant of two arbitrary differential polynomials

Finally we consider two arbitrary differential polynomials and we review Carra-Ferro's adaptation of the multipolynomial resultant to a pair of algebraic ordinary differential equations (AODEs) [2]. Such AODEs can be described by differential polynomials. We will deal with both homogeneous and non-homogeneous AODEs.

Suppose first we are given 2 homogeneous AODEs in the form of 2 homogeneous differential polynomial equations over a differential field $\mathcal{K}$. Observe that a homogeneous AODE of positive order always has the solution 0. So we are interested in determining whether such a pair of homogeneous AODEs has a non-trivial common solution. Denote the given homogeneous AODEs by:

$$\begin{aligned} F(x) &= 0, &\quad& \text{of order } m\,, \\ G(x) &= 0, && \text{of order } n\,. \end{aligned} \tag{9}$$

So the differential polynomial $F(x) \in \mathcal{K}\{x\}$ is of the form $F(x, x^{(1)}, \ldots, x^{(m)})$; and $G(x) \in \mathcal{K}\{x\}$ is of analogous form.

The system (9) has the same solution set as the system

$$\begin{array}{ll} F^{(n-1)}(x) = \cdots = F^{(1)} = F(x) = 0, & n \text{ equations}, \\ G^{(m-1)}(x) = \cdots = G^{(1)} = G(x) = 0, & m \text{ equations}. \end{array} \tag{10}$$

This system (10) contains the variables $x, x^{(1)}, \ldots, x^{(m+n-1)}$. So it is a system of $m + n$ homogeneous equations in $m + n$ variables. Considered as a system of homogeneous algebraic equations (with the $x^{(i)}$ considered as unrelated indeterminates), it has a multipolynomial resultant $\mathrm{res}(F^{(n-1)}, \ldots, F, G^{(m-1)}, \ldots, G)$ (defined in Subsection 4.4) whose vanishing gives a necessary and sufficient condition for the existence of a non-trivial solution over an extension of $\mathcal{K}$.

Definition 3. *For such homogeneous differential polynomials $F(x), G(x)$, we define the* (differential) resultant $\partial\mathrm{res}(F, G)$ *to be the multipolynomial resultant* $\mathrm{res}(F^{(n-1)}, \ldots, F, G^{(m-1)}, \ldots, G)$.

But, whereas a solution to the differential problem is also a solution to the algebraic problem, the converse is not true. So we do not expect the vanishing of this resultant to be a sufficient condition for the existence of a nontrivial solution to (9).

Example 3. Consider Example 4 in [2], p. 554.

$$\begin{array}{rcll} F(x) & = & xx^{(1)} - x^2 & = 0\,, \\ G(x) & = & xx^{(1)} & = 0\,. \end{array}$$

The corresponding system (10) would be the same in this case. Whereas the differential problem only has the trivial solution $x = x^{(1)} = 0$, the corresponding algebraic problem has also the non-trivial solutions $(0, a)$, for a in $\mathcal{K}$.

Indeed, in this case the differential resultant coincides with the Sylvester resultant: $\partial\mathrm{res}(F, G) = \mathrm{res}(F, G) = 0$. This reflects the fact that there are non-trivial algebraic solutions. But $x = 0$ does not lead to a non-trivial differential solution.

The following theorem follows from Proposition 6.

Theorem 7. *For such homogeneous differential polynomials $F(x), G(x)$, the vanishing of $\partial\mathrm{res}(F, G)$ is a necessary condition for the existence of a non-trivial common solution of the system $F(x) = 0, G(x) = 0$ in an extension of $\mathcal{K}$.*

Next we consider the more general case of a pair of non-homogeneous AODEs $f(x) = 0, g(x) = 0$ over $\mathcal{K}$, of orders m and n, respectively. This system has the same solution set as the system

$$\begin{array}{ll} f^{(n)}(x) = \cdots = f^{(1)} = f(x) = 0, & n+1 \text{ equations}, \\ g^{(m)}(x) = \cdots = g^{(1)} = g(x) = 0, & m+1 \text{ equations}. \end{array} \tag{11}$$

This system contains the variables $x, x^{(1)}, \ldots, x^{(m+n)}$. So it is a system of $m + n + 2$ non-homogeneous equations in $m + n + 1$ variables. Considered as a system of non-homogeneous algebraic equations (with the $x^{(i)}$ considered as unrelated indeterminates), it has a multipolynomial resultant $\mathrm{res}(f^{(n)}, \ldots, f, g^{(m)}, \ldots, g)$ (defined in Section 2) whose vanishing gives a necessary condition for the existence of a common solution to the system in an extension of $\mathcal{K}$.

Definition 4. *For such differential polynomials $f(x), g(x)$, we define the* (differential) resultant $\partial\mathrm{res}(f, g)$ *to be the multipolynomial resultant* $\mathrm{res}(f^{(n)}, \ldots, f, g^{(m)}, \ldots, g)$.

The following theorem follows from Proposition 7.

Theorem 8. *For such differential polynomials $f(x), g(x)$, the vanishing of $\partial\text{res}(f, g)$ is a necessary condition for the existence of a common solution of the system $f(x) = 0, g(x) = 0$ in an extension of $\mathcal{K}$.*

In the comprehensive resource [18] by Zwillinger on differential equations there appears a short section entitled "Differential Resultants". The concept of differential resultants is briefly introduced and an example of two coupled differential equations in two differential variables x and z is given, for which a differential resultant in only z is determined. However, no precise definition of the concept is given. Carra'-Ferro in [1] treats the same problem and arrives at the same answer. The treatment in both these papers is consistent with our Theorem 8.

5 Conclusion

We have reviewed formulations of different notions of resultants, for differential operators and differential polynomials. We have extended the definition of a resultant for linear homogeneous differential polynomials to the case of arbitrary differential polynomials. With this generalization, the formulations of Chardin and Carra'-Ferro are seen to be intimately related. The application of these ideas to the problem of Zwillinger for a system of coupled differential equations remains to be further investigated.

Acknowledgement
The second author wants to acknowledge partial support by the Austrian Science Fund (FWF): [P 31327-N32].

References

[1] Carra'-Ferro, G. A resultant theory for systems of linear PDEs. In *Proc. of Modern Group Analysis*, 1994.

[2] Carra'-Ferro, G. A resultant theory for the systems of two ordinary algebraic differential equations. *AAECC 8/6*, 539–560, 1997.

[3] Chardin, M. Differential resultants and subresultants. In *Proc. Fundamentals of Computation Theory 1991*. In *LNCS Vol. 529*, Springer-Verlag, 1991.

[4] Collins, G.E. The calculation of multivariate polynomial resultants. *J.ACM 18/4*, 515–532, 1971.

[5] Cox, D., Little, J., O'Shea, D. *Using Algebraic Geometry*, Springer, 1998.

[6] Gelfand, I., Kapranov, M., Zelevinsky, A. *Discriminants, Resultants and Multidimensional Determinants*, Birkhauser, Boston, 1994.

[7] Kaplansky, I. *An Introduction to Differential Algebra*, Hermann, 1957.

[8] Kolchin, E.R. Algebraic matric groups and the Picard-Vessiot theory of homogeneous linear ODEs. *Ann. of Math. 49*, 1–42, 1948.

[9] Kolchin, E.R. Existence theorems connected with the Picard-Vessiot theory of homogeneous linear ODEs. *Bull. Amer. Math. Soc. 54*, 927–932, 1948.

[10] Macaulay, F.S. *The Algebraic Theory of Modular Systems*, 1916. Reprinted in 1994 by Cambridge Univ. Press.

[11] McCallum, S. Factors of iterated resultants and discriminants. *J.Symb.Comp. 27*, 367–385, 1999.

[12] McCallum, S., Winkler, F. Resultants: Algebraic and Differential. Techn.Rep. RISC-18-08, J.Kepler University, Linz, Austria, 2018.

[13] Ritt, J. F. *Differential Equations from the Algebraic Standpoint.* AMS Coll. Publ. Vol. 14, New York, 1932.

[14] Rueda, S., Sendra, J.R. Linear complete differential resultants and the implicitization of linear DPPEs. *J.Symbolic Computation 45/3*, 324–341, 2010.

[15] Sturmfels, B. Introduction to resultants. In *Applications of Computational Algebraic Geometry.* Lecture Notes of the AMS, Short Course Series, 128–142, AMS, 1996.

[16] van der Waerden, B.L. *Modern Algebra, Vol. II* Ungar, New York, 1950.

[17] Zhang, Z.-Y., Yuan, C.-M., Gao, X.-S. Matrix formulae of differential resultant for first order generic ordinary differential polynomials. *Computer Mathematics*, R. Feng et al. (eds.), 479–503, 2014.

[18] Zwillinger, D. *Handbook of Differential Equations Third Edn.* Academic Press, 1998.

An algebraic method for quasi-linear first-order ODEs

Thieu N. Vo[1,*]

[1]Fractional Calculus, Optimization and Algebra Research Group, Faculty of Mathematics and Statistics, Ton Duc Thang University, Ho Chi Minh City, Vietnam

Abstract. In this paper, we consider the class of quasi-linear first-order ODEs of the form $y' = P(x, y)$, where P is a polynomial in y with coefficients in $\mathbb{C}(x)$, and study their algebraic solutions. Our method is intrinsically based on algebraic function field theory. We give an upper bound for the degrees of algebraic solutions which are in a quadratic extension of $\mathbb{C}(x)$.

1 Introduction

A quasi-linear first-order ODE is a differential equation of the form $y' = f(x, y)$ for some rational function $f \in \mathbb{C}(x, y)$. In this paper, we are interested in its algebraic solutions lying in a quadratic extension of $\mathbb{C}(x)$.

The study of quasi-linear first-order ODEs has a long history which can be traced back to the works of Fuchs [5] and Poincaré [7]. There are many solution methods for special classes of first-order ODEs. In case $f(x, y)$ is a polynomial of degree 2 in y, the differential equation is called a Riccati equation. Riccati equations form the simplest class of nonlinear first-order ODEs. In [6], Kovacic presents a complete algorithm for determining all algebraic solutions of a Riccati equation. The algorithm by Kovacic is based on a systematic study of differential Galois groups for second-order linear ordinary differential equations with rational function coefficients.

For general quasi-linear first-order ODEs, the problem of determining an algebraic solution still remains open. In case an upper bound for the degrees of the algebraic solutions is given, there are efficient algorithms to determine the solutions, such as [8] or [1]. The problem of determining an upper bound for the degrees of the algebraic solutions of a quasi-linear first-order ODE is one of the equivalent form of the Poincaré problem (see [1]). In [2], Carnicer gives an upper bound for the degrees of irreducible invariant algebraic curves of a planar polynomial vector field provided that the singularities of the curves are not too bad. This leads to a generic, but not complete, answer for the Poincaré problem.

In this paper, we modify a method developed in [10] for studying algebraic solutions of quasi-linear first-order ODEs. The method is based on the theory of algebraic function fields. In particular, we view the unknown function $y(x)$ and its derivative $y'(x)$ as elements in a suitable algebraic function field over $\mathbb{C}$. Standing from the algebraic function field theory point of view, since $y(x)$ and $y'(x)$ are algebraically dependent, certain relations between the poles of the principal divisors defined by them can be encoded (see [4]). In [10], the authors use this method to establish an algorithm for determining all solutions in the field $\mathbb{C}(x)$ of

*e-mail: vongocthieu@tdtu.edu.vn

rational functions, if there is any, of the differential equation. In this paper, we apply this method for a wider class of algebraic solutions, namely the class of algebraic solutions which are in a quadratic extension of $\mathbb{C}(x)$. We obtain a degree bound for algebraic solutions which are in a quadratic extension of $\mathbb{C}(x)$ of a quasi-linear first-order ODE $y' = P(x, y)$, where P is a polynomial of degree at least 5 in y with coefficients in $\mathbb{C}(x)$ (see Theorem 4).

The rest of the paper is organized as follows. In Section 2, we recall several definitions and basic facts from the theory of algebraic functions. Further study for a quadratic extension of $\mathbb{C}(x)$ is presented in Section 3. In Section 4, we apply facts presented in the previous sections to determine a degree bound for algebraic solutions which are in a quadratic extension of $\mathbb{C}(x)$ of a class of quasi-linear first-order ODEs.

2 Algebraic function fields

In this section, we recall basic notions and facts from algebraic function field theory. For more details, we refer the readers to [9].

We use the symbol $\mathbb{K}$ to denote an algebraically closed field of characteristic zero. In this paper, $\mathbb{K}$ can be the field $\mathbb{C}$ of complex numbers or the field $\overline{\mathbb{C}(x)}$ of algebraic functions in the variable x over $\mathbb{C}$. A field F containing $\mathbb{K}$ is called an *algebraic function field* over $\mathbb{K}$ if it has transcendence degree one over $\mathbb{K}$. For example, the field $\mathbb{K}(x)$ of rational functions in the variable x with coefficients in $\mathbb{K}$ is an algebraic function field over $\mathbb{K}$. Finite extensions of $\mathbb{K}(x)$ are also algebraic function fields over $\mathbb{K}$.

A *valuation ring* in F (over $\mathbb{K}$) is a ring O such that $\mathbb{K} \subsetneq O \subsetneq F$ and for every $y \in F$ either $y \in O$ or $y^{-1} \in O$. Since a valuation is a local ring, it admits a unique maximal ideal, which is the set of its non-invertible elements. Let P be the unique maximal ideal of O. Then P is principal, and every ideal of O is a power of P. Assume that P is generated by t, i.e. $P = tO$, then every element $y \in F \setminus \{0\}$ can be written uniquely as $y = u \cdot t^n$ for some unit u of O and $n \in \mathbb{Z}$. Note that the power n does not depend on how t is chosen. We call n the *valuation of y at P*, and denote it by $v_P(y)$. As a convention, we set $v_P(0) = \infty$.

A *prime divisor* of F (over $\mathbb{K}$) is defined to be the maximal ideal of a valuation ring of F (over $\mathbb{K}$). By $\mathbb{P}_{F/\mathbb{K}}$ we denote the set of all prime divisors of F. Due to [9, Theorem 1.1.13, p. 5], there is a one-to-one correspondence between valuation rings and prime divisors of F. Let $P \in \mathbb{P}_F$ be a prime divisor of F, and O_P the corresponding valuation ring. Since $\mathbb{K}$ is algebraically closed, the residue field O_P/P is isomorphic to $\mathbb{K}$. Therefore, there is a natural ring projection $O \to O/P \equiv \mathbb{K}$. We usually denote the image of $y \in O$ in $\mathbb{K}$ by $y(P)$. In case $F \subset \overline{\mathbb{K}(x)}$ is a finite extension of $\mathbb{K}(x)$, the projection map is exactly the usual evaluation map.

Let $y \in F$, and $P \in \mathbb{P}_{F/\mathbb{K}}$ a prime divisor of F. If $v_P(y) > 0$, we say that y has a zero of order $v_P(y)$ at P, and that P is a zero of y. In case $v_P(y) < 0$, we say that y has a pole of order $-v_P(y)$ at P, and that P is a pole of y. Every elements of F has only finitely many zeros and poles. Furthermore, the number of zeros of y is always equal to the number of poles of y, counting multiplicities.

Let $\mathbf{Div}_{\mathbb{K}}(F) = \bigoplus_{P \in \mathbb{P}_{F/\mathbb{K}}} \mathbb{Z}P$ be the set of all finitely many linear combinations of prime divisors of F. $\mathbf{Div}_{\mathbb{K}}(F)$ forms a free abelian group with the basis $\mathbb{P}_{F/\mathbb{K}}$. Elements of $\mathbf{Div}_{\mathbb{K}}(F)$ are called *divisors* of F (over $\mathbb{K}$). For a divisor $\delta = n_1P_1 + n_2P_2 + \ldots + n_rP_r$, we define $\deg \delta = n_1 + \ldots + n_r$ as the degree of δ. We define a partial order in $\mathbf{Div}_{\mathbb{K}}(F)$ as follows: for divisors δ, δ', we say $\delta \geq \delta'$ if all coefficients of $\delta - \delta'$ are non-negative. A divisor whose coefficients are all non-negative is called *effective*.

Principal divisors are essential examples of divisors. For each $x \in F^*$, the principal divisor of x in F is denoted by $(x) = \sum_{P \in \mathbb{P}_F} v_P(x)P$. Note that, since every element in F^* has only finitely many zeros and poles, the sum is always finite. It is important for us to gather information about poles of x. In order to do that, we set

$$(x)_\infty = -\sum_{\substack{P\in\mathbb{P}_F \\ v_P(x)<0}} v_P(x)P.$$

The simplest algebraic function field over $\mathbb{K}$ is the field $\mathbb{K}(x)$ of rational functions for some x transcendental over $\mathbb{K}$. In this case, there is a one-to-one correspondence between prime divisors of $\mathbb{K}(x)$ and the set $\mathbb{K}\cup\{\infty\}$. Therefore, we might identify $\mathbb{P}_{\mathbb{K}(x)}$ with $\mathbb{K}\cup\{\infty\}$. For a rational function $y(x)\in\mathbb{K}(x)$, an element $x_0\in\mathbb{K}$ is a pole of $y(x)$ if it is a root of the denominator of $y(x)$. The multiplicity of the root is the order of the pole. The infinity ∞ is a pole of $y(x)$ if and only if the degree of the denominator of $y(x)$ is larger than the degree of the numerator. In this case, the difference of the degrees is the order of the pole.

3 Quadratic extensions of $\mathbb{C}(x)$

In this section, we study the number of poles of an algebraic function in a quadratic extension of $\mathbb{C}(x)$. The result of this section is essential in the determination of an upper bound for algebraic solutions of nonlinear first-order differential equations in the next section. We also revisit a result in [10] which is a key step in the algebraic function field approach for differential equations.

Lemma 1. *Let $F\subset\overline{\mathbb{C}(x)}$ be a quadratic extension of $\mathbb{C}(x)$, and $y=y(x)\in F\setminus\mathbb{C}(x)$. Let $P(X,Y)\in\mathbb{C}[X,Y]$ be an irreducible polynomial such that $P(x,y)=0$. Then*

1. $\deg_Y P = 2$,
2. $\deg(y)_\infty = \deg_X P$, *and*
3. $\deg(\frac{dy}{dx})_\infty \le 4\deg_X P$.

Proof. 1. This item follows from the fact that F is an extension of degree 2 over $\mathbb{C}(x)$.

2. By [9, Theorem 1.4.11, p. 19], we have $\deg(y)_\infty = [F:\mathbb{C}(y)]$. Since $y\notin\mathbb{C}(x)$, $F=\mathbb{C}(x,y)$. Therefore $\deg(y)_\infty$ is equal to the degree of a minimal polynomial of x over $\mathbb{C}(y)$, which is $\deg_X P$.

3. Taking the derivative both sides of the equation $P(x,y)=0$, we obtain

$$\frac{\partial P}{\partial X}(x,y) + \frac{dy}{dx}\cdot\frac{\partial P}{\partial Y}(x,y) = 0.$$

Let $R(X,Z)$ be the resultant of $P(X,Y)$ and $\frac{\partial P}{\partial X}(X,Y) + Z\cdot\frac{\partial P}{\partial Y}(X,Y)$ with respect to Y. From the definition of the resultant, since $\deg_Y P = 2$, the polynomial $R(X,Z)$ is the determinant of a Sylvester matrix of size 4×4 (see [3, Definition 2, p. 162]). The entries of the matrix are coefficients of the polynomials $P(X,Y)$ and $\frac{\partial P}{\partial X}(X,Y) + Z\cdot\frac{\partial P}{\partial Y}(X,Y)$ which are treated as polynomials in Y with coefficients in $\mathbb{C}[X,Z]$. Therefore, $\deg_X R(X,Z)\le 4\deg_X P(X,Y)$. By [3, Proposition 5, p. 164], we have $R(x,\frac{dy}{dx})=0$. Therefore, the minimal polynomial of x over $\mathbb{C}\left(\frac{dy}{dx}\right)$ is a factor of R. Hence, the degree of the divisor $(\frac{dy}{dx})_\infty$, which is equal to the degree of the minimal polynomial of x over $\mathbb{C}\left(\frac{dy}{dx}\right)$, is at most $4\deg_X P$.

□

The following theorem is the key step in the computation of rational solutions of nonlinear first-order ODEs in [10]. A different version of it can be found in [4].

Theorem 2 (See [11]). *Let K be an algebraic function field over $\overline{\mathbb{C}(x)}$, and $u, v \in K$ such that $(u)_\infty \leq (v)_\infty$. Let $G(Y,Z) \in \overline{\mathbb{C}(x)}[Y,Z]$ be an irreducible polynomial such that $G(u,v) = 0$. Consider a finite extension $L \subset \overline{\mathbb{C}(x)}$ of $\mathbb{C}(x)$ containing coefficients of G. There exists an effective divisor $\delta \in \mathbf{Div}_{\mathbb{C}}(L)$, depending only on u and v, such that for every prime divisor $P \in \mathbb{P}_{K/\overline{\mathbb{C}(x)}}$, if $u(P)$ and $v(P)$ lie in L, then*

$$(u(P))_\infty \leq (v(P))_\infty + \delta,$$

as divisors in $\mathbf{Div}_{\mathbb{C}}(L)$.

In [10], the authors apply the rational function version of the above theorem to the problem of determining rational solutions of differential equations. The following is another consequence applying to the class of algebraic functions in a quadratic extension of $\mathbb{C}(x)$.

Corollary 3. *Let $u = u(x,y), v = v(x,y) \in \mathbb{C}(x,y)$ such that $(u)_\infty \leq (v)_\infty$ as divisors in $\mathbf{Div}_{\overline{\mathbb{C}(x)}}(\overline{\mathbb{C}(x)}(y))$. Let $F \subset \overline{\mathbb{C}(x)}$ be a quadratic extension of $\mathbb{C}(x)$. Then there exists a constant C, depending only on u and v, such that for every $y(x) \in F$, if $u(x, y(x))$ and $v(x, y(x))$ are defined then*

$$\deg(u(x, y(x)))_\infty \leq \deg(v(x, y(x)))_\infty + C.$$

4 A class of nonlinear first-order ODEs and their algebraic solutions in a quadratic extension of $\mathbb{C}(x)$

In this section, we consider a class of nonlinear first-order ODEs and study their solutions lying in a quadratic extension of $\mathbb{C}(x)$. We associate for each algebraic function $y(x) \in \overline{\mathbb{C}(x)}$ an irreducible polynomial $P(X,Y) \in \mathbb{C}[X,Y]$ such that $P(x, y(x)) = 0$. The polynomial P will be called the *annihilating polynomial* of $y(x)$. Note that the annihilating polynomial of an algebraic function is unique up to multiplication by a nonzero complex number. In this case, we shall call the degree of P in X to be the degree of $y(x)$, i.e. $\deg y(x) := \deg_X P$.

Theorem 4. *Consider the differential equation*

$$y' = P(x, y), \tag{1}$$

where $P(x,y) \in \mathbb{C}[x,y]$ and $\deg_y P \geq 5$. Then there exists a constant C, depending only on P, such that for every solution $y(x)$ lying in a quadratic extension of $\mathbb{C}(x)$, we have $\deg y(x) \leq C$.

Proof. Let $n = \deg_x P$ and consider two rational functions $u(x,y) = y^n$ and $v(x,y) = P(x,y)$ in $\mathbb{C}(x,y)$. As elements in the algebraic function field $\overline{\mathbb{C}(x)}(y)$ over $\overline{\mathbb{C}(x)}$, u and v have pole only at infinity with the same degree. Therefore they meet the requirement of Corollary 3. Let us fix a quadratic extension $F \subset \overline{\mathbb{C}(x)}$ of $\mathbb{C}(x)$. Then F is an algebraic function field over $\mathbb{C}$. Due to Corollary 3, there exists a constant $\tilde{C}$, depending only on P, such that for every algebraic function $y(x) \in F$, we have

$$\deg(u(x, y(x)))_\infty \leq \deg(v(x, y(x)))_\infty + \tilde{C}.$$

Or equivalently,

$$n \cdot \deg(y(x))_\infty \leq \deg(P(x, y(x)))_\infty + \tilde{C}. \tag{2}$$

If, furthermore, $y(x)$ is a solution of the differential equation (1), then the right hand side of the above inequality is equal to $\deg(\frac{dy(x)}{dx})_\infty + \tilde{C}$. By Lemma 1, it can be upper bounded by $4 \cdot \deg(y(x))_\infty + \tilde{C}$. Hence, from (2), we obtain

$$\deg(y(x))_\infty \leq \frac{\tilde{C}}{n-4}. \tag{3}$$

In case $y(x) \in \mathbb{C}(x)$, we have

$$\deg(y(x))_\infty = [F : \mathbb{C}(y(x))] = [F : \mathbb{C}(x)] \cdot [\mathbb{C}(x) : \mathbb{C}(y(x))] = 2 \deg y(x).$$

Otherwise, Lemma 1 yields $\deg(y(x))_\infty = \deg y(x)$. In any case, we can choose the constant $C = \frac{\tilde{C}}{n-4}$. □

Remark 4.1. In the proof of Theorem 4, the constant $\tilde{C}$ can be determined effectively (see [10]). Then so is the constant C. Theorem 4 provides an upper bound for the degree of the annihilating polynomial of an algebraic solution which belongs to a quadratic extension of $\mathbb{C}(x)$. As a consequence, there is an algorithm for computing all algebraic solutions lying in a quadratic extension of $\mathbb{C}(x)$ of the differential equation (1). Note that, an algorithm for determining all rational solutions of the given differential equation is provided in [10].

Example 4.1. Consider the differential equation

$$y' = -xy^5 + y^3. \tag{4}$$

We view $u = y^5$ and $v = -xy^5 + y^3$ as elements in the algebraic function field $\overline{\mathbb{C}(x)}(y)$ over $\overline{\mathbb{C}(x)}$. Then we see that they only have pole at infinity with the same order. By following the lines of the proof of [10, Theorem 4.3], we can choose $\tilde{C} = 2$. This means that for every quadratic extension $F \subset \overline{\mathbb{C}(x)}$ of $\mathbb{C}(x)$ and for every $y(x) \in F$, we always have

$$\deg(u(x, y(x)))_\infty \leq \deg(v(x, y(x)))_\infty + 2. \tag{5}$$

If furthermore $y(x)$ is a solution of the given differential equation, then we have

$$\deg(v(x, y(x)))_\infty = \deg(y'(x))_\infty \leq 4 \deg(y(x))_\infty.$$

By combining with (5), we obtain $\deg(y(x))_\infty \leq 2$. Hence, the minimal polynomial of $y(x)$ over $\mathbb{C}(x)$ must be a polynomial of degree at most 2 in x and at most 2 in y. Using algorithms in [1], we find all algebraic solutions of the given differential equation which are lying in a quadratic extension of $\mathbb{C}(x)$. They are $y(x) = 0$ and $\pm\frac{1}{\sqrt{2x}}$.

References

[1] A. Bostan, G. Chèze, T. Cluzeau, and J.-A. Weil. Efficient algorithms for computing rational first integrals and darboux polynomials of planar polynomial vector fields. *Mathematics of Computation*, 85(299):1393–1425, 2016.

[2] M. M. Carnicer. The Poincaré problem in the nondicritical case. *Annals of Mathematics. Second Series*, 140(2):289–294, 1994.

[3] David Cox, John Little, and Donal O'shea. *Ideals, varieties, and algorithms*, volume 3. Springer, 2007.

[4] A. Eremenko. Rational solutions of first-order differential equations. *Annales Academiae Scientiarum Fennicae. Mathematica*, 23(1):181–190, 1998.

[5] L. Fuchs. Über Differentialgleichungen, deren Integrale feste Verzweigungspunkte besitzen. *Sitzungsberichte Acad.*, 11(3):251–273, 1884.

[6] J. J. Kovacic. An algorithm for solving second order linear homogeneous differential equations. *Journal of Symbolic Computation*, 2(1):3–43, 1986.

[7] H. Poincaré. Sur un théorème de M. Fuchs. *Acta Math.*, 7:1–32, 1885.

[8] M.J. Prelle and M.F. Singer. Elementary first integrals of differential equations. *Transactions of the American Mathematical Society*, 279(1):215–229, 1983.

[9] H. Stichtenoth. *Algebraic Function Fields and Codes*. Springer Publishing Company, Incorporated, 2nd edition, 2008.

[10] N.T. Vo, G. Grasegger, and F. Winkler. Computation of all rational solutions of first-order algebraic odes. *Advances in Applied Mathematics*, 98:1–24, 2018.

[11] N. T. Vo, G. Grasegger, and F. Winkler. Deciding the Existence of Rational General Solutions for First-Order Algebraic ODEs. *Journal of Symbolic Computation*, 87:127–139, 2018.

A Presentation of the Kähler Differential Module for a Fat Point Scheme in $\mathbb{P}^{n_1} \times \cdots \times \mathbb{P}^{n_k}$

Tran N. K. Linh[1,3,*], *Elena* Guardo[2,**], and *Le Ngoc* Long[3,***]

[1]Department of Mathematics, Hue University of Education, 34 Le Loi, Hue, Vietnam
[2]Dipartimento di Matematica e Informatica, Viale A. Doria 6, 95100 - Catania, Italy
[3]Fakultät für Informatik und Mathematik, Universität Passau, D-94030 Passau, Germany

Abstract. Let $\mathbb{Y}$ be a fat point scheme in $\mathbb{P}^{n_1} \times \cdots \times \mathbb{P}^{n_k}$ over a field K of characteristic zero. In this paper we introduce the multi-graded Kähler differential module for $\mathbb{Y}$ and we establish a short exact sequence of this module in terms of the thickening of $\mathbb{Y}$.

1 Introduction

In [1], G. de Dominicis and M. Kreuzer introduced some methods using algebraic differential forms into the study of finite sets of points in the projective n-space $\mathbb{P}^n$ over a field K of characteristic zero. Explicitly, given a finite set of points $\mathbb{X}$ in $\mathbb{P}^n$ with homogeneous vanishing ideal $I_\mathbb{X}$ in $R = K[X_0, \dots, X_n]$ and homogeneous coordinate ring $R_\mathbb{X} = R/I_\mathbb{X}$, the *Kähler differential module* for $\mathbb{X}$ is the $R_\mathbb{X}$-module $\Omega^1_{R_\mathbb{X}/K} = J/J^2$ where J is the kernel of the multiplication map $\mu : R_\mathbb{X} \otimes_K R_\mathbb{X} \to R_\mathbb{X}$. One of interesting results in [1] is the canonical exact sequence for the Kähler differential module

$$0 \to I^{(2)}_\mathbb{X}/I^2_\mathbb{X} \to I_\mathbb{X}/I^2_\mathbb{X} \to R^{n+1}_\mathbb{X}(-1) \to \Omega^1_{R_\mathbb{X}/K} \to 0.$$

Based on this exact sequence, the structure of this module can be precisely described in several special cases, for instance, if $\mathbb{X}$ is the complete intersection of hypersurfaces of degrees $d_1, \dots, d_n$ then it follows that the Hilbert function of $\Omega^1_{R_\mathbb{X}/K}$ is given by $\mathrm{HF}_{\Omega^1_{R_\mathbb{X}/K}}(i) = (n+1)\,\mathrm{HF}_\mathbb{X}(i-1) - \sum_{j=1}^{n} \mathrm{HF}_\mathbb{X}(i-d_j)$ for all $i \in \mathbb{Z}$. Later, in [2, 4], the differential algebra techniques were extended to fat point schemes in $\mathbb{P}^n$ and in $\mathbb{P}^1 \times \mathbb{P}^1$.

In this paper we will consider the natural question of whether these differential algebraic methods can be applied to study fat point schemes of a multiprojective space $\mathbb{P}^{n_1} \times \cdots \times \mathbb{P}^{n_k}$ and, especially, we look closely at the generalization of the above canonical exact sequence to fat point schemes in $\mathbb{P}^{n_1} \times \cdots \times \mathbb{P}^{n_k}$. Let $\mathbb{Y}$ be a fat point scheme in $\mathbb{P}^{n_1} \times \cdots \times \mathbb{P}^{n_k}$ with multi-homogeneous vanishing ideal $I_\mathbb{Y} \subseteq S = K[X_{10}, \dots, X_{1n_1}, \dots, X_{k0}, \dots, X_{kn_k}]$ and let $R_\mathbb{Y} = S/I_\mathbb{Y}$ be the multi-graded coordinate ring of $\mathbb{Y}$. We show that the Kähler differential module $\Omega^1_{R_\mathbb{Y}/K}$ admits the following exact sequence.

*e-mail: tnkhanhlinh141@gmail.com
**e-mail: guardo@dmi.unict.it
***e-mail: nglong16633@gmail.com

Theorem 1.1. *(Theorem 3.5) Let $\mathbb{Y} = m_1P_1+\cdots+m_sP_s$ be a fat point scheme in $\mathbb{P}^{n_1}\times\cdots\times\mathbb{P}^{n_k}$, and let $\mathbb{V} = (m_1 + 1)P_1 + \cdots + (m_s + 1)P_s$ be the thickening of $\mathbb{Y}$. Then we have the exact sequence of $\mathbb{Z}^k$-graded $R_{\mathbb{Y}}$-modules*

$$0 \longrightarrow I_{\mathbb{Y}}/I_{\mathbb{V}} \longrightarrow R_{\mathbb{Y}}^{n_1+1}(-e_1) \oplus \cdots \oplus R_{\mathbb{Y}}^{n_k+1}(-e_k) \longrightarrow \Omega^1_{R_{\mathbb{Y}}/K} \longrightarrow 0.$$

This result shows that one can compute the Hilbert function of the Kähler differential module for $\mathbb{Y}$ from the Hilbert functions of $\mathbb{Y}$ and $\mathbb{V}$, in particularly, to compute the Hilbert function $\mathrm{HF}_{\Omega^1_{R_{\mathbb{Y}}/K}}(\underline{i})$, we need to compute $\mathrm{HF}_{\Omega^1_{R_{\mathbb{Y}}/K}}(\underline{i})$ for only a finite number of $\underline{i} \in \mathbb{Z}^k$.

2 Basic Facts and Notation

Let K be a field of characteristic zero. Let $k \geq 2$ be a positive integer, let $\underline{i}$ denote the tuple $(i_1, \ldots, i_k) \in \mathbb{Z}^k$, and let $|\underline{i}| = \sum_l i_l$. We write $\underline{i} \leq \underline{j}$ if $i_l \leq j_l$ for every $l = 1, \ldots, k$. Also, let $\{e_1, \ldots, e_k\}$, $e_i = (0, \ldots, 1, \ldots, 0)$, be the canonical basis of $\mathbb{Z}^k$.

The multi-graded coordinate ring of $\mathbb{P}^{n_1} \times \cdots \times \mathbb{P}^{n_k}$ is the the polynomial ring $S = K[X_{10}, \ldots, X_{1n_1}, \ldots, X_{k0}, \ldots, X_{kn_k}]$ equipped with the $\mathbb{Z}^k$-grading defined by $\deg X_{i0} = \cdots = \deg X_{in_i} = e_i$ for $i = 1, \ldots, k$. For $\underline{i} = (i_1, ..., i_k) \in \mathbb{Z}^k$, we let $S_{\underline{i}}$ be the homogeneous component of degree $\underline{i}$ of S, i.e., the K-vector space with basis

$$\{X_{10}^{\alpha_{10}} \cdots X_{1n_1}^{\alpha_{1n_1}} \cdots X_{k0}^{\alpha_{k0}} \cdots X_{kn_k}^{\alpha_{kn_k}} \mid \textstyle\sum_{l=1}^{n_j} \alpha_{jl} = i_j,\ \alpha_{jl} \in \mathbb{Z}\}.$$

Given an ideal $I \subseteq S$, we set $I_{\underline{i}} := I \cap S_{\underline{i}}$ for all $\underline{i} \in \mathbb{Z}^k$. Clearly, $I_{\underline{i}}$ is a K-vector subspace of $S_{\underline{i}}$ and $I \supseteq \bigoplus_{\underline{i}\in\mathbb{Z}^k} I_{\underline{i}}$. The ideal I is called *$\mathbb{Z}^k$-homogeneous* if $I = \bigoplus_{\underline{i}\in\mathbb{Z}^k} I_{\underline{i}}$. If I is a $\mathbb{Z}^k$-homogeneous ideal of S then the quotient ring S/I also inherits the structure of a multi-graded ring via $(S/I)_{\underline{i}} := S_{\underline{i}}/I_{\underline{i}}$ for all $\underline{i} \in \mathbb{Z}^k$.

A finitely generated S-module M is a *$\mathbb{Z}^k$-graded S-module* if it has a direct sum decomposition

$$M = \bigoplus_{\underline{i}\in\mathbb{Z}^k} M_{\underline{i}}$$

with the property that $S_{\underline{i}}M_{\underline{j}} \subseteq M_{\underline{i}+\underline{j}}$ for all $\underline{i}, \underline{j} \in \mathbb{Z}^k$.

Definition 2.1. Let M be a finitely generated $\mathbb{Z}^k$-graded S-module. The *Hilbert function* of M is the numerical function $\mathrm{HF}_M : \mathbb{Z}^k \to \mathbb{Z}$ defined by

$$\mathrm{HF}_M(\underline{i}) := \dim_K M_{\underline{i}} \quad \text{for all } \underline{i} \in \mathbb{Z}^k.$$

In particular, for a $\mathbb{Z}^k$-homogeneous ideal I of S, the Hilbert function of S/I satisfies

$$\mathrm{HF}_{S/I}(\underline{i}) := \dim_k (S/I)_{\underline{i}} = \dim_k S_{\underline{i}} - \dim_k I_{\underline{i}} \quad \text{for all } \underline{i} \in \mathbb{Z}^k.$$

If M is a finitely generated $\mathbb{Z}^k$-graded S-module such that $\mathrm{HF}_M(\underline{i}) = 0$ for $\underline{i} \not\geq (0, \ldots, 0)$, we write the Hilbert function of M as an infinite matrix, where the initial row and column are indexed by 0.

A point in the space $\mathbb{P}^{n_1} \times \cdots \times \mathbb{P}^{n_k}$ has the form

$$P = [a_{10} : a_{11} : \cdots : a_{1n_1}] \times \cdots \times [a_{k0} : a_{k1} : \cdots : a_{kn_k}] \in \mathbb{P}^{n_1} \times \cdots \times \mathbb{P}^{n_k}$$

where $[a_{j0} : a_{j1} : \cdots : a_{jn_j}] \in \mathbb{P}^{n_j}$. Its vanishing ideal is the bihomogeneous prime ideal of the form

$$I_P = \langle L_{11}, \ldots, L_{1n_1}, \ldots, L_{k1}, \ldots, L_{kn_k} \rangle$$

where $L_{jl} = a_{jl}X_{j0} - a_{j0}X_{jl}$ and $\deg(L_{jl}) = e_j$ for $l = 1, \ldots, n_j$.

When $\mathbb{X} = \{P_1, \ldots, P_s\}$ is a set of s distinct points in $\mathbb{P}^{n_1} \times \cdots \times \mathbb{P}^{n_k}$, then the vanishing ideal of $\mathbb{X}$ is $I_{\mathbb{X}} = I_{P_1} \cap \cdots \cap I_{P_s}$ and its $\mathbb{Z}^k$-graded coordinate ring is $R_{\mathbb{X}} = S/I_{\mathbb{X}}$.

Definition 2.2. Let $\mathbb{X} = \{P_1, \ldots, P_s\}$ is a set of s distinct points in $\mathbb{P}^{n_1} \times \cdots \times \mathbb{P}^{n_k}$, and let $m_1, \ldots, m_s$ are positive integers. The ideal $I_{\mathbb{Y}} = I_{P_1}^{m_1} \cap \cdots \cap I_{P_s}^{m_s}$ defines a 0-dimensional subscheme $\mathbb{Y}$ of $\mathbb{P}^{n_1} \times \cdots \times \mathbb{P}^{n_k}$.

(a) The scheme $\mathbb{Y}$ is called a *fat point scheme* of $\mathbb{P}^{n_1} \times \cdots \times \mathbb{P}^{n_k}$ and is denoted by $\mathbb{Y} = m_1P_1 + \cdots + m_sP_s$.

(b) The number m_j is called the *multiplicity* of the point P_j.

(c) The fat point scheme $\mathbb{V} = (m_1 + 1)P_1 + \cdots + (m_s + 1)P_s$ in $\mathbb{P}^{n_1} \times \cdots \times \mathbb{P}^{n_k}$ is called the *thickening* of $\mathbb{Y}$.

The support of $\mathbb{Y}$ is $\mathrm{Supp}(\mathbb{Y}) = \mathbb{X}$, the vanishing ideal of $\mathbb{Y}$ is $I_{\mathbb{Y}}$ and its $\mathbb{Z}^k$-graded coordinate ring is the residue class ring $R_{\mathbb{Y}} = S/I_{\mathbb{Y}}$.

If $\mathbb{Y}$ is a fat point scheme in $\mathbb{P}^n$, then there exists a linear form that is a non-zerodivisor for the homogeneous coordinate ring of $\mathbb{Y}$. Using this property, the following lemma for fat point schemes in $\mathbb{P}^{n_1} \times \cdots \times \mathbb{P}^{n_k}$ follows similarly from [3, Lemma 1.2].

Lemma 2.3. *Let $\mathbb{Y}$ be a fat point scheme of $\mathbb{P}^{n_1} \times \cdots \times \mathbb{P}^{n_k}$. Then for $i \in \{1, \ldots, k\}$ there exists $L_i \in S_{e_i}$ such that the image of L_i in the residue class ring $R_{\mathbb{Y}} = S/I_{\mathbb{Y}}$ is a non-zerodivisor of $R_{\mathbb{Y}}$.*

Remark 2.4. After a suitable change of coordinates, we can assume that $L_i = X_{i0}$ for $i = 1, \ldots, k$. This also implies that X_{i0} does not vanish at any point of $\mathrm{Supp}(\mathbb{Y}) = \mathbb{X}$. In this case if x_{ij} denotes the image of X_{ij} in $R_{\mathbb{Y}}$, then $x_{10}, \ldots, x_{k0}$ are non-zerodivisors of $R_{\mathbb{Y}}$.

As a consequence of the lemma and [8, Proposition 1.9] we get several basis properties of the Hilbert function of $\mathbb{Y}$.

Theorem 2.5. *Let $\mathbb{Y} = m_1P_1 + \cdots + m_sP_s$ be a fat point scheme of $\mathbb{P}^{n_1} \times \cdots \times \mathbb{P}^{n_k}$ with Hilbert function $\mathrm{HF}_{\mathbb{Y}}$, and let $j \in \{1, ..., k\}$ and $N = n_1 + \cdots + n_k$.*

(a) For all $\underline{i} \in \mathbb{Z}^k$ we have $\mathrm{HF}_{\mathbb{Y}}(\underline{i}) \leq \mathrm{HF}_{\mathbb{Y}}(\underline{i} + e_j) \leq \sum_{k=1}^{s} \binom{N+m_k-1}{m_k-1}$.

(b) If $\mathrm{HF}_{\mathbb{Y}}(\underline{i}) = \mathrm{HF}_{\mathbb{Y}}(\underline{i} + e_j)$ then $\mathrm{HF}_{\mathbb{Y}}(\underline{i}) = \mathrm{HF}_{\mathbb{Y}}(\underline{i} + 2e_j)$.

3 A Presentation of the Kähler Differential Module

In the following we let $\mathbb{Y}$ be a fat point scheme in $\mathbb{P}^{n_1} \times \cdots \times \mathbb{P}^{n_k}$ supported at $\mathbb{X}$. We also denote the image of X_{ij} in $R_{\mathbb{Y}}$ by x_{ij}. According to Remark 2.4, we may always assume that $x_{10}, \ldots, x_{k0}$ are non-zerodivisors of $R_{\mathbb{Y}}$.

In the multi-graded algebra

$$R_{\mathbb{Y}} \otimes_K R_{\mathbb{Y}} = \bigoplus_{\underline{i} \in \mathbb{Z}^k} \Big(\bigoplus_{\underline{j}+\underline{h}=\underline{i}} (R_{\mathbb{Y}})_{\underline{j}} \otimes (R_{\mathbb{Y}})_{\underline{h}} \Big)$$

we have the $\mathbb{Z}^k$-homogeneous ideal $J = \ker(\mu)$, where $\mu : R_{\mathbb{Y}} \otimes_K R_{\mathbb{Y}} \to R_{\mathbb{Y}}$ is the multi-homogeneous $R_{\mathbb{Y}}$-linear map given by $\mu(f \otimes g) = fg$. A $\mathbb{Z}^k$-homogeneous system of generators of J is

$$\{ x_{ij_i} \otimes 1 - 1 \otimes x_{ij_i} \mid 1 \leq i \leq k, 0 \leq j_i \leq n_i \}.$$

Definition 3.1. The $\mathbb{Z}^k$-graded $R_{\mathbb{Y}}$-module $\Omega^1_{R_{\mathbb{Y}}/K} = J/J^2$ is called the *module of Kähler differentials* of $R_{\mathbb{Y}}/K$. The $\mathbb{Z}^k$-homogeneous K-linear map $d_{R_{\mathbb{Y}}/K} : R_{\mathbb{Y}} \to \Omega^1_{R_{\mathbb{Y}}/K}$ given by $f \mapsto f \otimes 1 - 1 \otimes f + J^2$ satisfies universal property. We call d the *universal derivation* of $R_{\mathbb{Y}}/K$.

More generally, for any $\mathbb{Z}^k$-graded K-algebra T/S we can define in the same way the Kähler differential module $\Omega^1_{T/S}$, and the universal derivation of T/S (cf. [6, Section 2]).

Remark 3.2. (a) The $\mathbb{Z}^k$-graded S-module $\Omega^1_{S/K}$ has the representation

$$\Omega^1_{S/K} = \bigoplus_{i=1}^{k} \bigoplus_{j=0}^{n_i} S\, dX_{ij} \cong S^{n_1+1}(-e_1) \oplus \cdots \oplus S^{n_k+1}(-e_k).$$

(b) If I_1, I_2 are $\mathbb{Z}^k$-homogeneous ideals of S, then

$$I_1\Omega^1_{S/K} \cap I_2\Omega^1_{S/K} = (I_1 \cap I_2)\Omega^1_{S/K}$$

(see [7, Chapter 3, §7, Theorem 7.4(i)]).

Some following properties of the module of Kähler differentials follows from [6, Propositions 4.12 and 4.13].

Proposition 3.3. *Let $dI_{\mathbb{Y}}$ be the submodule of $\Omega^1_{S/K}$ generated by $\{dF \mid F \in I_{\mathbb{Y}}\}$.*

(a) There is an isomorphism of $\mathbb{Z}^k$-graded $R_{\mathbb{Y}}$-modules

$$\Omega^1_{R_{\mathbb{Y}}/K} \cong \Omega^1_{S/K}/(dI_{\mathbb{Y}} + I_{\mathbb{Y}}\Omega^1_{S/K}).$$

(b) The set $\{dx_{ij_i} \mid 1 \le i \le k, 0 \le j_i \le n_i\}$ is a $\mathbb{Z}^k$-homogeneous system of generators of $\Omega^1_{R_{\mathbb{Y}}/K}$.

Example 3.4. Let $\mathbb{X} = \{P\}$ be the set of only one point in $\mathbb{P}^{n_1} \times \cdots \times \mathbb{P}^{n_k}$ where $P = [1 : 0 : \cdots : 0] \times \cdots \times [1 : 0 : \cdots : 0]$. We have $I_P = \langle X_{ij_i} \mid 1 \le i \le k, 1 \le j \le n_i \rangle$ and $R_{\mathbb{X}} = K[x_{10}, \ldots, x_{k0}]$. Also, we see that

$$I_P/I_P^2 \cong R_{\mathbb{X}}^{n_1}(-e_1) \oplus \cdots \oplus R_{\mathbb{X}}^{n_k}(-e_k)$$

and

$$\Omega^1_{R_{\mathbb{X}}/K} = \langle dx_{i0} \mid 1 \le i \le k \rangle \cong R_{\mathbb{X}}(-e_1) \oplus \cdots \oplus R_{\mathbb{X}}(-e_k).$$

Moreover, we get the exact sequence of $\mathbb{Z}^k$-graded $R_{\mathbb{X}}$-modules

$$0 \longrightarrow I_P/I_P^2 \longrightarrow R_{\mathbb{X}}^{n_1+1}(-e_1) \oplus \cdots \oplus R_{\mathbb{X}}^{n_k+1}(-e_k) \longrightarrow \Omega^1_{R_{\mathbb{X}}/K} \longrightarrow 0.$$

In this case the Hilbert function of $\Omega^1_{R_{\mathbb{X}}/K}$ satisfies

$$\mathrm{HF}_{\Omega^1_{R_{\mathbb{X}}/K}}(\underline{i}) = \mathrm{HF}_{\mathbb{X}}(\underline{i} - e_1) + \cdots + \mathrm{HF}_{\mathbb{X}}(\underline{i} - e_k).$$

In general, we have a presentation of module of Kähler differential as follows.

Theorem 3.5. *Let $\mathbb{Y} = m_1P_1 + \cdots + m_sP_s$ be a fat point scheme in $\mathbb{P}^{n_1} \times \cdots \times \mathbb{P}^{n_k}$, and let $\mathbb{V}$ be the thickening of $\mathbb{Y}$. Then we have the exact sequence of $\mathbb{Z}^k$-graded $R_{\mathbb{Y}}$-modules*

$$0 \longrightarrow I_{\mathbb{Y}}/I_{\mathbb{V}} \longrightarrow R_{\mathbb{Y}}^{n_1+1}(-e_1) \oplus \cdots \oplus R_{\mathbb{Y}}^{n_k+1}(-e_k) \longrightarrow \Omega^1_{R_{\mathbb{Y}}/K} \longrightarrow 0.$$

To prove Theorem 3.5, we will first require the following lemma.

Lemma 3.6. *Let $P = [1:0:\cdots:0]\times\cdots\times[1:0:\cdots:0]$ be a point of $\mathbb{P}^{n_1}\times\cdots\times\mathbb{P}^{n_k}$, let $m \geq 1$, and let $F \in I_P^m \setminus I_P^{m+1}$ be a $\mathbb{Z}^k$-homogeneous polynomial of degree $\underline{i} \in \mathbb{Z}^k$. Then we have $dF \notin I_P^m \Omega^1_{S/K}$.*

Proof. Observe that $I_P^{m+1} = \langle X_{ij_i} \mid 1 \leq i \leq k, 1 \leq j \leq n_i \rangle^{m+1}$. Let $N = n_1 + \cdots + n_k$. For $\underline{j} \in \mathbb{N}^k$ and $t \geq 1$, we set

$$\Gamma_{\underline{j},t} = \left\{ \alpha = (\alpha_1,\dots,\alpha_k) \in \mathbb{N}^N \;\middle|\; \begin{array}{l} |\alpha| = t,\ (|\alpha_1|,\dots,|\alpha_k|) \leq \underline{j}, \\ \alpha_l = (\alpha_{l1},\dots,\alpha_{ln_l}),\ 1 \leq l \leq k \end{array} \right\}$$

Since $F \in I_P^m$, the polynomial F can be written as

$$F = \sum_{\alpha \in \Gamma_{\underline{i},m+1}} X_{11}^{\alpha_{11}} \cdots X_{1n_1}^{\alpha_{1n_1}} \cdots X_{k1}^{\alpha_{k1}} \cdots X_{kn_k}^{\alpha_{kn_k}} F_\alpha + \sum_{\beta \in \Gamma_{\underline{i},m}} X_{11}^{\beta_{11}} \cdots X_{1n_1}^{\beta_{1n_1}} \cdots X_{k1}^{\beta_{k1}} \cdots X_{kn_k}^{\beta_{kn_k}} G_\beta$$

where $F_\alpha \in S_{\underline{i}-(|\alpha_1|,\dots,|\alpha_k|)}$, and $G_\beta = a_\beta X_{10}^{c_{\beta 1}} \cdots X_{k0}^{c_{\beta k}}$ with $(c_{\beta 1},\dots,c_{\beta k}) = \underline{i} - (|\beta_1|,\dots,|\beta_k|)$ and $a_\beta \in K$. It follows from $F \notin I_P^{m+1}$ that there exists $\widetilde{\beta} \in \Gamma_{\underline{i},m}$ such that $a_{\widetilde{\beta}} \neq 0$.

In order to prove $dF \notin I_P^m \Omega^1_{S/K}$, it suffices to show that

$$\omega = \sum_{\beta \in \Gamma_{\underline{i},m}} G_\beta d(X_{11}^{\beta_{11}} \cdots X_{1n_1}^{\beta_{1n_1}} \cdots X_{k1}^{\beta_{k1}} \cdots X_{kn_k}^{\beta_{kn_k}}) \notin I_P^m \Omega^1_{S/K}.$$

Write $G_{\widetilde{\beta}} = a_{\widetilde{\beta}} \cdot T$ with the term $T = X_{10}^{c_{\widetilde{\beta}1}} \cdots X_{k0}^{c_{\widetilde{\beta}k}}$, and set

$$\widetilde{\Gamma} = \{\beta \in \Gamma_{\underline{i},m} \mid a_\beta \neq 0,\ G_\beta = a_\beta T\}.$$

Clearly, $\widetilde{\beta} \in \widetilde{\Gamma} \neq \emptyset$. Let (i,j) be the smallest tuple w.r.t. $\mathtt{Lex}$ such that there exists $\beta \in \widetilde{\Gamma}$ such that $b_{ij} \neq 0$. Write $\widetilde{\Gamma} = \{\widetilde{\beta}^1,\dots,\widetilde{\beta}^s\}$ such that $\widetilde{\beta}^1 \leq_{\mathtt{Lex}} \cdots \leq_{\mathtt{Lex}} \widetilde{\beta}^s$. Then $\widetilde{\beta}^1_{ij} \leq \cdots \leq \widetilde{\beta}^s_{ij}$. Set $\varrho = \min\{\widetilde{\beta}^l_{ij} \mid \widetilde{\beta}^l_{ij} > 0, 1 \leq l \leq s\}$. We have

$$\begin{aligned}
\widetilde{\omega} :&= \sum_{\beta \in \widetilde{\Gamma}} a_\beta T d(X_{11}^{\beta_{11}} \cdots X_{1n_1}^{\beta_{1n_1}} \cdots X_{k1}^{\beta_{k1}} \cdots X_{kn_k}^{\beta_{kn_k}}) \\
&= \sum_{\beta \in \widetilde{\Gamma}} a_\beta T d(X_{ij}^{\beta_{ij}} \cdots X_{in_i}^{\beta_{in_i}} \cdots X_{k1}^{\beta_{k1}} \cdots X_{kn_k}^{\beta_{kn_k}}) \\
&= \sum_{\beta \in \widetilde{\Gamma}, \beta_{ij} \neq \varrho} a_\beta T d(X_{ij}^{\beta_{ij}} \cdots X_{in_i}^{\beta_{in_i}} \cdots X_{k1}^{\beta_{k1}} \cdots X_{kn_k}^{\beta_{kn_k}}) + \sum_{\beta \in \widetilde{\Gamma}, \beta_{ij} = \varrho} a_\beta T d(X_{ij}^{\varrho} \cdots X_{in_i}^{\beta_{in_i}} \cdots X_{k1}^{\beta_{k1}} \cdots X_{kn_k}^{\beta_{kn_k}}) \\
&\overset{(\star)}{=} \sum_{\beta \in \widetilde{\Gamma}, \beta_{ij} \neq \varrho} a_\beta T d(X_{ij}^{\beta_{ij}} \cdots X_{in_i}^{\beta_{in_i}} \cdots X_{k1}^{\beta_{k1}} \cdots X_{kn_k}^{\beta_{kn_k}}) \\
&\quad + \sum_{\beta \in \widetilde{\Gamma}, \beta_{ij} = \varrho} a_\beta T \varrho X_{ij}^{\varrho-1} X_{ij+1}^{\beta_{ij+1}} \cdots X_{in_i}^{\beta_{in_i}} \cdots X_{k1}^{\beta_{k1}} \cdots X_{kn_k}^{\beta_{kn_k}} dX_{ij} \\
&\quad + \sum_{\beta \in \widetilde{\Gamma}, \beta_{ij} = \varrho} a_\beta T X_{ij}^{\varrho} d(X_{ij+1}^{\beta_{ij+1}} \cdots X_{in_i}^{\beta_{in_i}} \cdots X_{k1}^{\beta_{k1}} \cdots X_{kn_k}^{\beta_{kn_k}}).
\end{aligned}$$

Note that if $\beta, \beta' \in \widetilde{\Gamma}$, $\beta \neq \beta'$ and $\beta_{ij} = \beta'_{ij} = \varrho$, then $(\beta_{ij+1},\dots,\beta_{kn_k}) \neq (\beta'_{ij+1},\dots,\beta'_{kn_k})$. By Macaulay's Basis Theorem (cf. [5, Theorem 1.5.7]), we have

$$\sum_{\beta \in \widetilde{\Gamma}, \beta_{ij} = \varrho} a_\beta T \varrho X_{ij}^{\varrho-1} X_{ij+1}^{\beta_{ij+1}} \cdots X_{in_i}^{\beta_{in_i}} \cdots X_{k1}^{\beta_{k1}} \cdots X_{kn_k}^{\beta_{kn_k}} dX_{ij} \notin I_P^m \Omega^1_{S/K}.$$

It follows that $\widetilde{\omega} \notin I_P^m \Omega^1_{S/K}$ (since in the right-hand side of ($\star$) the first summand has $\beta_{ij} \neq \varrho$ and the last summand does not contain dX_{ij}). Therefore we get $\omega \notin I_P^m \Omega^1_{S/K}$, as we wanted to show. □

We are now ready to state and prove the main result of this section.

Proof. Let $\varphi : I_{\mathbb{Y}}/I_{\mathbb{V}} \to \Omega^1_{S/K}/I_{\mathbb{Y}}\Omega^1_{S/K}$ be the map given by $\varphi(F + I_{\mathbb{V}}) = dF + I_{\mathbb{Y}}\Omega^1_{S/K}$ for all $F \in I_{\mathbb{Y}}$. It is easy to check that the map φ is well-defined, $\mathbb{Z}^k$-homogeneous of degree $(0, \dots, 0)$, and $R_{\mathbb{Y}}$-linear. For any $\mathbb{Z}^k$-homogeneous element $F \in I_{\mathbb{Y}} \setminus I_{\mathbb{V}}$, we have $F \in I_{P_j}^{m_j} \setminus I_{P_j}^{m_j+1}$ for some $j \in \{1, \dots, s\}$. W.l.o.g. we may assume that $P_j = [1 : 0 : \cdots : 0] \times \cdots \times [1 : 0 : \cdots : 0]$. So, Lemma 3.6 yields $dF \notin I_{P_j}^{m_j}\Omega^1_{S/K}$. Since $I_{\mathbb{Y}}\Omega^1_{S/K} = \bigcap_{j=1}^{s} I_{P_j}^{m_j}\Omega^1_{S/K}$ by Remark 3.2(b), we obtain $dF \notin I_{\mathbb{Y}}\Omega^1_{S/K}$. Consequently, the map φ is injective. Moreover, by Remark 3.2(a) and Proposition 3.3, we see that

$$\Omega^1_{S/K}/I_{\mathbb{Y}}\Omega^1_{S/K} \cong R_{\mathbb{Y}}^{n_1+1}(-e_1) \oplus \cdots \oplus R_{\mathbb{Y}}^{n_k+1}(-e_k)$$

and

$$(\Omega^1_{S/K}/I_{\mathbb{Y}}\Omega^1_{S/K})/\operatorname{im}(\varphi) \cong \Omega^1_{S/K}/(dI_{\mathbb{Y}} + I_{\mathbb{Y}}\Omega^1_{S/K}) \cong \Omega^1_{R_{\mathbb{Y}}/K}.$$

Therefore the conclusion follows. □

The following relation between the Hilbert function of $\Omega^1_{R_{\mathbb{Y}}/K}$ and of $\mathbb{Y}$ and $\mathbb{V}$ is an immediate consequence of Theorem 3.5

Corollary 3.7. *$\mathbb{Y} = m_1P_1 + \cdots + m_sP_s$ be a fat point scheme in $\mathbb{P}^{n_1} \times \cdots \times \mathbb{P}^{n_k}$, and let $\mathbb{V}$ be the thickening of $\mathbb{Y}$. Then the Hilbert function of $\Omega^1_{R_{\mathbb{Y}}/K}$ satisfies*

$$\mathrm{HF}_{\Omega^1_{R_{\mathbb{Y}}/K}}(\underline{i}) = (n_1 + 1)\,\mathrm{HF}_{\mathbb{Y}}(\underline{i} - e_1) + \cdots + (n_k + 1)\,\mathrm{HF}_{\mathbb{Y}}(\underline{i} - e_k) + \mathrm{HF}_{\mathbb{Y}}(\underline{i}) - \mathrm{HF}_{\mathbb{V}}(\underline{i})$$

for all $\underline{i} \in \mathbb{Z}^k$.

References

[1] G. de Dominicis and M. Kreuzer, Kähler differentials for points in $\mathbb{P}^n$, J. Pure Appl. Alg. **141**, 153–173 (1999).

[2] E. Guardo, M. Kreuzer, T. N. K. Linh, and L.N. Long, Kähler differentials for fat point schemes in $\mathbb{P}^1 \times \mathbb{P}^1$, Submitted (2018).

[3] E. Guardo and A. Van Tuyl, Fat Points in $\mathbb{P}^1 \times \mathbb{P}^1$ and their Hilbert functions, Canad. J. Math. **56** , no. 4, 716–741 (2004).

[4] M. Kreuzer, T. N. K. Linh, and L. N. Long, Kähler differentials and Kähler differents for fat point schemes, J. Pure Appl. Algebra **219**, 4479–4509 (2015).

[5] M. Kreuzer and L. Robbiano, *Computational Commutative Algebra 1*, (Springer Verlag, Heildelberg, 2000).

[6] E. Kunz, *Kähler Differentials* (Adv. Lectures Math., Wieweg Verlag, Braunschweig, 1986).

[7] H. Matsumura, *Commutative Ring Theory* (Cambridge University Press, Cambrige, 1986).

[8] J. Sidman and A. Van Tuyl, Multigraded regularity: Syzygies and fat points, Beiträge zur Algebra und Geometrie **47**, 1–22 (2006).

Relationships between quantized algebras and their semi-classical limits

Sei-Qwon Oh[1,*]

[1]Department of Mathematics, Chungnam National University, 99 Daehak-ro, Yuseong-gu, Daejeon 34134, Korea

Abstract. A Poisson $\mathbb{C}$-algebra R appears in classical mechanical system and its quantized algebra appearing in quantum mechanical system is a $\mathbb{C}[[\hbar]]$-algebra $Q = R[[\hbar]]$ with star product $*$ such that for any $a, b \in R \subseteq Q$,

$$a * b = ab + B_1(a,b)\hbar + B_2(a,b)\hbar^2 + \ldots$$

subject to

$$\{a, b\} = \hbar^{-1}(a * b - b * a)|_{\hbar=0}, \qquad \cdots \quad (**)$$

where $B_i : R \times R \longrightarrow R$ are bilinear products. The given Poisson algebra R is recovered from its quantized algebra Q by $R = Q/\hbar Q$ with Poisson bracket $(**)$, which is called its semiclassical limit. But it seems that the star product in Q is complicate and that Q is difficult to understand at an algebraic point of view since it is too big. For instance, if λ is a nonzero element of $\mathbb{C}$ then $\hbar - \lambda$ is a unit in Q and thus a so-called deformation of R, $Q/(\hbar - \lambda)Q$, is trivial. Hence it seems that we need an appropriate $\mathbb{F}$-subalgebra A of Q such that A contains all generators of Q, $\hbar \in A$ and A is understandable at an algebraic point of view, where $\mathbb{F}$ is a subring of $\mathbb{C}[[\hbar]]$.
Here we discuss how to find nontrivial deformations from quantized algebras and the natural map in [6] from a class of infinite deformations onto its semi-classical limit. The results are illustrated by examples.

1 Motivation and Quantization

1.1 Star product

A commutative $\mathbb{C}$-algebra R is said to be a *Poisson algebra* if there exists a bilinear product $\{-,-\} : R \times R \to R$, called a *Poisson bracket*, such that $(R, \{-,-\})$ is a Lie algebra satisfying Leibniz's rule $\{ab, c\} = a\{b, c\} + \{a, c\}b$ for all $a, b, c \in R$. A quantization of R is an associative $\mathbb{C}[[\hbar]]$-algebra $R[[\hbar]]$ equipped with a star product $* : R[[\hbar]] \times R[[\hbar]] \to R[[\hbar]]$ such that for all $a, b \in R$,

$$a * b = ab + B_1(a,b)\hbar + B_2(a,b)\hbar^2 + \ldots$$

subject to

$$\{a, b\} = \hbar^{-1}(a * b - b * a)|_{\hbar=0}, \tag{1}$$

*e-mail: sqoh@cnu.ac.kr

where $B_i : R \times R \longrightarrow R$ are bilinear products. Denote by $Q = (R[[\hbar]], *)$. We can recover the Poisson algebra R from its quantization Q. That is, $\hbar Q$ is a nontrivial ideal such that $Q/\hbar Q \cong R$ as a commutative $\mathbb{C}$-algebra and the Poisson bracket $\{-,-\}$ in R is obtained by (1). But if λ is a nonzero element of $\mathbb{C}$ then $\hbar - \lambda$ is a unit in Q and thus $Q/(\hbar - \lambda)Q$ is trivial. Moreover, we do not know what $Q_\lambda = \{f|_{\hbar=\lambda} \mid f \in Q\}$ means mathematically. Let us see the following example.

1.2 Poisson Weyl algebra

The Poisson Weyl algebra is the $\mathbb{C}$-algebra $R = \mathbb{C}[x, y]$ with Poisson bracket

$$\{f, g\} = -\frac{\partial f}{\partial x}\frac{\partial g}{\partial y} + \frac{\partial g}{\partial x}\frac{\partial f}{\partial y} \tag{2}$$

for all $f, g \in R$, namely $\{y, x\} = 1$.

Define a multiplication on the set $R[[\hbar]]$ of formal power series over $R = \mathbb{C}[x, y]$ by

$$yx = xy + \hbar$$

and $\mathbb{C}[[\hbar]]$ is central in $R[[\hbar]]$. Then $Q = R[[\hbar]]$ is an associative $\mathbb{C}[[\hbar]]$-algebra and $\hbar$ is a nonzero, nonunit, non-zero-divisor and central element such that $Q/\hbar Q$ is commutative. Hence $Q/\hbar Q \cong R$ is a Poisson $\mathbb{C}$-algebra with Poisson bracket

$$\{\overline{f}, \overline{g}\} := \overline{\hbar^{-1}(fg - gf)}$$

for all $f, g \in Q$. In particular, $\{\overline{y}, \overline{x}\} = 1$.

For $\lambda \in \mathbb{C}$, let Q_λ be the set of formal elements $f|_{\hbar=\lambda}$ for all $f \in Q$. For the case $\lambda = 0$, observe that $Q_0 = R$. For $0 \neq \lambda \in \mathbb{C}$ and $f = 1 + \hbar x + \hbar^2 + \hbar^3 + \ldots \in Q$, we do not know mathematical meaning of the formal form

$$f|_{\hbar=\lambda} = 1 + \lambda x + \lambda^2 + \lambda^3 + \ldots \in Q_\lambda.$$

In particular, we should observe that $Q/(\hbar - \lambda)Q = 0$, since $\hbar - \lambda$ is a unit in Q, and thus $Q_\lambda \neq Q/(\hbar - \lambda)Q$.

The *Weyl algebra* W is the $\mathbb{C}$-algebra generated by x, y subject to the relation

$$yx = xy + 1.$$

Let A be a $\mathbb{C}[\hbar]$-algebra of Q generated by x and y. Then A is a $\mathbb{C}[\hbar]$-algebra generated by x, y subject to the relation $yx - xy = \hbar$. Note that the element $\hbar \in A$ satisfies the following:

- $\hbar$ is a nonzero, nonunit, non-zero-divisor and central element.
- $A/\hbar A$ is commutative.
- $(\hbar - \lambda)A \neq A$ for all $0 \neq \lambda \in \mathbb{C}$.

Hence the bilinear product $\{-,-\}$ on $A/\hbar A \cong R$ defined by

$$\{\overline{f}, \overline{g}\} := \overline{\hbar^{-1}(fg - gf)} \quad (f, g \in A)$$

is well defined and thus $A/\hbar A$ is a Poisson algebra isomorphic to the Poisson Weyl algebra and for each $0 \neq \lambda \in \mathbb{C}$, $A/(\hbar - \lambda)A$ is the $\mathbb{C}$-algebra generated by x, y subject to the relation $yx = xy + \lambda$, which is isomorphic to the Weyl algebra W. In particular,

$$A_\lambda := \{f|_{\hbar=\lambda} \mid f \in A\}$$

makes sense mathematically and is isomorphic to $A/(\hbar - \lambda)A$ as a $\mathbb{C}$-algebra.

2 Deformation and semiclassical limit

2.1 Semiclassical limit

Let A be a $\mathbb{C}$-algebra. An element $\hbar \in A$ is said to be a *regular* element if it is a nonzero, nonunit, non-zero-divisor and central element such that $A/\hbar A$ is commutative.

Let $\hbar$ be a regular element. Then $0 \neq \hbar A$ is a proper ideal such that the factor $\overline{A} := A/\hbar A$ is a Poisson algebra with Poisson bracket

$$\{\overline{a}, \overline{b}\} = \overline{\hbar^{-1}(ab - ba)} \tag{3}$$

for $\overline{a}, \overline{b} \in A/\hbar A$. The Poisson algebra $\overline{A} = A/\hbar A$ is called a *semiclassical limit* of A.

Let R and Q be as in §1.1. Then $\hbar$ is a regular element of Q and the semiclassical limit $Q/\hbar Q$ is isomorphic to R as a Poisson algebra. In §1.2, $\hbar$ is a regular element of A and the semiclassical limit $A/\hbar A$ is isomorphic to the Poisson Weyl algebra $\mathbb{C}[x, y]$ as a Poisson algebra.

2.2 Deformations

Retain the notations in §2.1. Suppose that there is an element $0 \neq \lambda \in \mathbb{C}$ such that $\hbar - \lambda$ is a nonunit in A. Then $(\hbar - \lambda)A$ is a proper ideal of A and thus the factor $A_\lambda := A/(\hbar - \lambda)A$ is a nontrivial $\mathbb{C}$-algebra such that its multiplication is induced by that of A. The factor A_λ is called a *deformation* of $\overline{A}$. For instance, the algebra A_λ in §1.2 is a deformation of the Poisson Weyl algebra.

An algorithm to obtain a deformation is given as follows. Let $\mathbb{F}$ be a subring of $\mathbb{C}[[\hbar]]$ containing $\mathbb{C}[\hbar]$ and let A be an $\mathbb{F}$-algebra generated by $x_1, \dots, x_n$ subject to relations $f_1, \dots, f_r$. For $\lambda \in \mathbb{C}$, assume that $f_i|_{\hbar=\lambda}$ makes sense mathematically for each $i = 1, \dots, r$. The $\mathbb{C}$-algebra generated by $x_1, \dots, x_n$ subject to the relations $f_1|_{\hbar=\lambda}, \dots, f_r|_{\hbar=\lambda}$, still denoted by A_λ, is deeply related to A. If $g|_{\hbar=\lambda}$ makes sense mathematically for all $g \in A$ then the evaluation map

$$\varphi_\lambda : A \longrightarrow A_\lambda, \quad g \mapsto g|_{\hbar=\lambda}$$

is a $\mathbb{C}$-algebra epimorphism and thus $A_\lambda \cong A/\ker \varphi_\lambda$ and the multiplication of A_λ is induced by that of A. If $\hbar$ is a regular element then A_λ is a deformation of $\overline{A}$. An example of this algorithm is given in §3.

2.3 Weyl algebra

As shown in §1.2, let A be the $\mathbb{C}[\hbar]$-subalgebra of $Q = R[[\hbar]]$ generated by x and y. Then $\hbar \in A$ is a regular element and its semiclassical limit $\overline{A} = A/\hbar A$ is the Poisson Weyl algebra $R = \mathbb{C}[x, y]$ with the Poisson bracket $\{y, x\} = 1$. For each $0 \neq \lambda \in \mathbb{C}$, deformation $A_\lambda = A/(\hbar - \lambda)A$ is the $\mathbb{C}$-algebra generated by x and y subject to the relation

$$yx - xy = \lambda.$$

Note that every deformation A_λ is isomorphic to the Weyl algebra W by [4, Proposition 3.4].

2.4 Another deformation of Poisson Weyl algebra

Here we recall [5, Example 3.8]. Let Q be a $\mathbb{C}[[\hbar]]$-algebra generated by x, y subject to the relation

$$yx = (\cos\hbar)xy + \sin\hbar.$$

Denote by A the $\mathbb{C}[\hbar]$-subalgebra of Q generated by $\cos\hbar, \sin\hbar$ and x, y. Consider the composition of two natural maps

$$\phi : A \longrightarrow Q \longrightarrow Q/\hbar Q \cong R = \mathbb{C}[x, y].$$

Then $\hbar$ is a regular element in A, since $\hbar$ is regular in Q, and $A/\ker\phi \cong \mathbb{C}[x, y]$ as a $\mathbb{C}$-algebra. Give a Poisson bracket in $A/\ker\phi$ by (3). Then the Poisson bracket in $A/\ker\phi$ is

$$\{\overline{y}, \overline{x}\} = \overline{-(\sin\hbar|_{\hbar=0})xy + \cos\hbar|_{\hbar=0}} = 1$$

and thus $A/\ker\phi \cong \mathbb{C}[x, y]$, the Poisson Weyl algebra as a Poisson $\mathbb{C}$-algebra.

For each $0 \neq \lambda \in \mathbb{C}$, let A_λ be the $\mathbb{C}$-algebra generated by $\cos\lambda, \sin\lambda, x, y$ subject to the relation

$$yx = (\cos\lambda)xy + \sin\lambda.$$

Then the evaluation map

$$\varphi_\lambda : A \longrightarrow A_\lambda, \quad g \mapsto g|_{\hbar=\lambda}$$

is an algebra ephimorphism and thus $A/\ker\varphi_\lambda \cong A_\lambda$. In particular, the $\mathbb{C}$-algebra A_π for the case $\lambda = \pi$ is generated by x, y subject to the relation

$$yx + xy = 0,$$

which can be considered as a deformation of the Poisson Weyl algebra.

2.5 Poisson affine 2-space

Let $\mathbb{F} = \mathbb{C}[\hbar, \hbar^{-1}]$ and let A be the $\mathbb{F}$-algebra generated by x, y subject to the relation

$$yx = \hbar xy.$$

Note that $\hbar - 1$ is a regular element in A and the semiclassical limit $\overline{A} = A/(\hbar - 1)$ is the Poisson algebra $\mathbb{C}[x, y]$ with Poisson bracket

$$\{y, x\} = xy.$$

Then for each $0, \pm 1 \neq q \in \mathbb{C}$, the deformation $A_q = A/(\hbar - q)A$ is the $\mathbb{C}$-algebra generated by x, y subject to the relation

$$yx = qxy$$

and thus A_q is the so-called coordinate ring $O_q(\mathbb{C}^2)$ of affine 2-space.

3 A natural map from a class of infinite deformations onto its semiclassical limit

We assume the following conditions (i)-(vii) in [6, Notation 1.1].

(i) Assume that $\mathbf{K}$ is an infinite subset of the set $\mathbb{C} \setminus \{0\}$.

(ii) Assume that $\mathbb{F}$ is a subring of the ring of regular functions on $\mathbf{K} \cup \{0\}$ containing $\mathbb{C}[\hbar]$. That is,

$$\mathbb{C}[\hbar] \subseteq \mathbb{F} \subseteq \{f/g \in \mathbb{C}(\hbar) | f, g \in \mathbb{C}[\hbar] \text{ such that } g|_{\hbar=\lambda} \neq 0 \ \forall \lambda \in \mathbf{K} \cup \{0\}\}. \tag{4}$$

(iii) Let $\mathbb{F}\langle x_1, \ldots, x_n \rangle$ be the free $\mathbb{F}$-algebra on the set $\{x_1, \ldots, x_n\}$. A finite product $\mathbf{x}$ of x_i's (repetitions allowed) is called a monomial. For each $i = 1, \ldots, r$, let f_i be an $\mathbb{F}$-linear combination of monomials

$$f_i = \sum_{\mathbf{x}} a^i_{\mathbf{x}}(\hbar)\mathbf{x}, \quad a^i_{\mathbf{x}}(\hbar) \in \mathbb{F}.$$

Set $A = \mathbb{F}\langle x_1, \ldots, x_n \rangle / I$, where I is the ideal of $\mathbb{F}\langle x_1, \ldots, x_n \rangle$ generated by $f_1, \ldots, f_r$. That is, A is the $\mathbb{F}$-algebra generated by $x_1, \ldots, x_n$ subject to the relations

$$f_1, \ldots, f_r.$$

(iv) Assume that $\hbar$ is a regular element. Hence there exists the semiclassical limit $A/\hbar A$. Denote by φ_0 the canonical projection

$$\varphi_0 : A \longrightarrow A/\hbar A, \quad f \mapsto f|_{\hbar=0}.$$

(v) For each $\lambda \in \mathbf{K}$, let A_λ be the $\mathbb{C}$-algebra generated by $x_1, \ldots, x_n$ subject to the relations

$$f_1|_{\hbar=\lambda}, \ldots, f_r|_{\hbar=\lambda}.$$

Note that $a^i_{\mathbf{x}}(\lambda)$ is a well-defined element of $\mathbb{C}$ by (4) and that the evaluation map

$$\varphi_\lambda : A \longrightarrow A_\lambda, \quad f \mapsto f|_{\hbar=\lambda}$$

is an epimorphism of $\mathbb{C}$-algebras. In other words, A_λ is a deformation of the Poisson algebra $A/\hbar A$.

(vi) Let $\prod_{\lambda \in \mathbf{K}} A_\lambda$ be the product of infinite deformations A_λ and let φ be the homomorphism of $\mathbb{C}$-algebras defined by

$$\varphi : A \longrightarrow \prod_{\lambda \in \mathbf{K}} A_\lambda, \quad \varphi(a) = (\varphi_\lambda(a))_{\lambda \in \mathbf{K}}. \tag{5}$$

Note that $\varphi(\hbar)$ is an invertible element of $\prod_{\lambda \in \mathbf{K}} A_\lambda$ since $0 \notin \mathbf{K}$.

(vii) Assume that there exists an $\mathbb{F}$-basis $\{\xi_i \mid i \in I\}$ of A such that $\{\varphi_0(\xi_i) | i \in I\}$ and $\{\varphi_\lambda(\xi_i) | i \in I\}$ are $\mathbb{C}$-bases of $A/\hbar A$ and A_λ, respectively, for each $\lambda \in \mathbf{K}$. Hence every element $a \in A$ is expressed uniquely by

$$a = \sum_i a_i(\hbar)\xi_i, \quad a_i(\hbar) \in \mathbb{F}$$

and, for each $\lambda \in \mathbf{K}$,

$$\varphi_\lambda(a) = \textstyle\sum_i a_i(\lambda)\varphi_\lambda(\xi_i), \quad \varphi_0(a) = \sum_i a_i(0)\varphi_0(\xi_i).$$

Note that $a_i(\lambda)$ and $a_i(0)$ are well-defined elements of $\mathbb{C}$ by (4).

(viii) Let $\widehat{q}$ be the parameter taking values in $\mathbf{K}$. That is, $\widehat{q}$ is a function defined by

$$\widehat{q} : \mathbf{K} \longrightarrow \mathbb{C}, \ \ \widehat{q}(\lambda) = \lambda.$$

(iv) Let $A_{\widehat{q}}$ be the $\mathbb{C}$-algebra obtained by replacing $\hbar$ in A by $\widehat{q}$ and let $\widehat{\ }$ be the map defined by

$$\widehat{\ } : A_{\widehat{q}} \longrightarrow \prod_{\lambda \in \mathbf{K}} A_\lambda, \quad f \mapsto \widehat{f} = (f|_{\widehat{q}=\lambda})_{\lambda \in \mathbf{K}}.$$

3.1 Theorem [Oh, 2017]

Here we write a natural map from a class of infinite deformations onto its semiclassical limit in [6, §1]. Set $\widehat{A} = \widehat{\ }^{-1}(\varphi(A)) \subset A_{\widehat{q}}$, the inverse image of $\varphi(A)$ by the map $\widehat{\ }$. Since φ is a monomorphism of $\mathbb{C}$-algebras by [6, Lemma 1.2], there exists the composition

$$\gamma : \widehat{A} \longrightarrow \varphi(A)\left(\subseteq \prod_{\lambda \in \mathbf{K}} A_\lambda\right) \longrightarrow A \longrightarrow A/\hbar A, \quad f \mapsto (\varphi_0 \circ \varphi^{-1})(\widehat{f}), \tag{6}$$

which is an epimorphism of $\mathbb{C}$-algebras such that $\gamma(\widehat{q}) = 0$ and $\gamma(x_i) = \varphi_0(x_i)$ for $i = 1, \ldots, r$.

Here we write applications of the map γ in (6).

• One should observe [6, Example 1.6] in which the map γ in (6) induces a homeomorphism between the primitive spectrum of the coordinate ring $O_q(\mathbb{C}^2)$ of quantized affine 2-space and the Poisson primitive spectrum of its semiclassical limit $O(\mathbb{C}^2)$.

• One should observe [4, Theorem 4.2] in which the map γ in (6) induces a monomorphism from the group of automorphisms of Weyl algebra into the group of Poisson automorphisms of Poisson Weyl algebra.

Acknowledgment The author is supported by National Research Foundation of Korea, NRF-2017R1A2B4008388.

References

[1] Eun-Hee Cho and Sei-Qwon Oh, *Semiclassical limits of Ore extensions and a Poisson generalized Weyl algebra*, Lett. Math. Phys. **106** (2016), no. 7, 997–1009.

[2] K. R. Goodearl, *Semiclassical limits of quantized coordinate rings*, in Advances in Ring Theory (D. V. Huynh and S. R. Lopez-Permouth, Eds.) Basel Birkhäuser (2009), 165–204.

[3] M. Kontsevich, *Deformation quantization of Poisson manifolds*, arXiv:q-alg/9709040.

[4] No-Ho Myung and Sei-Qwon Oh, *Automorphism groups of Weyl algebras*, arXiv: 1710.00432v2, (2018)

[5] No-Ho Myung and Sei-Qwon Oh, *A construction of an iterated Ore extension*, arXiv: 1707.05160v2, (2018)

[6] Sei-Qwon Oh, *A natural map from a quantized space onto its semiclassical limit and a multi-parameter Poisson Weyl algebra*, Comm. Algebra **45** (2017), 60–75.

A numerical scheme for problems in fractional calculus

M. Razzaghi[1,*]

[1]Department of Mathematics and Statistics, Mississippi state University, Mississippi State, MS 39762, USA

Abstract. In this paper, a new numerical method for solving the fractional differential equations with boundary value problems is presented. The method is based upon hybrid functions approximation. The properties of hybrid functions consisting of block-pulse functions and Bernoulli polynomials are presented. The Riemann-Liouville fractional integral operator for hybrid functions is given. This operator is then utilized to reduce the solution of the boundary value problems for fractional differential equations to a system of algebraic equations. Illustrative examples are included to demonstrate the validity and applicability of the technique.

Keywords: Hybrid functions; fractional-order differential equations; block-pulse; Bernoulli polynomials; Caputo derivative; numerical solution

1 Introduction

Fractional differential equations (FDEs) are generalized from integer order ones, which are obtained by replacing integer order derivatives with fractional ones. A history of the development of fractional differential operators can be found in [1, 2].

FDEs have attracted considerable interest because of their ability to model complex phenomena such as continuum and statistical mechanics [3] visco-elastic materials [4], solid mechanics [5] and dynamics of interfaces between soft-nanoparticles and rough substrates [6]. In general, most of the FDEs do not have exact solutions. In particular, there is no known method for solving fractional boundary value problems exactly [7].

Due to the extensive applications of FDEs in engineering and science, research in this area has grown significantly, and there has been considerable interest in developing numerical schemes for their solution. These methods include Fourier transforms [8], eigenvector expansion [9], Laplace transforms [10], Adomian decomposition method [11], variational iteration method [12], finite difference method (FDM) [13], the power series method [14], fractional differential transform method (FDTM) [15], and homotopy analysis method [16].

Most of the work published to date concerning exact and numerical solutions for FDEs is devoted to the initial value problems. The theory of boundary value problems for FDEs has received attention quite recently [7].

The available sets of orthogonal functions can be divided into three classes. The first class includes sets of piecewise constant basis functions (e.g., block-pulse, Haar, Walsh, etc.). The second class consists of sets of orthogonal polynomials (e.g., Chebyshev, Laguerre, Legendre, etc.). The third class is the set of sine-cosine functions in the Fourier series. Orthogonal

*e-mail: razzaghi@math.msstate.edu

functions have been used when dealing with various problems of the dynamical systems. The main advantage, of using orthogonal functions is that they reduce the dynamical system problems to those of solving a system of algebraic equations by using the operational matrix of integration.

In recent years, the hybrid functions consisting of the combination of block-pulse functions with Legendre polynomials, Chebyshev polynomials, Taylor series, Lagrange polynomials or Bernoulli polynomials [17–22] have been shown to be a mathematical power tool for discretization of selected problems. Among these hybrid functions, the hybrid functions of block-pulse and Bernoulli polynomials have been shown to be computationally more effective [21–23]. In 2015, to the best of our knowledge, the Riemann-Liouville fractional integral operator for hybrid of block-pulse functions and Bernoulli polynomials was derived directly for the first time [24,25].

In the present paper, a new numerical method for solving the initial and boundary value problems for fractional order differential equations is presented. The method is based upon Hybrid functions approximation. These hybrid functions, which consist of the hybrid of block-pulse and Bernoulli polynomials are given. We then obtain the Riemann-Liouville fractional integral operator for hybrid of block-pulse functions and Bernoulli polynomials. This operator is then utilized to reduce the solution of the fractional order differential equations to the solution of algebraic equations.

2 Preliminaries and notations

2.1 The fractional derivative and integral

There are various definitions of fractional derivative and integration. The widely used definition of a fractional derivative is the Caputo definition and a fractional integration is the Riemann-Liouville definition.

Definition 1. Caputo's fractional derivative of order q is defined as [10]

$$(D^q f)(t) = \frac{1}{\Gamma(n-q)} \int_0^t \frac{f^{(n)}(s)}{(t-s)^{q+1-n}} ds, \qquad n-1 < q \leq n,\ n \in \mathbb{N},$$

where $q > 0$ is the order of the derivative and n is the smallest integer greater than q.

The Caputo fractional derivative is a linear operation, namely

$$D^q(\lambda f(t) + \mu g(t)) = \lambda D^q f(t) + \mu D^q g(t),$$

where λ and μ are constants.

Definition 2. The Riemann-Liouville fractional integral operator of order q is defined as [10]

$$I^q f(t) = \begin{cases} \frac{1}{\Gamma(q)} \int_0^t \frac{f(s)}{(t-s)^{1-q}} ds = \frac{1}{\Gamma(q)} t^{q-1} * f(t), & q > 0, \\ f(t), & q = 0. \end{cases} \tag{1}$$

where $t^{q-1} * f(t)$ is the convolution product of t^{q-1} and $f(t)$. For the Riemann-Liouville fractional integral we have

$$I^q t^\nu = \frac{\Gamma(\nu+1)}{\Gamma(\nu+1+q)} t^{\nu+q}, \qquad \nu > -1. \tag{2}$$

The Riemann-Liouville fractional integral is also a linear operation.

The Caputo derivative and Riemann-Liouville integral satisfy the following properties [10]

$$I^{\alpha}(D^{\alpha}h(t)) = h(t) - \sum_{k=0}^{n-1} h^{k}(0)\frac{t^k}{k!}, \tag{3}$$

if $\alpha \in R$, $n-1 < \alpha \leq n$, $n \in N$ then

$$D^{\alpha}(h(t)) = I^{n-\alpha}D^{n}h(t). \tag{4}$$

3 Hybrid of block-pulse functions and Bernoulli polynomials

Hybrid functions $b_{nm}(t)$, $n = 1, 2, \ldots, N$, $m = 0, 1, \ldots, M$ are defined on the interval $[0, t_f)$ as [24]

$$b_{nm}(t) = \begin{cases} \beta_m(\frac{N}{t_f}t - n + 1), & t \in [\frac{n-1}{N}t_f, \frac{n}{N}t_f), \\ 0, & \text{otherwise}, \end{cases} \tag{5}$$

where n and m are the order of block-pulse functions and Bernoulli polynomials, respectively. In Eq. (5), $\beta_m(t)$, $m = 0, 1, 2, ...$ are the Bernoulli polynomials of order m, which can be defined by [26]

$$\beta_m(t) = \sum_{k=0}^{m} \binom{m}{k} \alpha_{m-k} t^k, \tag{6}$$

where α_k, $k = 0, 1, ..., m$ are Bernoulli numbers [27]. These polynomials satisfy the following formula [27]

$$\beta_m(1 - x) = (-1)^m \beta_m(x). \tag{7}$$

3.1 Function Approximation

Let $f \in L^2[0, 1]$, the best approximation of f by using hybrid functions is given by

$$f \simeq P_M^N f = \sum_{m=0}^{M}\sum_{n=1}^{N} c_{nm} b_{nm}(t) = C^T B(t), \tag{8}$$

where

$$C^T = [c_{10}, c_{20}, ..., c_{N0}, c_{11}, c_{21}, ..., c_{N1}, ..., c_{1M}, c_{2M}, ..., c_{NM}], \tag{9}$$

and

$$B^T(t) = [b_{10}(t), b_{20}(t), ..., b_{N0}(t), b_{11}(t), b_{21}(t), ..., b_{N1}(t), ..., b_{1M}(t), b_{2M}(t), ..., b_{NM}(t)]. \tag{10}$$

4 Riemann-Liouville fractional integral operator for hybrid of block-pulse functions and Bernoulli polynomials

We now derive the operator I^{α} for $B(t)$ in Eq. (10) given by

$$I^{\alpha}B(t) = \overline{B}(t, \alpha), \tag{11}$$

where

$$\overline{B}(t, \alpha) = [I^{\alpha}b_{10}(t), ..., I^{\alpha}b_{N0}(t), I^{\alpha}b_{11}(t), ..., I^{\alpha}b_{N1}(t), ..., I^{\alpha}b_{1M}(t), I^{\alpha}b_{2M}(t), ..., I^{\alpha}b_{NM}(t)]^T. \tag{12}$$

To obtain $I^{\alpha}b_{nm}(t)$, we use the Laplace transform. By using Eq. (5), we have

$$b_{nm}(t) = \mu_{\frac{n-1}{N}}(t)\beta_m(Nt-n+1) - \mu_{\frac{n}{N}}(t)\beta_m(Nt-n+1), \tag{13}$$

where $\mu_c(t)$ is unit step function defined as

$$\mu_c(t) = \begin{cases} 1 & t \geq c, \\ 0, & t < c. \end{cases}$$

By taking the Laplace transform from Eq. (13) and using Eq. (7), we get

$$L[b_{nm}(t)] = e^{-\frac{n-1}{N}s}L[\beta_m(N(t+\frac{n-1}{N})-n+1) - e^{-\frac{n}{N}s}L[\beta_m(N(t+\frac{n}{N})-n+1)],$$

$$= e^{-\frac{n-1}{N}s}L[\beta_m(Nt)] - (-1)^m e^{-\frac{n}{N}s}L[\beta_m(-Nt)]. \tag{14}$$

From the definition of Bernoulli polynomials in Eq. (6), we have

$$L[b_{nm}(t)] = e^{-\frac{n-1}{N}s}L[\sum_{k=0}^{m}\begin{pmatrix} m \\ k \end{pmatrix}\alpha_{m-k}N^k t^k] - (-1)^m e^{-\frac{n}{N}s}L[\sum_{k=0}^{m}\begin{pmatrix} m \\ k \end{pmatrix}\alpha_{m-k}(-N)^k t^k]$$

$$= e^{-\frac{n-1}{N}s}\sum_{k=0}^{m}\begin{pmatrix} m \\ k \end{pmatrix}\alpha_{m-k}N^k\frac{\Gamma(k+1)}{s^{k+1}} - (-1)^m e^{-\frac{n}{N}s}\sum_{k=0}^{m}\alpha_{m-k}\begin{pmatrix} m \\ k \end{pmatrix}(-N)^k\frac{\Gamma(k+1)}{s^{k+1}}. \tag{15}$$

From Eq. (1),

$$L[I^{\alpha}b_{nm}(t)] = L[\frac{1}{\Gamma(\alpha)t^{1-\alpha}} * b_{nm}(t)] =$$

$$e^{-\frac{n-1}{N}s}\sum_{k=0}^{m}\begin{pmatrix} m \\ k \end{pmatrix}\alpha_{m-k}N^k\frac{\Gamma(k+1)}{s^{k+1+\alpha}} - (-1)^m e^{-\frac{n}{N}s}\sum_{k=0}^{m}\begin{pmatrix} m \\ k \end{pmatrix}\alpha_{m-k}(-N)^k\frac{\Gamma(k+1)}{s^{k+1+\alpha}}. \tag{16}$$

Taking the inverse Laplace transform of Eq. (16) yields [24,25]

$$I^{\alpha}b_{nm}(t) = \begin{cases} 0, & t \in (-\infty, \frac{n-1}{N}), \\ (t-\frac{n-1}{N})^{\alpha}d_{nm}(t) & t \in [\frac{n-1}{N}, \frac{n}{N}), \\ (t-\frac{n-1}{N})^{\alpha}d_{nm}(t) - (-1)^m(t-\frac{n}{N})^{\alpha}\overline{d}_{n+1m}(t) & t \in [\frac{n}{N}, \infty), \end{cases} \tag{17}$$

where

$$d_{nm}(t) = \sum_{k=0}^{m}\begin{pmatrix} m \\ k \end{pmatrix}N^k\alpha_{m-k}\frac{\Gamma(k+1)}{\Gamma(k+\alpha+1)}(t-\frac{n-1}{N})^k,$$

$$\overline{d}_{nm}(t) = \sum_{k=0}^{m}(-1)^k\begin{pmatrix} m \\ k \end{pmatrix}N^k\alpha_{m-k}\frac{\Gamma(k+1)}{\Gamma(k+\alpha+1)}(t-\frac{n}{N})^k.$$

5 Problem statement

We focus on the following problem:

Caputo fractional differential equation

$$D^{\alpha} f(t) = h(t, f(t), D^{\beta} f(t)), \quad 1 < \alpha \leq 2, \quad 0 \leq \beta \leq 1, \tag{18}$$

with the boundary conditions

$$f(0) = f_0, \qquad f(1) = f_1. \tag{19}$$

For this problem we have:

Lemma 1. Assume that $h : [0,1] \times R \times R \longrightarrow R$ is continuous. Then, $f \in C[0,1]$ is a solution of the boundary value problem in Eq. (18) and Eq. (19) if, and only if, $f(t)$ is the solution of [28]

$$\begin{aligned} f(t) &= I^{\alpha} h(t, f(t), D^{\beta} f(t)) - tI^{\alpha} h(1, f(1), D^{\beta} f(1)) + (f_1 - f_0)t + f_0 \\ &= \int_0^1 G(t,s) h(s, f(s), D^{\beta} f(s)) ds + (f_1 - f_0)t + f_0, \end{aligned} \tag{20}$$

where $G(t,s)$ is the Green function given by

$$G(t,s) = \frac{1}{\Gamma(\alpha)} \begin{cases} (t-s)^{\alpha-1} - t(1-s)^{\alpha-1}, & 0 \leq s \leq t, \\ -t(1-s)^{\alpha-1}, & t < s \leq 1. \end{cases}$$

The existence and uniqueness results for this problem are given in [28].

6 Numerical method

In this section, we use the hybrid of block-pulse functions and Bernoulli polynomials for solving problem given in Eqs. (18) and (19). Here, we expand $f(t)$ by the hybrid functions as

$$f(t) = A^T B(t), \tag{21}$$

by using Eq. (11), we get

$$I^{\alpha-\beta} f(t) = A^T \overline{B}(t, \alpha - \beta) \tag{22}$$

Substituting Eqs. (21) and (22) in Eq. (20), collocating the resulting equation at the Newton-cotes nodes t_i, given by

$$t_i = \frac{i+1}{2N(M+1)}, \quad i = 0, 1, ..., 2N(M+1) - 2. \tag{23}$$

These equations give $N(M+1)$ algebraic equations, which can be solved for the unknown vector A^T using Newton's iterative method.

7 Illustrative Example

In this section, two examples are given to demonstrate the applicability and accuracy of our method.

7.1 Example 1

Consider the boundary value problem for inhomogeneous linear fractional differential equation [7]

$$D^{\alpha} f(t) + af(t) = g(t), \quad t \in [0, 1], \tag{24}$$

$$f(0) = 0, \qquad f(1) = \frac{1}{\Gamma(\alpha + 2)}, \tag{25}$$

where $1 < \alpha \leq 2$, $a \in R$. For $g(t) = t + \frac{at^{\alpha}}{\Gamma(\alpha+2)}$, the exact solution of this problem is $f(t) = \frac{t^{\alpha+1}}{\Gamma(\alpha+2)}$.
Here, we solve this problem by using hybrid functions with $N = 1$ and $M = 1$. Let

$$D^{\alpha} f(t) = A^T B(t) = [a_{10}, a_{11}] \begin{bmatrix} b_{10}(t) \\ b_{11}(t) \end{bmatrix}. \tag{26}$$

Then, by using Eqs. (11) and (26), we have

$$f(t) = A^T \overline{B}(t, \alpha) + f'(0)t, \tag{27}$$

by substituting Eqs. (26) and (27) in Eq. (24), we get

$$A^T B(t) + a(A^T \overline{B}(t, \alpha) + f'(0)t) = g(t) \tag{28}$$

By collocating Eq. (28) at the Newton-cotes nodes given in Eq. (23), we get $N(M + 1) + 1$ unknown with $N(M + 1)$ equations. Also, by using Eqs. (25) and (27), we have

$$f(1) = A^T \overline{B}(1, \alpha) + f'(0), \tag{29}$$

From Eqs. (28) and (29), we get $a_{10} = 0.5, \ a_{11} = 1, \ and \ f'(0) = 0$. Then, by using Eq. (27), we have the exact solution.

7.2 Example 2

Consider boundary value problems for a class of fractional differential equations with three-point boundary conditions [7],

$$D^{\alpha} f(t) + af(t) = g(t), \quad t \in [0, 1], \tag{30}$$

$$f(0) = 0, \qquad f(1) = \xi f(\eta),$$

where $\eta \in (0, 1)$, a, $\xi \in R$ and $\Delta := 1 - \xi\eta \neq 0$.
Integrating Eq. (30) and using the boundary conditions, we have the following integral representation

$$f(t) = -aI^{\alpha} f(t) - \frac{at^{\alpha-1}}{\Delta}(\xi I^{\alpha} f(\eta) - I^{\alpha} f(1)) + F(t), \tag{31}$$

where

$$F(t) = I^{\alpha} g(t) + \frac{at^{\alpha-1}}{\Delta}(\xi I^{\alpha} g(\eta) - I^{\alpha} g(1)).$$

We solve this problem by using hybrid functions with $N = 1$ and $M = 5$. Let

$$f(t) = A_1^T B(t), \quad I^{\alpha} f(\eta) = A_2^T B(t), \quad I^{\alpha} f(1) = A_3^T B(t), \quad F(t) = K^T B(t), \tag{32}$$

then by using Eq. (11), we have

$$I^{\alpha} f(t) = A_1^T \overline{B}(t, \alpha). \tag{33}$$

Substituting Eqs. (32) and (33) in Eq. (31), we get

$$A_1^T B(t) = -aA_1^T \overline{B}(t, \alpha) - \frac{at^{\alpha-1}}{\Delta}(\xi A_2^T B(t) - A_3^T B(t)) + K^T B(t). \tag{34}$$

By collocating Eq. (34) at the Newton-cotes nodes given in Eq. (23), we get A_1. Taking

$$\alpha = \frac{3}{2},\ a = \frac{e^{-3\Pi}}{\sqrt{\Pi}},\ \xi = -\frac{125}{196},\ \eta = \frac{2}{5},\ f(1) = -\frac{1}{40}$$

and

$$g(t) = \frac{e^{-3\pi}}{\sqrt{\pi}}(t^5 - \frac{37}{20}t^3 + \frac{33}{40}t^2) + \frac{1}{\sqrt{\pi}}(\frac{128}{7}t^{\frac{7}{2}} - \frac{74}{5}t^{\frac{3}{2}} + \frac{33}{10}t^{\frac{1}{2}}),$$

the exact solution is

$$f(t) = (t^3 - \frac{37t}{20} + \frac{33}{40})t^2.$$

For these values, the exact solution is obtained by the present method.

8 Conclusion

A general formulation for the Riemann-Liouville fractional integral operator for hybrid of block-pulse functions and Bernoulli polynomials has been given. This operator is used to approximate numerical solution of FDEs. Our numerical findings are compared with exact solutions and with the solutions obtained by some other numerical methods. The solution obtained using the present method shows that this approach can solve the problem effectively.

References

[1] Oldham K.B, Spanier J. The Fractional Calculus, Academic Press: New York. 1974.

[2] Miller K.S, Ross B. An Introduction to the Fractional Calculus and Fractional Differential Equations, Wiley: New York. 1993.

[3] Mainardi F. Fractional calculus: Some basic problems in continuum and statistical mechanics', in: A. Carpinteri, F. Mainardi (Eds.), Fractals and Fractional Calculus in Continuum Mechanics, Springer Verlag: New York. 1997; 291–348.

[4] Bagley R.L, Torvik P.J. Fractional calculus in the transient analysis of viscoelastically damped structures, AIAA J. 1985; 23: 918–925.

[5] Rossikhin Y.A, Shitikova M.V. Applications of fractional calculus to dynamic problems of linear and nonlinear hereditary mechanics of solids, Appl.Mech. Rev. 1997; 50: 15–67.

[6] Chow T.S. Fractional dynamics of interfaces between soft-nanoparticles and rough substrates, Phys. Lett. A. 2005; 342: 148–155.

[7] Rehman M.U, Ali khan R. A numerical method for solving boundary value problems for fractional differential equations, Appl. Math. Model. 2012; 36: 894–907.

[8] Gaul L, Klein P, Kemple S. Damping description involving fractional operators, Mech. Syst. Signal. Pr. 1991; 5: 81–88.

[9] Suarez L, Shokooh A. An eigenvector expansion method for the solution of motion containing fractional derivatives, J. Appl. Mech. 1997; 64: 629–735.

[10] Podlubny I. Fractional Differential Equations: An Introduction to Fractional Derivatives, Fractional Differential Equations, to Methods of Their Solution and Some of Their Applications, Academic Press, New York. 1998.

[11] Momani S, Al-Khaled K. Numerical solutions for systems of fractional differential equations by the decomposition method, Appl. Math. Comput. 2005; 162: 1351–1365.

[12] Odibat Z, Momani S. Application of variational iteration method to nonlinear differential equations of fractional order, Int. J. Nonlinear Sci. Numer. Simul. 2006; 7: 27–34.

[13] Meerschaert M, Tadjeran C. Finite difference approximations for two-sided space-fractional partial differential equations, Appl. Numer. Math. 2006; 56: 80–90.

[14] Odibat Z, Shawagfeh N. Generalized Taylor's formula, Appl. Math. Comput. 2007; 186: 286–293.

[15] Arikoglu A, Ozkol I. Solution of fractional integro-differential equations by using fractional differential transform method, Chaos Solitons Fract 2009; 40: 521–529.

[16] Hashim I, Abdulaziz O, Momani S. Homotopy analysis method for fractional IVPs, Commun. Nonlinear Sci. Numer. Simul. 2009; 14: 674–684.

[17] Razzaghi M, Marzban H. Hybrid analysis direct method in the calculus of variations, Int. J. Comput. Math. 2000; 75: 259–269.

[18] Razzaghi M, Marzban H. Direct method for variational problems via hybrid of block-pulse and Chebyshev functions, Math. Prob. Eng. 2000; 6: 85-97.

[19] Marzban H, Razzaghi M. Solution of multi-delay systems using hybrid of block-pulse functions and Taylor series, J. Soun. Vibr. 2006; 292: 954–963.

[20] Tabrizidooz H.R, Marzban H, Razzaghi M. A composite collocation method for the nonlinear mixed Volterra-Fredholm-Hammerstein integral equations, Commun. Nonlin. Sci. Numeric. Simul. 2011; 16: 1186–1194.

[21] Haddadi N, Ordokhani Y, Razzaghi M. Optimal Control of Delay Systems by Using a Hybrid Functions Approximation, J. Optimi. Theo. Appli. 2012; 153: 338–356.

[22] Mashayekhi S, Ordokhani Y, Razzaghi M. Hybrid functions approach for nonlinear constrained optimal control problems, Commun. Nonlin. Sci. Numer. Simulat. 2012; 17: 1831–1843.

[23] Mashayekhi S, Ordokhani Y, Razzaghi M. Hybrid functions approach for optimal control of systems described by integro-differential equations, Appl. Math. Model. 2013; 37: 3355–3368.

[24] Mashayekhi, S., and Razzaghi, M., 2015, "Numerical solution of nonlinear fractional integro-differential equations by hybrid functions," Eng. Analysis Bound. Elem., 56, pp. 81–89.

[25] Mashayekhi, S., and Razzaghi, M., 2016, "Numerical solution of distributed order fractional differential equations by hybrid functions," J. Comput. Physics., 315, pp. 169–181.

[26] Costabile F, Dellaccio F, Gualtieri M.I. A new approach to Bernoulli polynomials, Rendiconti di Matematica, Serie VII. 2006; 26: 1–12.

[27] Arfken G. Mathematical Methods for Physicists, Third edition, Academic Press: San Diego. 1985.

[28] Mujeeb U.R, Ali Rahmat K. The Legendre wavelet method for solving fractional differential equations, Commun Nonlinear Sci. Numer. Simulat. 2011; 16: 4163–4173.

Positive solutions for a one-dimensional Sturm-Liouville semipositone superlinear p-Laplacian problem

K. D. Chu[1,*] and *D. D.* Hai[2,**]

[1]Faculty of Mathematics and Statistics, Ton Duc Thang University, Ho chi Minh City, Vietnam
[2]Department of Mathematics and Statistics, Mississippi state University, Mississippi State, MS 39762, USA

Abstract. We prove the existence of a positive classical solution for the p-Laplacian equation

$$-(r(t)\phi(u'))' = -\lambda h(u) + f(t,u),\ t \in (0,1)$$

with Sturm-Liouville boundary conditions, where $\phi(s) = |s|^{p-2}s, p > 1, r : [0,1] \to (0,\infty), f : (0,1) \times [0,\infty) \to \mathbb{R}$ is a Carathéodory function satisfying a superlinear condition at 0 and ∞ involving the principal eigenvalue of $-(r(t)\phi(u'))'$, $h : (0,\infty) \to (0,\infty)$ is allowed to have infinite semipositone structure at 0, and $\lambda \geq 0$ is a small parameter.

1 Introduction

Consider the one-dimensional p-Laplacian problem

$$\begin{cases} -(r(t)\phi(u'))' = -\lambda g(u) + f(t,u),\ t \in (0,1), \\ au(0) - b\phi^{-1}(r(0))u'(0) = 0,\ cu(1) + d\phi^{-1}(r(1))u'(1) = 0, \end{cases} \tag{1.1}$$

where $\phi(s) = |s|^{p-2}s, p > 1, a, b, c, d$ are nonnegative constants with $\delta < 1, ac + ad + bc > 0,\ r : [0,1] \to (0,\infty),\ f : (0,1) \times [0,\infty) \to \mathbb{R},\ g : (0,\infty) \to (0,\infty)$, and λ is a nonnegative parameter.

We are interested in positive classical solution of (1.1) in the superlinear case involving with the principal eigenvalue of the corresponding operator.

When $\lambda = 0$ and $r \equiv 1$, Manásevich, Njoku, and Zanolin [14] used time-mapping estimates to prove the existence of a classical positive solution to (1.1) under Dirichlet boundary conditions when $f(t,z)$ satisfies

$$\limsup_{z\to 0^+} \frac{f(t,z)}{z^{p-1}} < \lambda_1 < \liminf_{z\to\infty} \frac{f(t,z)}{z^{p-1}} \leq \infty \tag{1.2}$$

and $\liminf_{z\to 0^+} \frac{f(t,z)}{z^{p-1}} > -\infty$ uniformly for a.e. $t \in (0,1)$, where $\lambda_1 = 2^p(p-1)\left(\int_0^1 \frac{ds}{(1-s^p)^{1/p}}\right)^p$ is the principal eigenvalue of $-(\phi(u'))'$ with zero boundary conditions (see [6,7]). Their result improved a previous work by Kaper, Knapp, and Kwong [12] where the stronger condition

$$\lim_{z\to 0^+} \frac{f(t,z)}{z^{p-1}} = l \leq 0 \text{ and } \lim_{z\to\infty} \frac{f(t,z)}{z^{p-1}} = \infty$$

*e-mail: chuduckhanh@tdt.edu.vn
**e-mail: dang@math.msstate.edu

uniformly for $t \in (0, 1)$ had been used. Related results were obtained by Webb and Lan [15] when $p = 2$, in which they gave a general method that covered many boundary conditions including nonlocal ones for both sublinear and superlinear cases involving with the principal eigenvalue of an operator. In the PDE case, the existence of a positive solution to the problem

$$-\Delta u = f(u) \text{ in } \Omega,\ u = 0 \text{ on } \partial\Omega$$

with $f \geq 0$ satisfying (1.2) was obtained in [10], while similar results for the p-Laplacian

$$-\Delta_p u = f(u) \text{ in } \Omega,\ u = 0 \text{ on } \partial\Omega$$

with $p > 1$ were obtained in [5] using Granas fixed point index [8]. The existence result in [14] was generalized to the Sturm-Liouville boundary value problem (1.1) with $\lambda = 0$ in [3] and to the singular problem (1.1) with $\lambda \geq 0$ small and $g(u) = u^{-\delta}$ for some $\delta \in [0, 1)$ under Dirichlet boundary condition in [4]. In this note, we shall extend the result in [4] to the Sturm-Liouville boundary condition with a more general $g(u)$ allowing the challenging infinite semipositone case i.e. $\lim_{u\to 0^+} g(u) = -\infty$ (see [13]). We refer to [9,11,16] and the references therein for the literature on existence results in the superlinear/sublinear cases not involving with the principal eigenvalue of the corresponding operator.

Let λ_1 be the principal eigenvalue of $-(r(t)\phi(u'))'$ on $(0, 1)$ with Sturm-Liouville boundary condition in (1.1) and let $\phi_1 > 0$ be the corresponding eigenfunction with $\|\phi_1\|_\infty = 1$ (see [2]).

We shall assume the following conditions:

(A1) $r : [0, 1] \to (0, \infty)$ is continuous and $g : (0, \infty) \to (0, \infty)$ is continuous, integrable on $(0, 1)$ and decreasing.

(A2) $f : (0, 1) \times [0, \infty)$ is a Carathéodory function, that is $f(., z)$ is measurable for each $z \geq 0$ and $f(t, .)$ is continuous for a.e. $t \in (0, 1)$.

(A3) For each $k > 0$, there exists $\gamma_k \in L^1(0, 1)$ such that

$$|f(t, z)| \leq \gamma_k(t)$$

for a.e. $t \in (0, 1)$ and all $z \in [0, k]$.

(A4) There exists $\gamma \in L^1(0, 1)$ with $\gamma \geq 0$ such that

$$f(t, z) + \gamma(t)z^{p-1} \geq 0$$

for a.e. $t \in (0, 1)$ and all $z \geq 0$.

(A5) There exists a subset $B \subset [0, 1]$ of measure 0 such that

$$\limsup_{z\to 0^+} \frac{f(t, z)}{z^{p-1}} < \lambda_1 < \liminf_{z\to\infty} \frac{f(t, z)}{z^{p-1}}$$

uniformly for $t \in [0, 1]\backslash B$.

By a positive classical solution of (1.1), we mean a function $u \in C^1[0, 1]$ with $u > 0$ on $(0, 1)$, $r(t)\phi(u')$ absolutely continuous on $[0, 1]$ and satisfying (1.1).

Our main result is

Theorem 1.1. *Let (A1)-(A5) hold. Then there exists a constant* $\lambda_0 > 0$ *such that for* $\lambda < \lambda_0$*,(1.1) has a positive classical solution u with* $\inf_{(0,1)} \frac{u}{\omega} > 0$*, where* $\omega(t) = \min(at+b, d+c(1-t))$.

In particular, when f is independent of t, we obtain

Corollary 1.1. *Let r satisfy (A1) and let $f : [0, \infty) \to R$ be continuous with*

$$-\infty < \lim_{z\to 0^+} \frac{f(z)}{z^{p-1}} < \lambda_1 < \lim_{z\to\infty} \frac{f(z)}{z^{p-1}} \leq \infty.$$

Then there exists a constant $\lambda_0 > 0$ such that for $\lambda < \lambda_0$, (1.1) has a positive classical solution u with $\inf\limits_{(0,1)} \frac{u}{p} > 0$

2. Preliminary results

We shall denote the norm in $L^q(0, 1)$ and $C^1[0, 1]$ by $\|.\|_q$ and $|.|_{C^1}$ respectively. We recall the following fixed point theorem of Krasnoselskii type in a Banach space (see Amann [1, Theorem 12.3]).

Theorem A. [1, Theorem 12.3] *Let E be a Banach space and $A : E \to E$ be a completely continuous operator. Suppose there exist $h \in E, h \neq 0$ and positive constants r, R with $r \neq R$ such that*
(a) If $y \in E$ satisfies $y = \theta Ay$ for some $\theta \in (0, 1]$ then $\|y\| \neq r$,
(b) If $y \in E$ satisfies $y = Ay + \xi h$ for some $\xi \geq 0$ then $\|y\| \neq R$.
Then A has a fixed point $y \in E$ with $\min(r, R) < \|y\| < \max(r, R)$.

Lemma 2.1. *Let $\gamma, h \in L^1(0, 1)$ with $\gamma, h \geq 0$ and let $\lambda > 0, u \in C^1[0, 1]$ satisfy*

$$\begin{cases} -(r(t)(\phi(u'))' + \gamma(t)\phi(u) \geq -\lambda h \ \ on\ (0, 1), \\ au(0) - b\phi^{-1}(r(0))u'(0) \geq 0,\ cu(1) + d\phi^{-1}(r(1))u'(1) \geq 0. \end{cases}$$

Then there exist constants $\kappa, \sigma > 0$ independent of u, λ, h such that

$$u(t) \geq \left(\kappa\|u\|_\infty - \sigma\,(\lambda\|h\|_1)^{\frac{1}{p-1}}\right) p(t)$$

for $t \in [0, 1]$.

Proof. Let $v \in C^1[0, 1]$ satisfy

$$\begin{cases} -(r(t)(\phi(v'))' + \gamma(t)\phi(v) = -\lambda h \ \text{ on } (0, 1), \\ av(0) - b\phi^{-1}(r(0))v'(0) = 0,\ cv(1) + d\phi^{-1}(r(1))v'(1) = 0. \end{cases}$$

By [3, Lemma 2.4], $v \leq 0$ and $|v|_{C^1} \leq M\,(\lambda\|h\|_1)^{\frac{1}{p-1}}$, where $M > 0$ is independent of u and λ. By [3, Lemma 2.5], $u \geq v$ on $[0, 1]$. Since there exists a constant $k_0 > 0$ such that $|v(t)| \leq k_0|v|_{C^1}\omega(t)$ for all $t \in [0, 1]$ and $v \in C^1[0, 1]$ satisfying the Sturm-Liouville boundary conditions in (1.1), it follows that $v(t) \geq -Mk_0\,(\lambda\|h\|_1)^{\frac{1}{p-1}}\,\omega(t)$ for $t \in [0, 1]$. If $\|u\|_\infty \leq C\,(\lambda\|h\|_1)^{\frac{1}{p-1}}$ for some $C > 0$ then

$$u(t) \geq -Mk_0\,(\lambda\|h\|_1)^{\frac{1}{p-1}}\,\omega(t) \geq \left(\|u\|_\infty - (C + Mk_0)\,(\lambda\|h\|_1)^{\frac{1}{p-1}}\right)\omega(t) \tag{2.1}$$

for $t \in [0, 1$. Suppose $\|u\|_\infty > C\,(\lambda\|h\|_1)^{\frac{1}{p-1}}$, where $C > M$. Let $\|u\|_\infty = |u(\tau)|$ for some $\tau \in [0, 1]$. Then $u(\tau) > 0$ for otherwise $\|u\|_\infty = -u(\tau) \leq -v(\tau) \leq M\,(\lambda\|h\|_1)^{\frac{1}{p-1}}$, a contradiction.

Let $z \in C^1[0, \tau]$ be the solution of

$$\begin{cases} -(r(t)\phi(z'))' + \gamma(t)\phi(z) = -\lambda h \ \text{ on } (0, \tau), \\ az(0) - b\phi^{-1}(r(0))z'(0) = 0,\ \ z(\tau) = \|u\|_\infty \end{cases}$$

Note that $u \geq z \geq v \geq -M(\lambda\|h\|_1)^{\frac{1}{p-1}}$ on $[0, \tau]$. Let $z_1(t) = z(t) + M(\lambda\|h\|_1)^{\frac{1}{p-1}}$. Then $z_1 \geq 0$ on $[0, 1]$ and

$$z_1(t) = z_1(0) + \int_0^t \phi^{-1}\left(\frac{r(0)\phi(z_1'(0) + \int_0^s(\gamma(\xi)\phi(z) + \lambda h)d\xi}{r(s)}\right). \tag{2.2}$$

If $b = 0$ then $z(0) = v(0) = 0$ and so $z_1'(0) = z'(0) \geq v'(0) \geq -M(\lambda\|h\|_1)^{\frac{1}{p-1}}$. On the other hand, if $b \neq 0$ then $z_1'(0) = \frac{a}{b\phi^{-1}(r(0))}z(0) \geq -\frac{Ma}{b\phi^{-1}(r(0))}(\lambda\|h\|_1)^{\frac{1}{p-1}}$. Hence

$$z_1'(0) \geq -M_1(\lambda\|h\|_1)^{\frac{1}{p-1}}, \tag{2.3}$$

where $M_1 = M$ if $b = 0$ and $M_1 = \frac{Ma}{b\phi^{-1}(r(0))}$ if $b \neq 0$. Since $\phi(z_1'(0)) \leq \phi\left(z_1'(0) + M_1(\lambda\|h\|_1)^{\frac{1}{p-1}}\right)$ and

$$(x + y)^q \leq 2^{(q-1)^+}(x^q + y^q) \text{ for } x, y \geq 0, q > 0, \tag{2.4}$$

it follows from (2.2) that

$$z_1(t) \leq m_1\left(z_1(0) + z_1'(0) + M_1(\lambda\|h\|_1)^{\frac{1}{p-1}} + \phi^{-1}\left(\int_0^t(\gamma(\xi)\phi(z_1) + \lambda h)d\xi\right)\right),$$

where $m_1 > 0$ is a constant depending only on p and r. Hence using (2.4) again and the inequality $\phi(x) + \phi(y) \leq 2\phi(x + y)$ for $x, y \geq 0$, we deduce that

$$\phi(z_1(t)) \leq m_2\left(\phi\left(z_1(0) + z_1'(0) + M_1(\lambda\|h\|_1)^{\frac{1}{p-1}}\right) + \lambda\|h\|_1 + \int_0^t \gamma(s)\phi(z_1)ds\right)$$

$$\leq 2m_2\phi\left(z_1(0) + z_1'(0) + M_2(\lambda\|h\|_1)^{\frac{1}{p-1}}\right) + m_2\int_0^t \gamma(s)\phi(z_1)ds$$

for $t \in [0, \tau]$, where $M_2 = M_1 + 1$ and $m_2 = 2^{(p-2)^+}\phi(m_1)$.

By Gronwall's inequality,

$$\phi(z_1(t)) \leq 2m_2\phi\left(z_1(0) + z_1'(0) + M_2(\lambda\|h\|_1)^{\frac{1}{p-1}}\right)e^{m_2\int_0^t \gamma(s)ds}$$

for $t \in [0, \tau]$. In particular when $t = \tau$, we obtain

$$z(0) + z'(0) + (M_2 + M)(\lambda\|h\|_1)^{\frac{1}{p-1}} \geq \kappa_0\|u\|_\infty, \tag{2.5}$$

where $\kappa_0 = \left(e^{-m_2\|\gamma\|_1}/2m_2\right)^{1/(p-1)}$. Since

$$(r(t)\phi(z'))' = \gamma(t)\phi(z) + \lambda h \geq -\gamma(t)\lambda M^{p-1}\|h\|_1 \text{ on } [0, \tau],$$

it follows upon integrating on $[0, t]$ that

$$r(t)\phi(z'(t)) \geq r(0)\phi(z'(0)) - \lambda M_3\|h\|_1,$$

where $M_3 = M^{p-1}\|\gamma\|_1$, which implies

$$z'(t) \geq \phi^{-1}\left(\frac{r(0)\phi(z'(0)) - \lambda M_3\|h\|_1}{r(t)}\right) \tag{2.6}$$

for $t \in [0, \tau]$. Using the inequality $\psi(x - y) \geq c_1\psi(x) - c_2\psi(y)$ for $x, y \geq 0$ and $\psi \in \{\phi, \phi^{-1}\}$, where $c_1 = \psi(1/2), c_2 = 2\psi(2)$, we deduce from (2.3) and (2.6) that

$$z'(t) \geq M_4 z'(0) - M_5(\lambda\|h\|_1)^{\frac{1}{p-1}} \geq -M_6(\lambda\|h\|_1)^{\frac{1}{p-1}} \tag{2.7}$$

where M_4, M_5 are positive constants depending only on p and r, and $M_6 = M_1M_4 + M_5$. If $b = 0$ then $z(0) = 0$ and (2.5) and (2.7) give

$$\begin{aligned} z(t) = \int_0^t z' &\geq \left(M_4\left((\kappa_0\|u\|_\infty - (M + M_2)(\lambda\|h\|_1)^{\frac{1}{p-1}}\right) - M_5(\lambda\|h\|_1)^{\frac{1}{p-1}}\right)t \\ &= \left(\kappa_1\|u\|_\infty - \sigma_1(\lambda\|h\|_1)^{\frac{1}{p-1}}\right)(at + b) \end{aligned} \tag{2.8}$$

for $t \in [0, \tau]$, where $\kappa_1 = a^{-1}M_4\kappa_0, \sigma_1 = a^{-1}\left(M_4(M + M_2) + M_5\right)$. On the other hand, if $b > 0$ then $z'(0) = \frac{a}{b\phi^{-1}(r(0))}z(0)$ from which (2.5) becomes

$$z(0) \geq \tilde{\kappa}_1\|u\|_\infty - M_7(\lambda\|h\|_1)^{\frac{1}{p-1}},$$

where $\tilde{\kappa}_1 = \kappa_0\left(1 + \frac{a}{b\phi^{-1}(r(0))}\right)^{-1}$, $M_7 = (M + M_2)\left(1 + \frac{a}{b\phi^{-1}(r(0))}\right)^{-1}$. Hence it follows upon integrating (2.7) that

$$\begin{aligned} z(t) \geq z(0) - M_6(\lambda\|h\|_1)^{\frac{1}{p-1}}t &\geq \tilde{\kappa}_1\|u\|_\infty - (M_6 + M_7)(\lambda\|h\|_1)^{\frac{1}{p-1}} \\ &\geq \left(\kappa_1\|u\|_\infty - \sigma_1(\lambda\|h\|_1)^{\frac{1}{p-1}}\right)(at + b) \end{aligned} \tag{2.9}$$

for $t \in [0, \tau]$, where $\kappa_1 = \frac{\tilde{\kappa}_1}{a+b}$, $\sigma_1 = \frac{M_6+M_7}{b}$. Combining (2.8) and (2.9), we see that in any case there exist constants κ_1 and $\sigma_1 > 0$ independent of u, z, λ, h such that

$$u(t) \geq z(t) \geq \left(\kappa_1\|u\|_\infty - \sigma_1(\lambda\|h\|_1)^{\frac{1}{p-1}}\right)(at + b) \tag{2.10}$$

for $t \in [0, \tau]$. Next, let $w \in C^1[\tau, 1]$ be the unique solution of

$$\begin{cases} -(r(t)\phi(w'))' + \gamma(t)\phi(w) = -\lambda h \ \text{ on } (\tau, 1), \\ w(\tau) = \|u\|_\infty, \ \ cw(1) + d\phi^{-1}(r(1))w'(1) = 0. \end{cases}$$

Then $u \geq w \geq v$ on $[\tau, 1]$. Let $w_1 = w + M(\lambda\|h\|_1)^{\frac{1}{p-1}}$. Then $w_1 \geq 0$ and by using the integral formula

$$w_1(t) = w_1(1) + \int_t^1 \phi^{-1}\left(\frac{-r(1)\phi(w_1'(1) + \int_s^1(\gamma(\xi)\phi(w) + \lambda h)d\xi}{r(s)}\right)ds$$

for $t \in [\tau, 1]$ with the same arguments as above, we obtain

$$u(t) \geq w(t) \geq \left(\kappa_2\|u\|_\infty - \sigma_2(\lambda\|h\|_1)^{\frac{1}{p-1}}\right)(d + c(1 - t)) \tag{2.11}$$

for $t \in [\tau, 1]$, where $\kappa_2 > 0$ is a constant independent of u, w, λ, h. Hence if $C > \max(\sigma_1/\kappa_1, \sigma_2/\kappa_2, M)$ then it follows from (2.1),(2.10), and (2.11) that $u(t) \geq \left(\kappa\|u\|_\infty - \sigma(\lambda\|h\|_1)^{\frac{1}{p-1}}\right)\omega(t)$ for $t \in [0, 1]$, where $\kappa = \min(\kappa_1, \kappa_2, 1), \sigma = \max(\sigma_1, \sigma_2, C + Mk_0)$, which completes the proof of Lemma 2.1.

3. Proof of the main result

Proof of Theorem 1.1. Since $\limsup_{z\to 0^+} \frac{f(t,z)}{z^{p-1}} < \lambda_1$, there exist constants $r, \bar{\lambda} > 0$ with $\bar{\lambda} < \lambda_1$ such that

$$f(t,z) \le \bar{\lambda} z^{p-1} \tag{3.1}$$

for $z \le r$ and a.e. $t \in (0,1)$. Let κ, σ be given by Lemma 2.1 with $h(t) = g(r_0\omega(t))$, where $r_0 = \kappa r/2$, and let $\lambda > 0$ be small enough so that $\sigma(\lambda\|h\|_1)^{\frac{1}{p-1}} < r_0$. For $v \in E = C[0,1]$, we have $0 \le f(t,|v|) + \gamma(t)|v|^{p-1} \in L^1(0,1)$ in view of (A2) and (A4). Hence the problem

$$\begin{cases} -(r(t)\phi(u'))' + \gamma(t)\phi(u) = -\lambda g(\max(v, r_0\omega)) + f(t,|v|) + \gamma(t)|v|^{p-1} \text{ on } (0,1), \\ au(0) - b\phi^{-1}(r(0))u'(0) = 0, \ cu(1) + d\phi^{-1}(r(1))u'(1) = 0, \end{cases}$$

has a unique solution $u = Av \in C^1[0,1]$. Since $A = T_0 \circ S_0$, where $S_0 : C[0,1] \to L^1(0,1)$ is defined by $(S_0 v)(t) = -\lambda g(\max(v, r_0\omega)) + f(t,|v|) + \gamma(t)|v|^{p-1}$ and T_0 is defined in Lemma A with $\alpha = \beta = 0$, we see that $A : E \to E$ is completely continuous. We shall verify that

(i) $u = \theta Au$, $\theta \in (0,1] \Longrightarrow \|u\|_\infty \ne r$.

Indeed, let $u \in E$ satisfy $u = \theta Au$ for some $\theta \in (0,1]$ and suppose $\|u\|_\infty = r$. Then $u \in C^1[0,1]$ and

$$\begin{aligned} -(r(t)\phi(u'))' + \gamma(t)\phi(u) &= \theta^{p-1}\left(-\lambda g(\max(u, r_0\omega)) + f(t,|u|) + \gamma(t)|u|^{p-1}\right) \\ &\ge -\lambda h(t) \end{aligned}$$

on $(0,1)$, from which Lemma 2.1 gives

$$u(t) \ge \left(\kappa\|u\|_\infty - \sigma(\lambda\|h\|_1)^{\frac{1}{p-1}}\right)\omega(t) \ge r_0\omega(t) > 0 \tag{3.2}$$

for $t \in (0,1)$. Hence

$$-(r(t)\phi(u'))' \le \theta^{p-1} f(t,u) \tag{3.3}$$

on $(0,1)$. Let $\delta_0 = \sup\limits_{(0,1)} \frac{u}{\phi_1} \in (0,\infty)$. Then it follows from (3.1) and (3.3) that

$$-(r(t)\phi(u'))' \le \bar{\lambda} u^{p-1} \le \bar{\lambda}\delta_0^{p-1}\phi_1^{p-1} \text{ on } (0,1),$$

from which the weak comparison principle gives

$$u \le (\bar{\lambda}\delta_0^{p-1}/\lambda_1)^{\frac{1}{p-1}}\phi_1$$

on $[0,1]$, a contradiction with the definition of δ_0. Thus $\|u\|_\infty \ne r$ i.e. (i) holds.

Next, we claim that

(ii) There exists a constant $R > r$ such that $u = Au + \xi$, $\xi \ge 0 \Longrightarrow \|u\|_\infty \ne R$.

Let $u \in E$ satisfy $u = Au + \xi$ for some $\xi \in [0,\infty)$. Then $u - \xi = Au$ and therefore

$$-(r(t)\phi(u'))' + \gamma(t)\phi(u-\xi) = -\lambda g(\max(u, r_0\omega)) + f(t,|u|) + \gamma(t)|u|^{p-1} \tag{3.4}$$

on $(0,1)$, which implies

$$-(r(t)\phi(u'))' + \gamma(t)\phi(u) \ge -\lambda h(t) \tag{3.5}$$

on $(0,1)$. Since $\liminf\limits_{z\to\infty} \frac{f(t,z)-\lambda g(z)}{z^{p-1}} > \lambda_1$ uniformly for a.e. $t \in (0,1)$, there exist positive constants $L, \tilde{\lambda}$ with $\tilde{\lambda} > \lambda_1$ such that $f(t,z) - \lambda g(z) \ge \tilde{\lambda} z^{p-1}$ for a.e. $t \in (0,1)$ and $z > L$.

For $z \le L$, (A2) gives $f(t,z) - \lambda g(\max(z, r_0\omega(t))) \ge -\tilde{\gamma}_L(t)$ for a.e. $t \in (0,1)$ and all $z \ge 0$, where $\tilde{\gamma}_L = \gamma_L + \lambda h$. Suppose $\|u\|_\infty = R > r$. Then

$$u(t) \ge \left(\kappa R - \sigma(\lambda\|h\|_1)^{\frac{1}{p-1}}\right)p(t) \ge \frac{\kappa R}{2}\omega(t) \ge \frac{\kappa R c}{2}\phi_1(t)$$

for $t \in [0,1]$ in view of Lemma 2.1, where $c = \inf_{(0,1)} \frac{\omega}{\phi_1}$. In particular, $u(t) \geq r_0\omega(t)$ for $t \in [0,1]$. Let $\delta_1 = \inf_{(0,1)} \frac{u}{\phi_1}$. Then $\delta_1 \geq 1$ if $R > 2/\kappa c$, which we assume. Hence

$$-(r(t)\phi(u'))' \geq f(t,u) - \lambda g(u) \geq \begin{cases} \tilde{\lambda} u^{p-1} & \text{if } u > L \\ -\tilde{\gamma}_L & \text{if } u \leq L \end{cases}.$$

Since $u(t) > L$ if $\omega(t) > 2L/\kappa R$, it follows that

$$-\left(r(t)\phi\left(\frac{u'}{\delta_1}\right)\right)' \geq \begin{cases} \tilde{\lambda}\phi_1^{p-1} & \text{in } D_R, \\ -\tilde{\gamma}_L & \text{in } (0,1)\backslash D_R \end{cases} \equiv h_{1R},$$

where $D_R = \{t : \omega(t) > 2L/\kappa R\}$. Let $u_{1R}, u_2 \in C^1[0,1]$ satisfy

$$-\left(r(t)\phi\left(u'_{1R}\right)\right)' = h_{1R} \text{ on } (0,1),$$

and

$$-(r(t)\phi(u_2'))' = \tilde{\lambda}{\phi_1}^{p-1} \equiv h_2 \text{ on } (0,1).$$

with Sturm-Liouville boundary condition in (1.1). Note that $u_2 = (\tilde{\lambda}/\lambda_1)^{\frac{1}{p-1}}\phi_1$. Since

$$\|h_{1R} - h_2\|_1 = \int_{(0,1)\backslash D_R} (\tilde{\lambda}\phi_1^{p-1} + \tilde{\gamma}_L) \to 0$$

as $R \to \infty$, it follows that $|u_{1R} - u_2|_{C^1} \to 0$ as $R \to \infty$ (see e.g. [11, Lemma 3.1]). Let $\varepsilon > 0$ be such that $\left(\tilde{\lambda}/\lambda_1\right)^{\frac{1}{p-1}} - \varepsilon c_0 \equiv c_1 > 1$, where $c_0 = \sup_{(0,1)} \frac{\omega}{\phi_1} > 0$. Then, if R is large enough (independent of u), we get

$$u_{1R} \geq u_2 - \varepsilon\omega \geq u_2 - \varepsilon c_0\phi_1 = c_1\phi_1 \text{ on } (0,1),$$

and consequently, $u \geq \delta_1 u_{1R} \geq \delta_1 c_1 \phi_1$ on $(0,1)$, a contradiction with the definition of δ_1. Thus $\|u\|_\infty \neq R$ if R is large enough i.e. (ii) holds.

By Lemma A, A has a fixed point $u \in E$ with $\|u\|_\infty > r$, which is a classical positive solution of (1.1) in view of (3.2) and Lemmas 2.2. This completes the proof of Theorem 1.1.

References

[1] H. Amann, Fixed point equations and nonlinear eigenvalue problems in ordered Banach Spaces, *SIAM Rev.* **18** (1976), 620-709.

[2] P. Binding and P. Drabek, Sturm-Liouville theory for the p-Laplacian, *Studia Sci. Math. Hunga.* **40** (2003), no. 4, 375-396.

[3] K. D. Chu and D. D. Hai, Positive solutions for the one-dimensional Sturm-Liouville superlinear p-Laplacian problem, *Electron. J. Differential Equations* (2018), No. 92, pp. 1-14.

[4] K. D. Chu and D. D. Hai, Positive solutions for the one-dimensional singular superlinear p-Laplacian (submitted).

[5] A. Cwiszewski and M. Maciejewski, Positive stationary solutions for p-Laplacian problems with nonpositive perturbation, *J. Differential Equations* **254** (2013), 1120-1136.

[6] M. Del Pino, M. Elgueta, and R. Manasevich, A homotopic deformation along p of a Leray-Schauder degree result and existence for $(|u'|^{p-2}u')' + f(t,u) = 0, u(0) = u(T) = 0, p > 1$. *J. Differential Equations* **80** (1989), no. 1, 1–13.

[7] P. Drabek, Ranges of a -homogeneous operators and their perturbations, *Casopis Pest. Mat.* **105** (1980), 167-183.

[8] J. Dugundji and A. Granas, Fixed Point Theory, Springer-Verlag, 2004.

[9] L. Erbe and H. Wang, On the existence of positive solutions of ordinary differential equations. *Proc. Amer. Math. Soc.* **120** (1994), no. 3, 743–748.

[10] D. de Figueiredo, P. L. Lions, and R. D. Nusbaum, Estimations a priori pour les solutions positives de problèmes elliptiques superlinéaires. (French) *C. R. Acad. Sci. Paris Sér. A-B* **290** (1980), no. 5, 217–220.

[11] D. D. Hai, On singular Sturm-Liouville boundary-value problems. *Proc. Roy. Soc. Edinburgh Sect. A* **140** (2010), no. 1, 49–63.

[12] H. G. Kaper, M. Knaap, and M. K. Kwong, Existence theorems for second order boundary value problems. *Differential Integral Equations* **4** (1991), no. 3, 543–554.

[13] E. Lee, R. Shivaji, and J. Ye, Subsolutions: a journey from positone to infinite semipositone problems, *Electron. J. Differ. Equ. Conf.* **17** (2009), 123-131.

[14] R. Manásevich, F. Njoku and Zanolin, F, Positive solutions for the one-dimensional p-Laplacian. *Differential Integral Equations* **8** (1995), 213–222.

[15] J. R. L. Webb and K. Q. Lan, Eigenvalue criteria for existence of multiple positive solutions of nonlinear boundary vale problems of local and nonlocal types, *Topol. Methods Nonlinear Anal.* **27** (2006), 91-116.

[16] J. Wang, The existence of positive solutions for the one-dimensional p-Laplacian, *Proc. Amer. Math. Soc.* **125** (1997), 2275-2283.

A note on inverse problem for strongly damped wave equation with Gaussian white noise

Duc Khanh Chu[1,*], *Hoang Luc* Nguyen[2,**], *Van* Phan[3,***], and *Huy Tuan* Nguyen[4,****]

[1]Faculty of Mathematics and Statistics, Ton Duc Thang University, Ho Chi Minh City, Viet Nam
[2]Applied Analysis Research Group, Faculty of Mathematics and Statistics, Ton Duc Thang University, Ho Chi Minh City, Viet Nam
[3]Faculty of Applied Science, University of Technology, Ho Chi Minh City, Viet Nam
[4]Applied Analysis Research Group, Faculty of Mathematics and Statistics, Ton Duc Thang University, Ho Chi Minh City, Viet Nam

Abstract. In this paper, we study for the first time the inverse initial problem for the one-dimensional strongly damped wave with Gaussian white noise data. Under some a priori assumptions on the true solution, we propose the Fourier truncation method for stabilizing the ill-posed problem. Error estimates are given in both the L^2- and H^p-norms.
Keywords and phrases: Fourier regularization method; final value problem; strongly damped wave ; Gaussian white noise.
Mathematics subject Classification 2000: 35K05, 35K99, 47J06, 47H10

1 Introduction

Let T be a positive number and $D = (0, \pi)$. We are interested in the problem of recovering the initial state $u(x, 0)$, $x \in D$, for the following strongly damped wave equation

$$u_{tt} - \alpha u_{xxt} - u_{xx} = 0, \quad (x, t) \in D \times (0, T), \tag{1.1}$$

subject to the final conditions

$$\begin{cases} u(x, T) = g(x), & (x, t) \in D \times (0, T), \\ u_t(x, T) = 0, & (x, t) \in D \times (0, T), \end{cases} \tag{1.2}$$

and the boundary condition

$$u(0, t) = u(\pi, t) = 0, \quad t \in (0, T), \tag{1.3}$$

where $\alpha > 2$ is a fixed positive constant. In physics, the homogeneous Dirichlet boundary condition (1.3) expresses that two ends of the string are fixed. We can consider other types of

[*]e-mail: chuduckhanh@tdtu.edu.vn
[**]e-mail: hoangluctt@gmail.com (corresponding author)
[***]e-mail: khanhvanphan@hcmut.edu.vn
[****]e-mail: nguyenhuytuan@tdtu.edu.vn

boundary condition instead. In practice, we cannot measure g exactly, but we observe with the presence of a Gaussian white noise process ξ

$$g_\epsilon^{obs}(x) = g(x) + \epsilon\xi(x), \tag{1.4}$$

where $\epsilon > 0$ is the amplitude of the noise. Moreover, it can only be observed in discretized form:

$$\langle g_\epsilon^{obs}, \phi_j \rangle = \langle g, \phi_j \rangle + \epsilon\langle \xi, \phi_j \rangle, \quad j = \overline{1, N}, \tag{1.5}$$

where the natural number N is the number of steps of discrete observation and ϕ_j, $j = \overline{1, N}$, are defined in the next section. For more details on this random model, we refer to [1, 4].

This problem is well-known to be severely ill-posed and regularization methods for it are required. Strongly damped wave equation (SDWE) occurs in a wide range of applications such as modeling motion of viscoelastic materials [2, 7, 8]. From both the theoretical and numerical points of view, the initial value problem has been extensively studied (see e.g., [3, 5, 9]). However, to the best of our knowledge, the final value problem (1.1)-(1.3) with Gaussian white noise data has not been studied. In [6], Lions and Lattes introduced Problem (1.1)-(1.3), but regularization methods haven't been mentioned. The major object of this paper is to propose a stable regularized solution for the problem (1.1)-(1.3) using the Fourier truncation method.

2 Preliminaries

Let us recall that the problem

$$\begin{cases} -\phi_{xx} = \lambda\phi, & x \in D, \\ \phi(0) = \phi(\pi) = 0, \end{cases} \tag{2.6}$$

admits the eigenvalues $\lambda_j = j^2, j = \overline{1, \infty}$, and $\phi_j(x) = \sqrt{\frac{2}{\pi}}\sin(jx)$ are the corresponding eigenfunctions, which form an orthonormal basis of $L^2(D)$.

Definition 2.1 *Let H be a Hilbert space. The stochastic error is a Hilbert space process, i.e. a bounded linear operator $\xi : H \to L^2(\Omega, \mathcal{A}, \mathcal{P})$, where $(\Omega, \mathcal{A}, \mathcal{P})$ is the underlying probability space and $L^2(\cdot)$ is the space of all square integrable measurable functions.*

Remark 2.1 *Let $g \in H$. It follows from the representation* (1.4) *that*

$$\langle g_\epsilon, \chi \rangle = \langle g, \chi \rangle + \epsilon\langle \xi, \chi \rangle, \quad \forall\chi \in H.$$

Since $\xi(x)$ is a Gaussian white noise process, we have that $\langle \xi, \chi \rangle \sim N\left(0, \|\chi\|_H^2\right)$, and for any $\chi_1, \chi_2 \in H$, we have

$$\mathbf{E}\left(\langle \xi, \chi_1 \rangle \langle \xi, \chi_2 \rangle\right) = \langle \chi_1, \chi_2 \rangle.$$

In (1.5)*, we consider the test functions $\chi = \phi_j$, $j = \overline{1, N}$, - the eigenfunctions of the problem* (2.6)*, and $H = L^2(D)$. Then, $\xi_i := \langle \xi, \phi_j \rangle \sim N(0, 1)$ are i.i.d standard Gaussian random variables.*

For $p > 0$, we denote by $H^p(D)$ the closed subspace of $L^2(D)$, given by

$$H^p(D) = \left\{ u \in L^2(D) : \sum_{j=1}^{\infty} \lambda_j^p \langle u, \phi_j \rangle^2 < +\infty \right\},$$

and equipped with the norm

$$\|u\|_{H^p(D)} = \sqrt{\sum_{j=1}^{\infty} \lambda_j^p \langle u, \phi_j \rangle^2},$$

where $\langle \cdot, \cdot \rangle$ denotes the inner product in $L^2(D)$. Throughout this paper, denote by $\| \cdot \|$ the norm in $L^2(D)$.

Lemma 2.1 *Let $\overline{G}_{\epsilon,N(\epsilon)} \in L^2(D)$ be such that*

$$\overline{G}_{\epsilon,N(\epsilon)} = \sum_{j=1}^{N(\epsilon)} \left\langle g_\epsilon^{obs}, \phi_j \right\rangle \phi_j.$$

Assume that $g \in H^\gamma(D)$ for $\gamma > 0$. Then, we have the following estimate

$$\mathbf{E} \left\| \overline{G}_{\epsilon,N(\epsilon)} - g \right\|^2 \leq \epsilon^2 N(\epsilon) + \frac{1}{N(\epsilon)^{2\gamma}} \|g\|^2_{H^\gamma(D)}.$$

Here, N depends on ϵ and satisfies that $\lim_{\epsilon \to 0} N(\epsilon) = +\infty$.

Proof. By the usual MISE decomposition which involves a variance term and a bias term, we get

$$\begin{aligned}
\mathbf{E} \left\| \overline{G}_{\epsilon,N(\epsilon)} - g \right\|^2 &= \mathbf{E} \left(\sum_{j=1}^{N(\epsilon)} \left\langle g_\epsilon^{obs} - g, \phi_j \right\rangle^2 \right) + \sum_{j \geq N(\epsilon)+1} \left\langle g, \phi_j \right\rangle^2 \\
&= \epsilon^2 \mathbf{E} \left(\sum_{j=1}^{N(\epsilon)} \xi_j^2 \right) + \sum_{j \geq N(\epsilon)+1} \lambda_j^{-\gamma} \lambda_j^{\gamma} \langle g, \phi_j \rangle^2.
\end{aligned}$$

Since $\xi_j = \langle \xi, \phi_j \rangle \sim \mathcal{N}(0, 1)$, it follows that $\mathbf{E}\xi_j^2 = 1$. Therefore,

$$\mathbf{E} \left\| \overline{G}_{\epsilon,N(\epsilon)} - g \right\|^2 \leq \epsilon^2 N(\epsilon) + \frac{1}{N^{2\gamma}(\epsilon)} \|g\|^2_{H^\gamma(D)}.$$

3 Main results

Assume that Problem (1.1)-(1.3) has a unique true solution, we first find its Fourier series representation. From (1.1)-(1.3), we can derive the following ODE with given data at $t = T$

$$\begin{cases}
\dfrac{d^2}{dt^2} u_j(t) + \alpha \lambda_j \dfrac{d}{dt} u_j(t) + \lambda_j u_j(t) = 0, & t \in (0, T), \\
u_j(T) = g_j, \quad \dfrac{d}{dt} u_j(T) = 0,
\end{cases}$$

where $u_j(t) = \int_0^\pi u(x,t)\phi_j(x)dx$, $g_j = \int_0^\pi g(x)\phi_j(x)dx$.
For fixed damping $\alpha > 2$, the solution of the latter equation is given by

$$u_j(t) = \underbrace{\frac{\frac{\alpha j^2 + \sqrt{\alpha^2 j^4 - 4j^2}}{2}\exp\left((T-t)\frac{\alpha j^2 - \sqrt{\alpha^2 j^4 - 4j^2}}{2}\right) - \frac{\alpha j^2 - \sqrt{\alpha^2 j^4 - 4j^2}}{2}\exp\left((T-t)\frac{\alpha j^2 + \sqrt{\alpha^2 j^4 - 4j^2}}{2}\right)}{\sqrt{\alpha^2 j^4 - 4j^2}}}_{\mathcal{A}_j(T-t)} g_j. \tag{3.7}$$

Hence, the solution of problem (1.1)-(1.3) can be represented as

$$u(x,t) = \sum_{j=1}^{\infty} u_j(t)\phi_j(x).$$

For fixed t and $\beta_{N(\epsilon)}$, let us define the operator $Q_{\beta_{N(\epsilon)}}(t) : L_2(D) \to L_2(D)$

$$Q_{\beta_{N(\epsilon)}}(t)\varphi = \sum_{\lambda_j \le \beta_{N(\epsilon)}} \mathcal{A}_j(t)\left\langle \varphi, \phi_j \right\rangle \phi_j.$$

Now, we sate and prove the following lemma, which is needed for our analysis.

Lemma 3.1 *For fixed $t \in [0,T]$, we have*

$$\left\| Q_{\beta_{N(\epsilon)}}(t)\varphi \right\| \le \sqrt{2 + \frac{8T^2}{\alpha^2} e^{\alpha t \beta_{N(\epsilon)}}} \, \|\varphi\|. \tag{3.8}$$

Moreover,

$$\left\| Q_{\beta_{N(\epsilon)}}(t)\varphi \right\|_{H^p(D)} \le \sqrt{\beta_{N(\epsilon)}^p \left(2 + \frac{8T^2}{\alpha^2}\right) e^{\alpha t \beta_{N(\epsilon)}}} \, \|\varphi\|. \tag{3.9}$$

Proof. Using Parseval's equality, we obtain

$$\left\| Q_{\beta_{N(\epsilon)}}(t)\varphi \right\|^2 = \sum_{\lambda_j \le \beta_{N(\epsilon)}} \mathcal{A}_j^2(t)\left\langle \varphi, \phi_j \right\rangle^2. \tag{3.10}$$

From the equality

$$\frac{\alpha j^2 + \sqrt{\alpha^2 j^4 - 4j^2}}{2} = \sqrt{\alpha^2 j^4 - 4j^2} + \frac{\alpha j^2 - \sqrt{\alpha^2 j^4 - 4j^2}}{2},$$

$\mathcal{A}_j(t)$ can be rewritten as

$$\begin{aligned}
\mathcal{A}_j(t) &= \frac{\frac{\alpha j^2 + \sqrt{\alpha^2 j^4 - 4j^2}}{2}\exp\left(t\frac{\alpha j^2 - \sqrt{\alpha^2 j^4 - 4j^2}}{2}\right) - \frac{\alpha j^2 - \sqrt{\alpha^2 j^4 - 4j^2}}{2}\exp\left(t\frac{\alpha j^2 + \sqrt{\alpha^2 j^4 - 4j^2}}{2}\right)}{\sqrt{\alpha^2 j^4 - 4j^2}} \\
&= \exp\left(t\frac{\alpha j^2 - \sqrt{\alpha^2 j^4 - 4j^2}}{2}\right) \\
&\quad + \left(\alpha j^2 - \sqrt{\alpha^2 j^4 - 4j^2}\right)\frac{\exp\left(t\frac{\alpha j^2 - \sqrt{\alpha^2 j^4 - 4j^2}}{2}\right) - \exp\left(t\frac{\alpha j^2 + \sqrt{\alpha^2 j^4 - 4j^2}}{2}\right)}{2\sqrt{\alpha^2 j^4 - 4j^2}}.
\end{aligned}$$

Applying the inequalities $(a+b)^2 \le 2a^2 + 2b^2$ and $|e^a - e^b| \le \max\{e^a, e^b\}|a-b|$, we obtain

$$\begin{aligned}
\mathcal{A}_j^2(t) &\le 2\exp\left[t\left(\alpha j^2 - \sqrt{\alpha^2 j^4 - 4j^2}\right)\right] + \frac{\left(\alpha j^2 - \sqrt{\alpha^2 j^4 - 4j^2}\right)^2 t^2 \exp\left[t\left(\alpha j^2 + \sqrt{\alpha^2 j^4 - 4j^2}\right)\right]}{2} \\
&\le \exp\left(2\alpha j^2 t\right)\left(2 + \frac{8 j^4 t^2}{\left(\alpha j^2 + \sqrt{\alpha^2 j^4 - 4j^2}\right)^2}\right) \\
&\le \exp\left(2\alpha\beta_{N(\epsilon)} t\right)\left(2 + \frac{8T^2}{\alpha^2}\right). \qquad (3.11)
\end{aligned}$$

From (3.10) and (3.11), we can deduce (3.8).
For the second statement of Lemma 3.1, from (3.8), we have

$$\left\|Q_{\beta_{N(\epsilon)}}(t)\varphi\right\|^2_{H^p(D)} = \sum_{\lambda_j \le \beta_{N(\epsilon)}} \lambda_j^p \mathcal{A}_j^2(t)\left\langle \varphi, \phi_j\right\rangle^2 \le \beta^p_{N(\epsilon)} \exp\left(2\alpha\beta_{N(\epsilon)} t\right)\left(2 + \frac{8T^2}{\alpha^2}\right).$$

The proof of Lemma 3.1 is complete.

Define the truncation operator $\mathbf{I}_{\beta_N(\epsilon)} : L^2(D) \to L^2(D)$ such that

$$\mathbf{I}_{\beta_N(\epsilon)} v = \sum_{\lambda_j \le \beta_{N(\epsilon)}} \langle v, \phi_j\rangle \phi_j, \quad \text{for all v} \in L^2(D).$$

Next, we formulate a regularized problem for Problem (1.1)-(1.3) as follows

$$\begin{cases}
W_{tt}^{\epsilon,N(\epsilon)} - \alpha W_{xxt}^{\epsilon,N(\epsilon)} - W_{xx}^{\epsilon,N(\epsilon)} = 0, & (x,t) \in D \times (0,T), \\
W^{\epsilon,N(\epsilon)}(0,t) = W^{\epsilon,N(\epsilon)}(\pi,t) = 0, & t \in (0,T), \\
W^{\epsilon,N(\epsilon)}(x,T) = \mathbf{I}_{\beta_{N(\epsilon)}} \overline{G}_{\epsilon,N(\epsilon)}(x), & (x,t) \in D \times (0,T), \\
W_t^{\epsilon,N(\epsilon)}(x,T) = 0, & (x,t) \in D \times (0,T),
\end{cases} \qquad (3.12)$$

where $\beta_{N(\epsilon)}$ plays the role of regularization parameter. Using the same techniques as the beginning of this section for the problem (1.1)-(1.3), we can easily show that Problem (3.12) has a unique solution $W^{\epsilon,N(\epsilon)} \in C([0,T], L^2(D))$ given by

$$W^{\epsilon,N(\epsilon)}(x,t) = Q_{\beta_{N(\epsilon)}}(T-t)\overline{G}_{\epsilon,N(\epsilon)}(x).$$

We are now in position to state the main results of this paper.

Theorem 3.1 *Let us choose $N(\epsilon)$ and $\beta_{N(\epsilon)}$ such that*

$$\lim_{\epsilon\to 0} N(\epsilon) = +\infty, \quad \lim_{\epsilon\to 0} \beta_{N(\epsilon)} = +\infty, \quad \lim_{\epsilon\to 0} \frac{e^{\alpha T \beta_{N(\epsilon)}}}{N^\gamma(\epsilon)} = 0, \quad \lim_{\epsilon\to 0} e^{\alpha T\beta_{N(\epsilon)}} \epsilon \sqrt{N(\epsilon)} = 0. \qquad (3.13)$$

Assume that Problem (1.1)-(1.3) has a unique solution $u \in H^{2\theta}(D)$ and $g \in H^\gamma(D)$ for $\theta, \gamma > 0$. Then, there exists $C_0 = C_0(\alpha, T) > 0$ such that

$$\mathbf{E}\left\|u(\cdot,t) - W^{\epsilon,N(\epsilon)}(\cdot,t)\right\|^2 \le C_0 e^{2\alpha T\beta_{N(\epsilon)}}\left(\epsilon^2 N(\epsilon) + \frac{1}{N^{2\gamma}(\epsilon)}\|g\|^2_{H^\gamma(D)}\right) + \frac{2\|u(\cdot,t)\|^2_{H^{2\theta}(D)}}{\beta^{2\theta}_{N(\epsilon)}}. \qquad (3.14)$$

Remark 3.1 *We now give an example for the choice of $N(\epsilon)$ and $\beta_{N(\epsilon)}$, which satisfy the condition (3.13). Since $\lambda_N(\epsilon) = N^2(\epsilon)$, we can choose $\beta_{N(\epsilon)}$ such that $e^{\alpha T \beta_{N(\epsilon)}} = N(\epsilon)^a$ for $0 < a < \gamma$. Then, we have $\beta_{N(\epsilon)} = \frac{a}{\alpha T} \ln(N(\epsilon))$. The number $N(\epsilon)$ is chosen as follows*

$$N(\epsilon) = \left(\frac{1}{\epsilon}\right)^{\frac{2b}{2a+1}}$$

for $0 < b < 1$. With $N(\epsilon)$ chosen as above, $\mathbf{E}\left\|u(\cdot,t) - U^{\epsilon,N(\epsilon)}(\cdot,t)\right\|$ is of order

$$\frac{1}{\ln^{\theta}\left(\frac{1}{\epsilon}\right)}.$$

Proof of Theorem 3.1. It follows from (3.7) that

$$V^{\epsilon,N(\epsilon)}(x,t) := \mathbf{I}_{\beta_N(\epsilon)} u(x,t) = \sum_{\lambda_j \le \beta_{N(\epsilon)}} u_j(t)\phi_j(x) = Q_{\beta_{N(\epsilon)}}(T-t)g(x).$$

Applying the inequality $(a+b)^2 \le 2a^2 + 2b^2$, Parseval's equality, from (3.8) and Lemma 2.1, we have

$$\begin{aligned}
\mathbf{E}\left\|u(\cdot,t) - W^{\epsilon,N(\epsilon)}(\cdot,t)\right\|^2 &\le 2\left\|u(\cdot,t) - V^{\epsilon,N(\epsilon)}(\cdot,t)\right\|^2 + 2\mathbf{E}\left\|W^{\epsilon,N(\epsilon)}(\cdot,t) - V^{\epsilon,N(\epsilon)}(\cdot,t)\right\|^2 \\
&= 2\sum_{\lambda_j > \beta_{N(\epsilon)}} u_j^2(t) + 2\mathbf{E}\left\|Q_{\beta_{N(\epsilon)}}(T-t)\left(\overline{G}_{\epsilon,N(\epsilon)} - g\right)\right\|^2 \\
&\le 2\sum_{\lambda_j > \beta_{N(\epsilon)}} \lambda_j^{-2\theta}\lambda_j^{2\theta}u_j^2(t) + \left(4 + \frac{16T^2}{\alpha^2}\right)e^{2\alpha(T-t)\beta_{N(\epsilon)}}\mathbf{E}\left\|\overline{G}_{\epsilon,N(\epsilon)} - g\right\|^2 \\
&\le 2\beta_{N(\epsilon)}^{-2\theta}\|u(\cdot,t)\|_{H^{2\theta}(D)}^2 + \left(4 + \frac{16T^2}{\alpha^2}\right)e^{2\alpha(T-t)\beta_{N(\epsilon)}}\left(\epsilon^2 N(\epsilon) + \frac{1}{N^{2\gamma}(\epsilon)}\|g\|_{H^\gamma(D)}^2\right).
\end{aligned}$$

From the above inequality, we can imply (3.14). This completes the proof of Theorem 3.1. The next theorem gives the error estimate in the $H^p(D)$−norm.

Theorem 3.2 *Let us choose $N(\epsilon)$ and $\beta_{N(\epsilon)}$ such that*

$$\lim_{\epsilon\to 0} N(\epsilon) = +\infty, \quad \lim_{\epsilon\to 0}\beta_{N(\epsilon)} = +\infty, \quad \lim_{\epsilon\to 0}\frac{\sqrt{\beta_{N(\epsilon)}^p}\,e^{\alpha T\beta_{N(\epsilon)}}}{N^\gamma(\epsilon)} = 0, \quad \lim_{\epsilon\to 0} e^{\alpha T\beta_{N(\epsilon)}}\epsilon\sqrt{\beta_{N(\epsilon)}^p N(\epsilon)} = 0.$$

Suppose that the problem (1.1)-(1.3) has a unique solution $u \in H^{p+2\theta}(D)$ and $g \in H^\gamma(D)$ for $p, \gamma > 0$. Then,

$$\begin{aligned}
&\mathbf{E}\left\|W^{\epsilon,N(\epsilon)}(\cdot,t) - u(\cdot,t)\right\|_{H^p(D)}^2 \\
&\le 2\beta_{N(\epsilon)}^{-2\theta}\|u(\cdot,t)\|_{H^{p+2\theta}(D)}^2 + C_0\beta_{N(\epsilon)}^p e^{2\alpha T\beta_{N(\epsilon)}}\left(\epsilon^2 N(\epsilon) + \frac{1}{N^{2\gamma}(\epsilon)}\|g\|_{H^\gamma(D)}^2\right). \qquad (3.15)
\end{aligned}$$

Remark 3.2 *We give one example for the choice of $N(\epsilon)$. Choose $\beta_{N(\epsilon)}$ such that $\beta_{N(\epsilon)} = \frac{a}{\alpha T}\ln(N(\epsilon))$ for $0 < a < \gamma$. The number $N(\epsilon)$ is chosen as follows*

$$N(\epsilon) = \left(\frac{1}{\epsilon}\right)^{\frac{2b}{2a+1}}$$

for $0 < b < 1$. Then, $\mathbf{E}\left\|W^{\epsilon,N(\epsilon)}(\cdot,t) - u(\cdot,t)\right\|_{H^p(D)}^2$ is of order

$$\frac{1}{\ln^{\theta}\left(\frac{1}{\epsilon}\right)}.$$

Proof of Theorem 3.2. Indeed, using (3.9) and Lemma 2.1, we have

$$\begin{aligned}
&\mathbf{E}\left\|W^{\epsilon,N(\epsilon)}(\cdot,t)-u(\cdot,t)\right\|_{H^p(D)}^2 \\
&\le 2\left\|u(\cdot,t)-V^{\epsilon,N(\epsilon)}(\cdot,t)\right\|_{H^p(D)}^2+2\mathbf{E}\left\|W^{\epsilon,N(\epsilon)}(\cdot,t)-V^{\epsilon,N(\epsilon)}(\cdot,t)\right\|_{H^p(D)}^2 \\
&=2\sum_{\lambda_j>\beta_{N(\epsilon)}}\lambda_j^p u_j^2(t)+2\mathbf{E}\left\|Q_{\beta_{N(\epsilon)}}(T-t)\left(\overline{G}_{\epsilon,N(\epsilon)}-g\right)\right\|_{H^p(D)}^2 \\
&\le 2\sum_{\lambda_j>\beta_{N(\epsilon)}}\lambda_j^{-2\theta}\lambda_j^{2\theta+p}u_j^2(t)+\left(4+\frac{16T^2}{\alpha^2}\right)\beta_{N(\epsilon)}^p e^{2\alpha(T-t)\beta_{N(\epsilon)}}\mathbf{E}\left\|\overline{G}_{\epsilon,N(\epsilon)}-g\right\|^2 \\
&\le 2\beta_{N(\epsilon)}^{-2\theta}\|u(\cdot,t)\|_{H^{p+2\theta}(D)}^2+\left(4+\frac{16T^2}{\alpha^2}\right)\beta_{N(\epsilon)}^p e^{2\alpha(T-t)\beta_{N(\epsilon)}}\left(\epsilon^2N(\epsilon)+\frac{1}{N^{2\gamma}(\epsilon)}\|g\|_{H^\gamma(D)}^2\right),
\end{aligned}$$

from which we can deduce (3.15). This completes the proof.

References

[1] L. Cavalier, *Inverse problems in statistics. Inverse problems and high-dimensional estimation*, Lect. Notes Stat. Proc., 203, Springer, Heidelberg (2011).

[2] W. Fitzgibbon, *Strongly damped quasilinear evolution equations,* Journal of Mathematical Analysis and Applications 79, 536–550 (1981).

[3] V. Kalantarov and S. Zelik, *Finite-dimensional attractors for the quasi-linear strongly-damped wave equation,* Journal of Differential Equations 247, 1120–1155 (2009).

[4] H. Kekkonen, M. Lassas, and S. Siltanen, *Analysis of regularized inversion of data corrupted by white Gaussian noise*, Inverse Problems 30, p. 045009 (2014).

[5] S. Larsson, V. Thomée, and L. B. Wahlbin, *Finite-element methods for a strongly damped wave equation,* IMA journal of numerical analysis 11, 115–142 (1991).

[6] R. Lattès, J.-L. Lions, *Méthode de Quasi-réversibilité et Applications*, Dunod, Paris (1967).

[7] P. Massatt, *Limiting behavior for strongly damped nonlinear wave equations,* Journal of Differential Equations 48, 334–349 (1983).

[8] V. Pata and M. Squassina , *On the strongly damped wave equation,* Communications in mathematical physics 253, 511–533 (2005).

[9] V. Thomée and L. Wahlbin, *Maximum-norm estimates for finite-element methods for a strongly damped wave equation,* BIT Numerical Mathematics 44, 165–179 (2004).

An unconditionally stable implicit difference scheme for 2D porous medium equations using four-point NEGMSOR iterative method

Jackel *Chew* Vui Lung[1] and *Jumat* Sulaiman[1,*]

[1]Faculty of Science and Natural Resources, Universiti Malaysia Sabah, 88400, Kota Kinabalu, Sabah, Malaysia

Abstract. In this paper, a numerical method has been proposed for solving several two-dimensional porous medium equations (2D PME). The method combines Newton and Explicit Group MSOR (EGMSOR) iterative method namely four-point NEGMSOR. Throughout this paper, an initial-boundary value problem of 2D PME is discretized by using the implicit finite difference scheme in order to form a nonlinear approximation equation. By taking a fixed number of grid points in a solution domain, the formulated nonlinear approximation equation produces a large nonlinear system which is solved using the Newton iterative method. The solution vector of the sparse linearized system is then computed iteratively by the application of the four-point EGMSOR method. For the numerical experiments, three examples of 2D PME are used to illustrate the efficiency of the NEGMSOR. The numerical result reveals that the NEGMSOR has a better efficiency in terms of number of iterations, computation time and maximum absolute error compared to the tested NGS, NEG and NEGSOR iterative methods. The stability analysis of the implicit finite difference scheme used on 2D PME is also provided.

1 Introduction

This paper considers the numerical solution of the following initial-boundary value problem of the two-dimensional porous medium equation (2D PME):

$$\frac{\partial u}{\partial t}=\alpha\left[\frac{\partial}{\partial x}\left(u^{m}\frac{\partial u}{\partial x}\right)+\frac{\partial}{\partial y}\left(u^{m}\frac{\partial u}{\partial y}\right)\right], \tag{1}$$

that is subjected to the initial condition:

$$u(x,y,0)=u_0(x,y),\ 0\le x,y\le 1, \tag{2}$$

and the boundary conditions:

[*] Corresponding author: jumat@ums.edu.my

$$u(0,y,t) = g_1(y,t),\ u(1,y,t) = g_2(y,t),$$
$$u(x,0,t) = g_3(x,t),\ u(x,1,t) = g_4(x,t),\ 0 < t \le 1, \tag{3}$$

with some real parameters: $m \ge 1$ and $\alpha > 0$.

Equation (1) can be known as the generalized form of the heat equation. By referring to equation (1), the heat equation can be obtained by taking $m = 1$. In a particular case like $m = 2$, the Boussinesq equation can be obtained. Equation (1) is a type of partial differential equation in the nonlinear degenerate parabolic class which is widely used to describe models for gases in porous media, heat diffusion, underground infiltration, population dynamics as well as high energy physics [1]. The presence of a nonlinear term u^m in equation (1) makes the exact solution difficult to be obtained.

Generally, the numerical method can be considered as an alternative option in solving the partial differential equation especially for the case where the exact solution is almost impossible to be obtained. The numerical method yields an approximate solution that is differing from the exact solution by less than a specified tolerance. Furthermore, in terms of the implementation of the numerical method, efficiency is one of the important characteristics to be observed. The selected numerical method should be the one that requires less computation time and complexity in obtaining the desired result [2].

The numerical solution of multidimensional PME has received considerable interest from a number of researchers. For instance, Borana et al. [3] studied the numerical solution of the one-dimensional (1D) case of PME that is modelled based on the instability phenomenon in the double phase flow of two immiscible fluids through the inclined homogeneous porous medium. The method used is a Crank-Nicolson finite difference scheme. Meanwhile, Pradhan et al. [4] applied the finite element method for solving the 1D PME model that representing the instability phenomenon in the recovery of oil from a reservoir. For the numerical solution of 2D PMEs, Ngo and Huang [5] studied the application of an adaptive moving mesh finite element method as an improvement to the existing moving mesh method. In addition, Chew and Sulaiman [6] started the investigation on the efficiency of a numerical method that combines Newton and MSOR (NMSOR) iterative method for solving 1D PMEs. Then, Chew and Sulaiman [7] extended their study with the implementation of the Explicit Group (EG) method [8] to improve the efficiency of the NMSOR iterative method in solving the 1D PMEs.

The aim of this paper is to study the efficiency of the Newton and Explicit Group MSOR iterative method or NEGMSOR for solving several examples of 2D PME. The outline of this paper is arranged as follows. Section 2 shows the discretization of 2D PME by using the implicit finite difference scheme into a nonlinear approximation equation and the formation of a sparse and large nonlinear system. Section 3 presents the solution of the nonlinear system by the implementation of the Newton and EGMSOR iterative methods. Section 4 illustrates the numerical experiments and Section 5 concludes the finding of this study.

2 Implicit finite difference approximation equation

To obtain the numerical solutions, we consider a rectangular domain that is marked off in the unit square using the grid lines as follows:

$$x_i = ih,\ y_j = jh,\ h = \frac{1}{M}. \tag{4}$$

and for the time step, we define

$$t_n = nk, k = \frac{1}{N} \tag{5}$$

The numerical solutions that are represented by the points of intersection between the two grid lines are denoted by $u(x_i, y_j, t_n)$. For the sack of simplicity, we use $u_{i,j,n} = u(x_i, y_j, t_n)$ with $1 \le i, j \le M-1$ and $1 \le n \le N$ By using the implicit finite difference scheme, we obtain the implicit finite difference approximation equation of equation (1) that can be written as

$$\begin{aligned} f(u_{i,j,n+1}) &= u_{i,j,n+1} - A_1\left(u^{m}_{i,j,n+1}u_{i+1,j,n+1} - 2u^{m+1}_{i,j,n+1} + u^{m}_{i,j,n+1}u_{i-1,j,n+1}\right) \\ &- A_2\left(mu^{m-1}_{i,j,n+1}u^2_{i+1,j,n+1} - 2mu^{m-1}_{i,j,n+1}u_{i+1,j,n+1}u_{i-1,j,n+1} + mu^{m-1}_{i,j,n+1}u^2_{i-1,j,n+1}\right) \\ &- A_3\left(u^{m}_{i,j,n+1}u_{i,j+1,n+1} - 2u^{m+1}_{i,j,n+1} + u^{m}_{i,j,n+1}u_{i,j-1,n+1}\right) \\ &- A_4\left(mu^{m-1}_{i,j,n+1}u^2_{i,j+1,n+1} - 2mu^{m-1}_{i,j,n+1}u_{i,j+1,n+1}u_{i,j-1,n+1} + mu^{m-1}_{i,j,n+1}u^2_{i,j-1,n+1}\right) - u_{i,j,n} \end{aligned} \tag{6}$$

where $A_1 = A_3 = \alpha k / h^2$, $A_2 = A_4 = \alpha k / 4h^2$.

Equation (6) is the function at each interior grid point over the solution domain that links with neighbourhood grid points to both x- and y-directions. As a result, a sparse and large nonlinear system is formed and generally written as

$$\underset{\sim}{F}\left(\underset{\sim}{U}_{n+1}\right) = 0 \tag{7}$$

where

$$\underset{\sim}{F}\left(\underset{\sim}{U}_{n+1}\right) = \left(f_{1,1}, f_{2,1}, \ldots, f_{1,2}, f_{2,2}, \ldots, f_{M-1,M-1}\right)^T,$$

$$\underset{\sim}{U}_{n+1} = \left(u_{1,1,n+1}, u_{2,1,n+1}, \ldots, u_{1,2,n+1}, u_{2,2,n+1}, \ldots, u_{M-1,M-1,n+1}\right).$$

3 Linear stability analysis

Generally, the application of the linear Fourier analysis to the nonlinear partial differential equation such as 2D PME cannot be rigorously justified. Nevertheless, it has been found to effective in practice [9 12]. To show the stability of the implicit finite difference scheme used on 2D PME, we assume that the solution function $u(x, y, t)$ is bounded in the spatiotemporal region of $0 \le x, y \le 1$ and $0 \le t \le 1$. Now, the nonlinear term u^m can be frozen at each grid point in the solution domain and gives

$$\frac{\partial u}{\partial t} = \rho\left(\frac{\partial}{\partial x}\left(v\frac{\partial u}{\partial x}\right) + \frac{\partial}{\partial y}\left(v\frac{\partial u}{\partial y}\right)\right) \tag{8}$$

where $v = max\left|u^m\right|$. Equation (8) can be rewritten into

$$\frac{\partial u}{\partial t} = \rho v\left(\frac{\partial^2 u}{\partial x^2} + \frac{\partial^2 u}{\partial y^2}\right) \tag{9}$$

From here, the stability of the implicit finite difference scheme used in this paper can be shown by implementing the von Neumann method over equation (9) as shown in Theorem 3.

Theorem 3.
The fully implicit finite difference scheme of equation (9) on the finite domain $0 \le x, y \le 1$ with zero boundary conditions for all $t \ge 0$ is unconditionally stable.

Proof. To begin, we define the corresponding implicit finite difference operators as follows.

$$D_{-t}U_{p,q,n+1} = \frac{U_{p,q,n+1} - U_{p,q,n}}{k} \tag{10}$$

$$D_{+x}D_{-x}U_{p,q,n+1} = \frac{U_{p+1,q,n+1} - 2U_{p,q,n+1} + U_{p-1,q,n+1}}{h^2} \tag{11}$$

and

$$D_{+y}D_{-y}U_{p,q,n+1} = \frac{U_{p,q+1,n+1} - 2U_{p,q,n+1} + U_{p,q-1,n+1}}{h^2} \tag{12}$$

By substituting equation (10) - (12) into equation (9) and after some rearranging, the corresponding equation obtained is

$$\begin{aligned} U_{p,q,n+1} - \lambda\left(U_{p+1,q,n+1} - 2U_{p,q,n+1} + U_{p-1,q,n+1}\right) \\ -\lambda\left(U_{p,q+1,n+1} - 2U_{p,q,n+1} + U_{p,q-1,n+1}\right) = U_{p,q,n}, \end{aligned} \tag{13}$$

where $\lambda = k\rho v / h^2$.

To apply the von Neumann method, let

$$U_{p,q,n} = \xi^n e^{i\alpha p} e^{i\beta q} \tag{14}$$

By using equation (14), equation (13) can be transformed into

$$\xi\left(1 - \lambda\left(e^{i\alpha} - 2 + e^{-i\alpha}\right) - \lambda\left(e^{i\beta} - 2 + e^{-i\beta}\right)\right) = 1. \tag{15}$$

Since $e^{i\alpha} - 2 + e^{-i\alpha} = -4\sin^2\frac{\alpha}{2}$ and $e^{i\beta} - 2 + e^{-i\beta} = -4\sin^2\frac{\beta}{2}$, we obtain the amplication factor

$$\xi = \frac{1}{1 + 4\lambda\sin^2\frac{\alpha}{2} + 4\lambda\sin^2\frac{\beta}{2}} \tag{16}$$

Based on equation (16), it can be said that the implicit finite difference scheme to 2D PME is unconditionally stable with $0 < \xi \le 1$ for all positive values of λ and for $\alpha, \beta \in [-\pi, \pi]$.

4 Derivation of the four-point NEGMSOR iterative method

In the context of reducing the computational complexity, the convergence rate of the solution process needs to be increased. There are several iterative methods that have been proposed in order to reduce the computational complexity and this paper considers the application of the Explicit Group (EG) method which was introduced by Evans [8]. This method uses the concept of breaking the original matrix into several small fixed-size groups of points. By this way, the computational complexity can be reduced by solving several small groups of points instead of a large linear system per iteration. The EG method has been used extensively together with some efficient iterative methods such as SOR, MSOR and AOR [13-15].

Before the EG method is applied, equation (17) can be transformed into a linear system by using the Newton method as follows:

$$J_F\left(\underset{\sim}{U}_{n+1}^{(\ell)}\right)\Delta \underset{\sim}{H}_{n+1}^{(\ell)} = -\underset{\sim}{F}\left(\underset{\sim}{U}_{n+1}^{(\ell)}\right),\ \ell = 1,2,\ldots \tag{17}$$

where $J_F\left(\underset{\sim}{U}_{n+1}^{(\ell)}\right)$ is a penta-diagonal Jacobian matrix with a dimension of $(M-1)^2 \times (M-1)^2$, $\Delta \underset{\sim}{H}_{n+1}^{(\ell)}$ and $\underset{\sim}{F}\left(\underset{\sim}{U}_{n+1}^{(\ell)}\right)$ are the column matrices with a dimension of $(M-1)^2 \times 1$, and for the iterative index, we denote it as (ℓ) and $n+1$ is considered to be the next time level. Then, equation (17) is divided into several completed groups of four points and some ungroup that can be treated as a group of two points and/or single point [8]. Based on any completed group of four grid points, the general form of the EG iterative method is given by

$$\begin{bmatrix} \Delta h_{i,j} \\ \Delta h_{i+1,j} \\ \Delta h_{i,j+1} \\ \Delta h_{i+1,j+1} \end{bmatrix}^{(\ell+1)} = \begin{bmatrix} c_{i,j} & d_{i,j} & & e_{i+1,j} \\ b_{i+1,j} & c_{i+1,j} & d_{i-1,j+1} & \\ & b_{i+2,j} & c_{i,j+1} & d_{i,j+1} \\ a_{i,j+1} & & b_{i+1,j+1} & c_{i+1,j+1} \end{bmatrix}^{-1} \begin{bmatrix} S_1 \\ S_2 \\ S_3 \\ S_4 \end{bmatrix} \tag{18}$$

where

$$S_1 = f_{i,j} - a_{i,j}\Delta h_{i,j-1}^{(\ell)} - b_{i,j}\Delta h_{i-1,j}^{(\ell)},\quad S_2 = f_{i+1,j} - a_{i+1,j}\Delta h_{i+1,j-1}^{(\ell)} - e_{i+1,j}\Delta h_{i+1,j+1}^{(\ell)}$$

$$S_3 = f_{i,j+1} - a_{i,j+1}\Delta h_{i,j}^{(\ell)} - e_{i,j+1}\Delta h_{i,j+2}^{(\ell)},$$

$$S_4 = f_{i+1,j+1} - d_{i+1,j+1}\Delta h_{i+2,j+1}^{(\ell)} - e_{i+1,j+1}\Delta h_{i+1,j+2}^{(\ell)}.$$

Besides that, Kincaid and Young [16] introduced Modified Successive Over-Relaxation (MSOR) method which can be considered as a point iterative method that has two relaxation parameters. The basic idea of the MSOR method is to associate the different parameter with each row and column of a sparse and large linear system. Therefore, to derive the NEGMSOR iterative method, two different relaxation parameters, $1 \le \omega_1 < 2$ and $1 \le \omega_2 < 2$, are added to equation (18) and forms

$$\begin{bmatrix} \Delta h_{i,j} \\ \Delta h_{i+1,j} \\ \Delta h_{i,j+1} \\ \Delta h_{i+1,j+1} \end{bmatrix}^{(\ell+1)} = (1-\omega_1)\begin{bmatrix} \Delta h_{i,j} \\ \Delta h_{i+1,j} \\ \Delta h_{i,j+1} \\ \Delta h_{i+1,j+1} \end{bmatrix}^{(\ell)} + \omega_1 \begin{bmatrix} c_{i,j} & d_{i,j} & & e_{i+1,j} \\ b_{i+1,j} & c_{i+1,j} & d_{i-1,j+1} & \\ & b_{i+2,j} & c_{i,j+1} & d_{i,j+1} \\ a_{i,j+1} & & b_{i+1,j+1} & c_{i+1,j+1} \end{bmatrix}^{-1} \begin{bmatrix} S_1 \\ S_2 \\ S_3 \\ S_4 \end{bmatrix} \tag{19}$$

$$\begin{bmatrix} \Delta h_{i,j} \\ \Delta h_{i+1,j} \\ \Delta h_{i,j+1} \\ \Delta h_{i+1,j+1} \end{bmatrix}^{(\ell+1)} = (1-\omega_2)\begin{bmatrix} \Delta h_{i,j} \\ \Delta h_{i+1,j} \\ \Delta h_{i,j+1} \\ \Delta h_{i+1,j+1} \end{bmatrix}^{(\ell)} + \omega_2 \begin{bmatrix} c_{i,j} & d_{i,j} & & e_{i+1,j} \\ b_{i+1,j} & c_{i+1,j} & d_{i-1,j+1} & \\ & b_{i+2,j} & c_{i,j+1} & d_{i,j+1} \\ a_{i,j+1} & & b_{i+1,j+1} & c_{i+1,j+1} \end{bmatrix}^{-1} \begin{bmatrix} S_1 \\ S_2 \\ S_3 \\ S_4 \end{bmatrix} \tag{20}$$

Based on equations (19) and (20), the NEG and NEGSOR iterative methods can be obtained by choosing $\omega_1 = \omega_2 = 1$ and $\omega_1 = \omega_2$ within the range of $[1,2)$ respectively. The implementation of the NEGMSOR iterative method can be shown in Figure 1.

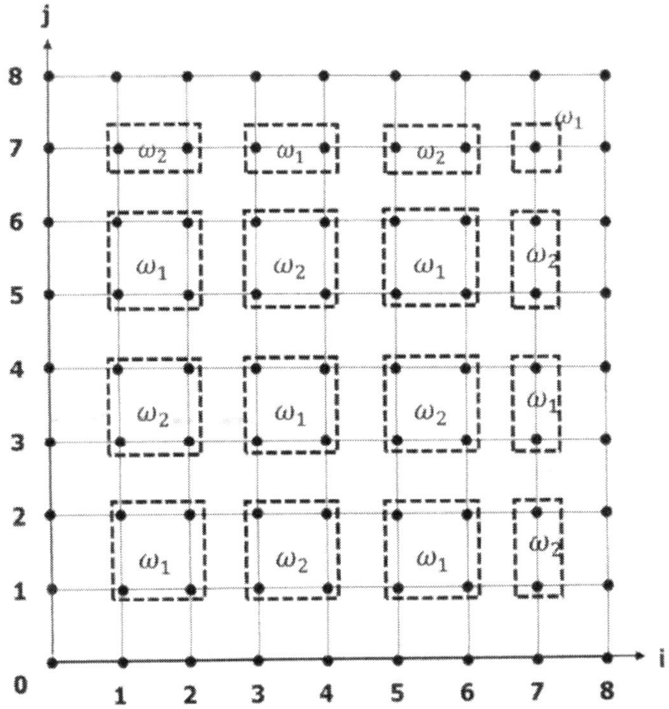

Figure 1. The implementation of NEGMSOR iterative method.

By referring to Figure 1, the numerical solution of 2D PME by using NEGMSOR iterative method can be summarized into Algorithm 1. The optimum values of ω_1 and ω_2 are determined when the least number of iterations is obtained after a few executions. During the executions, we consider a tolerance error of $\varepsilon = 1 \times 10^{-10}$.

Algorithm 1: NEGMSOR iterative method

i. Set up the values for A_1, A_2, A_3 , and A_4 ,

ii. Set the initial guess $\underset{\sim n+1}{U^{(0)}} = 1.0$,

iii. Formulation of the square matrix $J_F\left(\underset{\sim n+1}{U^{(\ell)}}\right)$ and the column matrix $-\underset{\sim}{F}\left(\underset{\sim n+1}{U^{(\ell)}}\right)$,

iv. Iteration of equations (19) and (20),

v. Convergence test 1: if $\left|\underset{\sim n+1}{\Delta H^{(\ell+1)}} - \underset{\sim n+1}{\Delta H^{(\ell)}}\right| < \varepsilon$, then $\underset{\sim n+1}{U^{(\ell+1)}} = \underset{\sim n+1}{\Delta H^{(\ell+1)}} + \underset{\sim n+1}{U^{(\ell)}}$,

vi. Convergence test 2: if $\left|\underset{\sim n+1}{U^{(\ell+1)}} - \underset{\sim n+1}{U^{(\ell)}}\right| < \varepsilon$, then go to the next time step,

vii. Display the approximate solutions.

5 Numerical experiments

As a comparative analysis, three examples of 2D PME are used. The efficiency of the NEGMSOR iterative method is investigated and is compared to the NGS, NEG and NEGSOR iterative methods. Number of iterations, execution time in seconds and maximum absolute error (error) become the subjects of comparison.

Example 1 [17]:

$$\frac{\partial u}{\partial t} = 0.2\left[\frac{\partial}{\partial x}\left(u\frac{\partial u}{\partial x}\right) + \frac{\partial}{\partial y}\left(u\frac{\partial u}{\partial y}\right)\right] \tag{21}$$

with exact solution $u(x,y,t) = x + y + (0.4)t$.

Example 2 [17]:

$$\frac{\partial u}{\partial t} = 0.2\left[\frac{\partial}{\partial x}\left(u^2\frac{\partial u}{\partial x}\right) + \frac{\partial}{\partial y}\left(u^2\frac{\partial u}{\partial y}\right)\right] \tag{22}$$

with exact solution $u(x,y,t) = (5(x+y+t))^{1/2}$.

Example 3 [18]:

$$\frac{\partial u}{\partial t} = \frac{\partial}{\partial x}\left(u^5\frac{\partial u}{\partial x}\right) + \frac{\partial}{\partial y}\left(u^5\frac{\partial u}{\partial y}\right) \tag{23}$$

with exact solution $u(x,y,t) = (4(x+y+2t)/5)^{1/4}$.

In the numerical experiments, we consider five different orders of the linear system, $(M-1)$ in which the different values of mesh size, M are 16, 32, 64, 128 and 256. The numerical results are indicated in Table 1 to 3 and the comparative analysis between the NEGMSOR iteration family against the NGS methods is done by calculating the reduction percentage in the number of iterations and the computation time as follows:

Against other iterative methods:

$$(\Delta\ell)\% = \left|\frac{\ell_{NGS} - \ell_{NEGMSOR}}{\ell_{NGS}}\right| \times 100\%, \quad (\Delta s)\% = \left|\frac{s_{NGS} - s_{NEGMSOR}}{s_{NGS}}\right| \times 100\% \tag{24}$$

where $(\Delta\ell)$ and (Δs) are reduction percentage of number of iterations and execution time for NEGMSOR method respectively.

Table 1. Comparison in terms of number of iterations based on all iterative methods for three examples.

Example	Method	Mesh Size				
		16	32	64	128	256
1	NGS	136	436	1525	5462	19404
	NEG	84	244	818	2903	10338
	NEGSOR	64	131	262	523	1020
	NEGMSOR	54	102	203	399	787
2	NGS	130	440	1380	4901	17458
	NEG	80	223	738	2596	9279
	NEGSOR	62	124	255	505	992
	NEGMSOR	58	104	202	394	779
3	NGS	739	2630	9478	34098	121649
	NEG	406	1393	4996	18034	64894
	NEGSOR	199	395	780	1525	2971
	NEGMSOR	181	349	680	1333	2616

Table 2. Comparison in terms of execution time (seconds) based on all iterative methods for three examples.

Example	Method	Mesh Size				
		16	32	64	128	256
1	NGS	0.76	2.93	19.59	360.89	4586.69
	NEG	0.44	2.64	8.55	214.01	3717.50
	NEGSOR	0.33	2.05	6.37	62.79	390.80
	NEGMSOR	0.29	1.73	5.60	40.01	263.99
2	NGS	0.97	2.70	18.26	248.82	4243.79
	NEG	0.62	1.93	8.56	187.24	2295.15
	NEGSOR	0.43	1.53	5.35	50.98	369.60
	NEGMSOR	0.39	1.28	4.90	38.32	225.95
3	NGS	0.98	9.51	113.63	1653.85	29234.80
	NEG	0.86	7.75	61.96	763.70	14890.79
	NEGSOR	0.59	2.05	12.89	118.29	921.40
	NEGMSOR	0.57	1.96	10.21	96.27	739.37

Table 3. Comparison in terms of maximum absolute error based on all iterative methods for three examples.

Example	Method	Mesh Size				
		16	32	64	128	256
1	NGS	8.86e-11	3.25e-10	1.90e-09	9.02e-09	3.86e-08
	NEG	1.05e-11	1.36e-10	6.59e-10	3.96e-09	1.80e-08
	NEGSOR	1.27e-12	7.44e-12	1.62e-11	2.60e-11	3.82e-11
	NEGMSOR	1.26e-12	7.98e-12	1.32e-11	2.90e-11	4.12e-11
2	NGS	7.57e-11	2.31e-09	1.31e-08	4.95e-08	1.75e-07
	NEG	2.09e-11	8.45e-10	6.39e-09	2.79e-08	9.92e-08
	NEGSOR	8.52e-13	2.11e-11	5.45e-11	3.69e-10	1.28e-09
	NEGMSOR	1.07e-12	9.03e-12	1.93e-11	3.91e-11	5.67e-11
3	NGS	1.10e-09	6.69e-09	3.56e-08	1.72e-07	7.78e-07
	NEG	1.55e-10	2.50e-09	1.51e-08	7.45e-08	3.56e-07
	NEGSOR	1.43e-11	2.89e-11	3.00e-11	4.06e-11	4.91e-11
	NEGMSOR	5.58e-12	3.34e-11	4.36e-11	3.96e-11	4.86e-11

From the numerical experiment results in Table 1-3, the reduction percentage of the NEGMSOR iteration family as compared the GS iterative method for three examples are summarized in Table 4.

Table 4. Reduction percentage of the number of iterations and the execution time for NMEGSOR iteration family compared to NGS method

Example	Methods	Number of Iterations (%)	Execution Time (%)
1	NEG	38.23-46.85%	9.89-56.36%
	NEGSOR	52.94-94.74%	30.03-91.48%
	NEGMSOR	60.29-95.94%	40.96-94.24%
2	NEG	38.46-49.32%	24.75-53.12%
	NEGSOR	52.31-94.32%	43.33-91.29%
	NEGMSOR	55.38-95.54%	52.59-94.68%
3	NEG	45.06-47.03%	12.24-53.82%
	NEGSOR	73.07-91.77%	39.80-96.84%
	NEGMSOR	75.51-96.09%	41.84-97.47%

Referring to the results of numerical experiments for three tested examples in Table 4, it can be pointed out that the percentage reduction of number of iterations for the NEGMSOR iterative method have declined approximately by 60.29-95.94%, 55.38-95.54%, and 75.51-96.09% respectively as compared with the NGS method. In fact, implementations of execution time for NEGMSOR iterative method are much faster about 40.96-94.24%, 52.59-94.68%, and 41.84-97.47% respectively than the NGS method.

6 Conclusion

This paper proposed the unconditionally stable implicit difference scheme together with the NEGMSOR iterative method for solving several examples of 2D PME. The implicit finite difference scheme is successfully applied to discretize the general form of 2D PME. In addition to that, the sparse and large linear system is solved using the NEGMSOR iterative method and the numerical results reveal that the proposed method is more efficient in terms of number of iterations, and execution time as compared to the tested NGS, NEG and NEGSOR iterative methods. Overall, in term of and the magnitude of maximum absolute error, the accuracy of the proposed NEGMSOR iterative method and the three tested numerical methods are comparable.

The authors gratefully acknowledge that this paper was partially supported by the research grant scheme (GUG0022-SG-M-1/2016) from Universiti Malaysia Sabah, Malaysia..

References

1. J.L. Vazquez. *The Mathematical Theories of Diffusion: Nonlinear and Fractional Diffusion*. Lecture Notes in Mathematics, 2186 (2016).
2. P.B. Patil, U.P. Verma. *Numerical Computational Methods*. Revised Ed. Oxford: Alpha Science International Ltd. (2009)
3. R.N. Borana, V.H. Pradhan, M.N. Mehta. Intern. J. of Innovative Research in Science, Engineering and Technology. **3**, 2 (2014)
4. V.H. Pradhan, M.N. Mehta, T. Patel. Intern. J. of Advanced Engineering Technology **2,** 1 (2011).
5. C. Ngo, W.Z. Huang. J. Comp. Physics. **331** (2017)
6. J.V.L. Chew. J. Sulaiman. *The 2nd International Conference on Mathematical Sciences and Statistics (ICMSS2016)* AIP Conf. Proc. **1739** (2016).
7. J.V.L. Chew. J. Sulaiman. Journal of Applied Mathematics and Computational Mechanics **15,** 2 (2016)
8. D.J. Evans. 1985. Intern. J. Comp Math. **17** (1985)
9. R.D. Richtmyer, K.W. Morton. *Difference methods for initial-value problems*. 2nd ed. New York: John Wiley & Sons, Inc. (1967)
10. B.F. Feng, T. Mitsui. J. Computational and Applied Mathematics. **90**, (1998)
11. D. Kim, W. Proskurowski. 2004. Intern. J. Math. & Mathematical Sciences. **17,** (2004)
12. J. Strikwerda. 2004. Finite difference schemes and partial differential equations. Philadelphia: SIAM. (2004)
13. J. Sulaiman, M.K. Hasan, M. Othman, S.A.A. Karim. Journal of Mathematics and System Science. **2,** (2012)
14. A. Saudi, J. Sulaiman. International Journal of Computer Applications. **53,** 16 (2012)

15. M.K.M. Akhir, J. Sulaiman. Applied Mathematical Sciences. **9,** 112 (2015
16. D.R. Kincaid, D.M. Young. 1972. Mathematics of Computation. **26,** 119 (1972)
17. A.D. Polyanin, V.F. Zaitsev. *Handbook of Nonlinear Partial Differential Equations.* Boca Raton: Chapman and Hall/CRC Press. (2004)
18. B.P. Sommeijer, P.J. van der Houwen. ACM Transactions on Mathematical Software. **10,** 4 (1984).

Regularity for the second order Riesz transforms associated with Schrödinger operators on weighted BMO type spaces

Nguyen Ngoc Trong[1,*], *Nguyen Anh* Dao[2,**], and *L. X.* Truong[3,***]

[1]Faculty of Mathematics and Computer Science, VUNHCM - University of Science, Ho Chi Minh city, Vietnam. Department of Primary Education, HoChiMinh City University of Education, Vietnam
[2]Applied Analysis Research Group, Faculty of Mathematics and Statistics, Ton Duc Thang University, Ho Chi Minh City, Viet Nam
[3]Division of Computational Mathematics and Engineering, Institute for Computational Science, Ton Duc Thang University, Ho Chi Minh City, Vietnam. Faculty of Mathematics and Statistics, Ton Duc Thang University, Ho Chi Minh City, Vietnam

Abstract. Let $\mathcal{L} = -\Delta + V$ be a Schrödinger operator on $\mathbb{R}^n$, where V is a nonnegative potential satisfying the suitable reverse Hölder's inequality. In this paper, we study the boundedness of the second order Riesz transforms such as $\mathcal{L}^{-1}\nabla^2$ on the spaces of BMO type for weighted case. We generalized the known results to the weighted case.

MSC: MSC43; MSC47

Keywords: Riesz transform, Schrödinger operator, BMO type space, weights.

1 Introduction and preliminaries

The study of both function spaces and the boundedness of singular integral operators associated with Schrödinger operators arose from practical applications in some mathematical fields, such as harmonic analysis and partial differential equations, and has become an active area of research in the last few years. These type of questions have been already considered by many authors and they are very important questions which appeal to various techniques from partial differential equations and harmonic analysis.

In recent years, many authors have been interested in the problems of harmonic analysis associated with Schrödinger operators on $\mathbb{R}^n$, see [1–3, 5, 6, 12, 14–17, 19, 21, 25]. More specifically, boundedness of singular integral versions on stratified Lie group was also attracted the attention, see [12, 14–16]. Very recently, D. Yang developed the theory of localized non-weighted BMO spaces associated to admissible functions in spaces of homogeneous type and eastablished the resuts for the boundedness of the singular integrals on these space, see [22–24].

*e-mail: trongnn37@gmail.com
**e-mail: daonguyenanh@tdtu.edu.vn
***e-mail: lxuantruong@gmail.com

Let $\mathcal{L} = -\Delta + V$ be a Schrödinger operator on $\mathbb{R}^n$, where the nonnegative potential V belongs to the reverse Hölder class RH_q for $q \geq \frac{n}{2}$, i.e., V satisfies the reverse Hölder's inequality: there exists $C > 0$ such that

$$\Big(\frac{1}{|B|}\int_B V(x)^q dx\Big)^{\frac{1}{q}} \leq \frac{C}{|B|}\int_B V(x)dx,$$

for all ball $B \subset \mathbb{R}^n$.

It is well-known that if $V \in RH_q$ for some $q > 1$, then there exists $\epsilon > 0$, such that $V \in RH_{q+\epsilon}$. We introduce the definition of the reverse Hölder index of V as $q_0 := \sup\{q : V \in RH_q\}$. Under assumption $V \in RH_m$ we may conclude $q_0 > m$.

We say $V \in RH_\infty$ if $V \in L^\infty_{loc}(\mathbb{R}^n)$ and

$$\|V\|_{L^\infty(B(x,r))} \leq \frac{C}{|B(x,r)|}\int_{B(x,r)} V(y)dy,$$

hold for every $x \in \mathbb{R}^n$ and $0 < r < \infty$. For every $1 < p < \infty$, it is easy to see that $RH_\infty \subset RH_p$.

For a weight ω, we shall mean ω is a nonnegative measurable and locally integrable function on $\mathbb{R}^n$. For $p \in (1, \infty)$, we say that ω belongs to the Muckenhoupt class A_p if the following holds: there exists $C > 0$ such that

$$\Big(\int_B \omega(x)dx\Big)\Big(\int_B \omega^{-\frac{1}{p-1}}(x)dx\Big)^{p-1} \leq C|B|^p, \tag{1.1}$$

for all balls B in $\mathbb{R}^n$.

For $p = 1$, we say that $\omega \in A_1$ if there is a positive constant C such that for every ball $B \subset \mathbb{R}^n$,

$$\frac{1}{|B|}\int_B \omega(y)dy \leq C\omega(x) \text{ for a.e. } x \in B.$$

We set $A_\infty = \cup_{p\geq 1} A_p$. For the ball $B = B(x, r)$ and $\lambda > 0$, we denote by λB the ball $B(x, \lambda r)$.

For $\alpha \geq 1$, we say that the weight ω satisfies the doubling property with the doubling order α if there exists a constant C such that

$$\omega(\lambda B) \leq C\lambda^{n\alpha}\omega(B),$$

for all balls $B \subset \mathbb{R}^n$ and all $\lambda > 1$, where $\omega(E) = \int_E \omega(x)dx$ for any measurable subset E of $\mathbb{R}^n$.

We then denote by D_α the set of all weights ω satisfying the doubling property with the doubling order α. It is important to note that if $\omega \in A_p$ for $p \in [1, \infty)$ then $\omega \in D_p$. Moreover, if $V \in RH_q$, then V belongs to a certain A_p class for some $p \in [1, \infty)$ and hence $V \in D_p$. See for example [20].

Associated to the potential V, the function of the critical radius is defined by

$$\rho(x) := \sup\Big\{r > 0 : \frac{1}{r^{n-2}}\int_{B(x,r)} V(y)dy \leq 1\Big\},$$

for all $x \in \mathbb{R}^n$, see [16].

Let $\rho_1(x)$ be the function of the critical radius associated to $|\nabla V|$. The auxillary functions $\rho(x), \rho_1(x)$ play an important role in the problem of harmonic analysis related to Schrödinger operators.

Definition 1.1. *Let $0 \le \beta < 1$ and w be a weight. The space $BMO^{\beta}_{\mathcal{L}}(\omega)$ is defined as the set of functions $f \in L^1_{loc}(\mathbb{R}^n)$ satisfying*

$$\int_B |f(y) - f_B|\, dy \le C\omega(B)|B|^{\beta/n}, \tag{1.2}$$

for all the balls $B \subset \mathbb{R}^n$, with $r < \rho(x)$, where $f_B = \frac{1}{|B|}\int_B f(y)dy$, and

$$\int_B |f(y)|dy \le C\omega(B)|B|^{\beta/n}, \tag{1.3}$$

for all balls $B = B(x, r)$, with $r \ge \rho(x)$, where C is a positive constant.

A norm in the space $BMO^{\beta}_{\mathcal{L}}(\omega)$ can be given by the maximum of the two infima of the constants in (1.2) and (1.3) respectively. This norm will be denoted by $\|\cdot\|_{BMO^{\beta}_{\mathcal{L}}(\omega)}$.

We now recall the definition of spaces $BMO^{\beta}(\omega)$ for $0 \le \beta < 1$,

$$BMO^{\beta}(\omega) = \Big\{f \in L^1_{\text{loc}} : \sup_B \frac{1}{|B|^{\beta/n}\omega(B)} \int_B |f(x) - f_B|dx < \infty\Big\},$$

with the supremum taken over all balls B and $f_B = \frac{1}{|B|}\int_B f(y)dy$.

For brevity, if $\omega \equiv 1$, then we will denote $BMO^{\beta}_{\mathcal{L}}, BMO^{\beta}$ by $BMO^{\beta}_{\mathcal{L}}(\omega), BMO^{\beta}(\omega)$ respectively . When $\beta = 0$, we write $BMO_{\mathcal{L}}, BMO$ instead of $BMO^{\beta}_{\mathcal{L}}, BMO^{\beta}$ respectively.

Inspired by the recent studies on the boundedness of the Riesz transforms in $\mathbb{R}^n$, see for example [1–4, 7, 8] and the references therein. In the case $\omega \equiv 1$, it is well-known that if $V \in RH_q$ then the integral operators $V\mathcal{L}^{-1}$ is bounded on $L^p(\mathbb{R}^n)$ for $1 \le p \le s$, and $\nabla^2\mathcal{L}^{-1}$ is bounded on $L^p(\mathbb{R}^n)$ for $1 < p \le s'$, in [19], where s, s' depend only on q, n. The ranges of p in these results are optimal. This seminal results of Z. Shen play the crucial role in the boundedness of the Riesz transform on the other types functional space. When $\omega \equiv 1$, and $\beta = 0$, J. Dong and Y. Liu [8] proved $\mathcal{L}^{-1}\nabla^2$ is bounded on $BMO^{\beta}_{\mathcal{L}}$.

In this paper, we improve the results of J Dong and Y. Liu in [8] by studying the boundedness of the second order Riesz transform $R^* = \mathcal{L}^{-1}\nabla^2$ on the spaces $BMO^{\beta}_{\mathcal{L}}(\omega), BMO^{\beta}(\omega)$. The next theorems are the main results of this paper.

Theorem 1.2. *Suppose that $V \in RH_q$ for some $q > n, |\nabla V| \in RH_{n/2}$ such that $q_1^* \le n, \rho \lesssim \rho_1 \lesssim 1$, where $q_1^* := \sup\{q : |\nabla V| \in RH_q\}$ is the reverse Hölder index of $|\nabla V|$. Let $w \in D_\alpha \cap \bigcup_{s>p'} \left(A_{p/s'} \cap RH_s\right)$ where $\frac{1}{p} = \frac{1}{q_1^*} - \frac{1}{n}$. For any $0 \le \beta < \varpi := \min\{1 - n/q, 2 - n/q_1^*\}$ and $1 \le \alpha < 1 + \frac{\varpi-\beta}{n}$, the operator $\mathcal{L}^{-1}\nabla^2$ is bounded from $BMO^{\beta}(\omega)$ into $BMO^{\beta}_{\mathcal{L}}(\omega)$.*

In [12], the authors gave some examples for the potential V which satisfy the assumption $\rho \lesssim \rho_1 \lesssim 1$. By Theorem 1.7 in [8], the authors proved that $\mathcal{L}^{-1}\nabla^2$ is bounded on the space $BMO_{\mathcal{L}}$. Note that, since $BMO^{\beta}_{\mathcal{L}}(\omega) \subset BMO^{\beta}(\omega)$, our results can be considered as improvements of those in [8], even in the case $\omega \equiv 1$.

The organization of this paper is as follows. In section 2, we give some preliminary results. Section 3 is devoted to the estimates on the size and smoothness of the kernels. Finally, the proof of the main theorems are also investigated in Section 4. Throughout this article, all the positive constants are signified as C although they may be different on the same line. Note that, ∇ means that we are taking all the partial derivaties with respect to the first variable. We write $A \sim B$ and $A \lesssim B$ if there exists some positive constants C, C' such that $C \leq \frac{A}{B} \leq C'$ and $A \leq C'B$, respectively.

2 Some preliminary results

Let $V \in RH_q$ where $q > n/2$. We begin with the results concerning the estimates on the potential V which are taken from [4, Lemma 1] and [19, Lemma 1.2], respectively.

Lemma 2.1. *The measure $V(x)dx$ satisfies the doubling condition, that is, there exists $C > 0$ such that*

$$\int\limits_{B(x,2r)} V(y)dy \leq C \int\limits_{B(x,r)} V(y)dy,$$

for all balls $B(x, r)$ in $\mathbb{R}^n$.

Lemma 2.2. *There exists $C > 0$ such that, for $0 < r < R < +\infty$,*

$$\frac{1}{r^{n-2}} \int\limits_{B(x,r)} V(y)dy \leq C\Big(\frac{r}{R}\Big)^{2-\frac{n}{q}} \frac{1}{R^{n-2}} \int\limits_{B(x,R)} V(y)dy.$$

Lemma 2.3. *If $r = \rho(x)$, then*

$$\frac{1}{r^{n-2}} \int\limits_{B(x,r)} V(y)dy = 1.$$

Moreover,

$$\frac{1}{r^{n-2}} \int\limits_{B(x,r)} V(y)dy \sim 1 \quad \textit{if and only if} \quad r \sim \rho(x).$$

Lemma 2.4. *There exist $C > 0$ and $l_0 > 0$ such that, for any x and y in G,*

$$\frac{1}{C}\Big(1 + \frac{|x-y|}{\rho(x)}\Big)^{-l_0} \leq \frac{\rho(y)}{\rho(x)} \leq C\Big(1 + \frac{|x-y|}{\rho(x)}\Big)^{\frac{l_0}{l_0+1}}.$$

In particular, $\rho(x) \sim \rho(y)$ if $|x-y| \lesssim \rho(x)$.

Lemma 2.5. *There exist $C > 0$ and $l_1 > 0$ such that*

$$\int\limits_{B(x,R)} \frac{V(y)}{|x-y|^{n-2}} dy \leq \frac{C}{R^{n-2}} \int\limits_{B(x,R)} V(y)dy \leq C\Big(1 + \frac{R}{\rho(x)}\Big)^{l_1}.$$

See [19] for the proofs of Lemmas 2.1 - 2.5.

Lemma 2.6. *Let $V \in RH_q$ with $q > n/2$ and $\epsilon > \frac{n}{q}$. Then, for any constant C_1, there exists a constant C_2 such that*

$$\int_{B(x,C_1 r)} \frac{V(u)}{|u-x|^{n-\epsilon}} du \le C_2 r^{\epsilon-2} \Big(\frac{r}{\rho(x)}\Big)^{2-n/q},$$

if $0 < r \le C_1\rho(x)$, and

$$\int_{B(x,C_1 r)} \frac{V(u)}{|u-x|^{n-\epsilon}} du \le C_2 r^{\epsilon-2} \Big(\frac{r}{\rho(x)}\Big)^{2+(\mu-1)n},$$

if $r > C_1\rho(x)$, where μ is such that $V \in D_\mu$.

The proof of this above lemma is similar to that of Lemma 1 in [4]. The following results give other characterization of the space $BMO^{\beta}_{\mathcal{L}}(\omega)$, the proof of this proposition is very similar with a small modification to that of [4, Proposition 2] and [18, Theorem 2.2], respectively.

Proposition 2.1. *Let $0 \le \beta < 1$ and a weight $\omega \in D_\alpha$ for some $\alpha \ge 1$. A function f belongs to $BMO^{\beta}_{\mathcal{L}}(\omega)$ if and only if condition (1.2) is satisfied for every ball $B = B(x,r)$ with $r < \rho(x)$ and*

$$\int_{B(x,\rho(x))} |f(y)| dy \le C\omega(B(x,\rho(x)))|\rho(x)|^{\beta}, \tag{2.4}$$

for all $x \in \mathbb{R}^n$.

It follows from [18, Theorem 2.2], we have the following lemma.

Lemma 2.7. ***(John-Nirenberg inequality)***
Let $1 \le p < \infty, 0 \le \beta < 1, \omega \in A_p, 1 \le r \le p'$, and $r < \infty$. Then $f \in BMO^{\beta}(\omega)$ if and only if

$$\sup_B \frac{1}{|B|^{\beta/n}} \Big(\frac{1}{\omega(B)} \int_B |f(x) - f_B|^r \omega^{1-r}(x) dx\Big)^{1/r} < \infty, \tag{2.5}$$

and, moreover, this quantity gives an equivalent norm.

In what follows we denote $\mathcal{I}_1 = (-\Delta)^{-1/2}$ the classical fractional integral of order 1. To prove the main theorems, we need the following technical lemma.

Lemma 2.8. *Let $V \in RH_q$ with $n/2 < q < n$ and $\omega \in RH_s \cap A_{p/s'}$ for some $s > p'$, where $\frac{1}{p} = \frac{1}{q} - \frac{1}{n}$. There exists $C > 0$ such that for any $k > 0, 0 \le \beta < 1, f \in BMO^{\beta}(\omega)$, and any ball $B(x,r)$,*

$$\begin{aligned}\int_{2^k B} |f(u) - f_B| \mathcal{I}_1(V\chi_{2^{k+2}B})(u) du &\le C(k+1) 2^{k(n\eta+\beta-1)} \\ &\quad \times \|f\|_{BMO^{\beta}(\omega)} \omega(B) r^{\beta-1} \Phi_{\beta,\eta}\Big(\frac{2^k r}{\rho(x)}\Big),\end{aligned} \tag{2.6}$$

where

$$\Phi_{\beta,\eta}(t) = \begin{cases} t^{2+\mu n - n} & \text{if } t \ge 1, \\ t^{n-n\eta+2-n/q} & \text{if } t < 1, \end{cases}$$

for η and μ being the exponent of the doubling property satisfied by ω and V respectively.

Proof. We first apply Hölder's inequality to estimate the right-hand side of (2.6) by

$$\|(f - f_B)\chi_{2^k B}\|_{p'} \|\mathcal{I}_1(V\chi_{2^{k+2}B})\|_p.$$

To bound the first factor we apply again Hölder's inequality with exponent σ such that $\sigma p' = (p/s')' = v$ to the functions $|f - f_B|^{p'}\omega^{\frac{1}{\sigma}-p'}$ and $\omega^{p'-\frac{1}{\sigma}}$. It is easy to check that $(p' - \frac{1}{\sigma})\sigma' = s$ and $\frac{1}{\sigma' p'} = \frac{s'}{sp}$. Therefore,

$$\begin{aligned}\|(f - f_B)\chi_{2^k B}\|_{p'} &\le \Big(\int_{2^k B} \omega^s(z)dz\Big)^{s'/sp}\Big(\int_{2^k B} |f(z) - f_B|^v \omega^{1-v}(z)dz\Big)^{1/v} \\ &\lesssim \frac{\omega(2^k B)^{s'/p}}{|2^k B|^{1/p}}\Big(\int_{2^k B} |f(z) - f_B|^v \omega^{1-v}(z)dz\Big)^{1/v}.\end{aligned}$$

By Lemma 2.7, we come up to

$$\begin{aligned}&\Big(\int_{2^k B} |f(z) - f_B|^v \omega^{1-v}(z)dz\Big)^{1/v} \\ &\le \Big(\int_{2^k B} |f(z) - f_{2^k B}|^v \omega^{1-v}(z)dz\Big)^{1/v} + (\omega^{1-v}(2^k B))^{1/v}\sum_{j=0}^{k-1}|f_{2^{j+1}B} - f_{2^j B}| \\ &\lesssim \omega(2^k B)^{1/v}|2^k B|^{\beta/n}\|f\|_{BMO^\beta(\omega)} + |2^k B|\omega(2^k B)^{-1/v'} \\ &\times \sum_{j=0}^{k-1}\frac{1}{|2^j B|}\int_{2^{j+1}B}|f(z) - f_{2^{j+1}B}|dz.\end{aligned}$$

We then obtain that

$$\begin{aligned}&\Big(\int_{2^k B} |f(z) - f_B|^v \omega^{1-v}(z)dz\Big)^{1/v} \\ &\lesssim \omega(2^k B)^{1/v}|2^k B|^{\beta/n}\|f\|_{BMO^\beta(\omega)} + |2^k B|\omega(2^k B)^{-1/v'}\sum_{j=0}^{k-1}\omega(2^{j+1}B) \\ &\times 2^{j+1}B|^{\beta/n-1}\|f\|_{BMO^\beta(\omega)}.\end{aligned}$$

This implies that

$$\begin{aligned}&\Big(\int_{2^k B} |f(z) - f_B|^v \omega^{1-v}(z)dz\Big)^{1/v} \\ &\lesssim \omega(2^k B)^{1/v}\|f\|_{BMO^\beta(\omega)}\Big[|2^k B|^{\beta/n} + |2^k B||B|^{\beta/n-1}\sum_{j=0}^{k-1}2^{j(\beta-n)}\Big] \\ &\lesssim \omega(2^k B)^{1/v}\|f\|_{BMO^\beta(\omega)}\Big[|2^k B|^{\beta/n} + k|2^k B||B|^{\beta/n-1}\Big] \\ &\lesssim (k+1)\omega(2^k B)^{1/v}2^{kn}|B|^{\beta/n}.\end{aligned}$$

By combining the above estimates, we obtain

$$\|(f - f_B)\chi_{2^k B}\|_{p'} \lesssim (k+1)2^{kn(1/p'+\eta)}r^{\beta-n/p}\omega(B)\|f\|_{BMO^\beta(\omega)}. \tag{2.7}$$

On the other hand, due to the boundedness of $\mathcal{I}_1$ and $V \in RH_q$, we have

$$\|\mathcal{I}_1(V\chi_{2^{k+2}B})\|_p \lesssim \|V\chi_{2^{k+2}B}\|_q \lesssim \frac{V(2^{k+2}B)}{|2^{k+2}B|^{1/q'}},$$

where $V(2^{k+2}B) = \int_{2^{k+2}B} V(y)dy$.

In the case $2^k r \geq \rho(x_0)$, we apply the second part of Lemma 2.6 to get

$$\|\mathcal{I}_1(V\chi_{2^{k+2}B})\|_p \lesssim (2^k r)^{n-2-n/q'}\Big(\frac{2^k r}{\rho(x)}\Big)^{2+(\mu-1)n}.$$

Combining the above estimates we arrive to (2.6). The case $2^k r < \rho(x)$, is handled similarly by using the first part of Lemma 2.6. □

We have the following simple result.

Lemma 2.9. *For $\alpha > 0$, we have* $\displaystyle\int_{B(x_0,r)} \frac{1}{|x_0 - z|^{n-\alpha}} dz \lesssim r^{\alpha}$.

Proof. Proof of this lemma is straightforward and we omit details here. □

3 Some estimates for the kernels

Let $\Gamma(g,h), \Gamma_0(g,h)$ denote the fundamental solution for the operator $-\Delta + V, -\Delta$ respectively. The following estimates of the fundamental solution for the Schrödinger operator on the nilpotent Lie group have been proved in [16].

Lemma 3.1. *Let $l > 0$ be an integer.*

1. *Suppose $V \in RH_{n/2}$. Then there exists $C_l > 0$ such that for $x \neq y$,*

$$|\Gamma(x,y)| \leq \frac{C_l}{(1+|x-y|)^l(1+|x-y|\rho(x)^{-1})^l}\frac{1}{|x-y|^{n-2}}.$$

2. *Suppose $V \in RH_n$. Then there exists $C_l > 0$ such that for $x \neq y$,*

(a) $$|\nabla\Gamma(x,y)| + |\nabla_2\Gamma(x,y)| \leq \frac{C_l}{(1+|x-y|)^l(1+|x-y|\rho(x)^{-1})^l}\frac{1}{|x-y|^{n-1}},$$

(b) $$|\nabla\nabla_2\Gamma(x,y)| \leq \frac{C_l}{(1+|x-y|)^l(1+|x-y|\rho(x)^{-1})^l}\frac{1}{|x-y|^{n}},$$

where ∇, ∇_2 means that we are taking all the partial derivaties with respect to the first and second variable, respectively.

Next, we recall the imbedding theorem of Morrey.

Lemma 3.2. *(The imbedding theorem of Morrey)*
Let the ball $B = B(a,R) \subset \mathbb{R}^n$ and $u \in W_0^{1,p}(B)$ with $p > n$. Then

$$|u(x) - u(y)| \leq CR^{\gamma}\|(\nabla u)\chi_B\|_p, \quad \text{for all} \quad x, y \in B,$$

where $C = C(n,p), \gamma = 1 - n/p$.

The following results give the estimate and the connection between the fundamental solutions of the Schrödinger operator $\mathcal{L}$ and the Laplacian $-\Delta$. The proofs of these lemmas are similar to that in [12, 13].

Lemma 3.3. *Suppose that $V \in RH_q$ for some $q > n$ and $|\nabla V| \in RH_{q_1}$ for some $q_1 > \frac{n}{2}$. There exists a constant $C_l > 0$ for each $l > 0$,*

$$|\nabla^2\Gamma(x,y)| \leq \frac{C_l}{\left(1+|x-y|\rho(x)^{-1}\right)^l}\frac{1}{|x-y|^{n-2}} \times\Big(\int_{B(x,2|x-y|)}\frac{|\nabla V(z)|}{|x-z|^{n-1}}dz+\frac{1}{|x-y|^2}\Big).$$

Lemma 3.4. *Suppose that $V \in RH_q$ for some $q > n$ and $|\nabla V| \in RH_{q_1}$ for some $q_1 > \frac{n}{2}$. Then*

$$|\nabla^2\Gamma(x,y)-\nabla^2\Gamma_0(x,y)| \lesssim \frac{1}{|x-y|^{n-2}}\int_{B(x,|x-y|)}\frac{|\nabla V(z)|}{|x-z|^{n-1}}dz + \frac{1}{|x-y|^{n-1}}\Big(\frac{|x-y|}{\rho_1(y)}\Big)^{\delta_1}+\frac{1}{|x-y|^{n}}\Big(\frac{|x-y|}{\rho(y)}\Big)^{\delta},$$

where $\delta = 2-n/q$ and $\delta_1 = 2-n/q_1$, if $|x-y| \leq A\rho_1(x)$ for some positive constant A.

Lemma 3.5. *Suppose that $V \in RH_q$ for some $q > n$ and $|\nabla V| \in RH_{q_1}$ for some $q_1 > \frac{n}{2}$. There exists a constant $C_l > 0$ for each $l > 0$,*

$$|\nabla^2\Gamma(x,y+h)-\nabla^2\Gamma(x,y)| \leq \frac{C_l}{\left(1+|x-y|\rho(x)^{-1}\right)^l}\frac{|h|^{\delta}}{|x-y|^{n-2+\delta}} \times\Big(\int_{B(x,|x-y|)}\frac{|\nabla V(z)|}{|x-z|^{n-1}}dz+\frac{1}{|x-y|^2}\Big),$$

where $\delta = 1-n/t$ for $t > n$.

Lemma 3.6. *Suppose $V \in RH_q$ for some $q > n/2$. Then, for any integer $k > 0$,*

$$|\Gamma(x,u)-\Gamma(y,u)| \lesssim \frac{|x-y|^{\delta}}{|x-u|^{n-2+\delta}}\Big(1+\frac{|x-u|}{\rho(u)}\Big)^{-k},$$

for $|x-y| < \frac{2}{3}|x-u|$ and $0 < \delta < \min\{1, 2-\frac{n}{q}\}$.

Lemma 3.7. *If $V \in RH_q$ for some $q > n$ then*

$$\|\nabla^2\Gamma(x,y)\|_{L^q(B(x_0,R))} \leq \frac{C_k}{R^{n-n/q}}\Big(1+\frac{R}{\rho(x_0)}\Big)^{k_0-k},$$

for any integer $k > 0$.

We will prove the following lemma.

Lemma 3.8. *Suppose $V \in RH_q$ for $q > n$. Then, for any integer $k > 0$,*

$$|\nabla\Gamma(x,u)-\nabla\Gamma(y,u)| \lesssim \frac{|x-y|^{\delta}}{|x-u|^{n-1+\delta}}\Big(1+\frac{|x-u|}{\rho(x)}\Big)^{-k},$$

for $|x-y| < c|x-u|$, $0 < \delta < \min\{1, 1-\frac{n}{q}\}$, and $c \in (0,1)$.

Proof. Fix $x, u \in \mathbb{R}^n$. Set $R = c|x-u|$. Then for every $h \in \mathbb{R}^n$ ($|h| < c|x-u|$), it follows from the imbedding theorem of Morrey and Lemma 3.7 that

$$\begin{aligned}|\nabla\Gamma(x+h,u) - \nabla\Gamma(x,u)| &\leq C|h|^{1-n/q}\Big(\int\limits_{B(x,R)} |\nabla^2\Gamma(z,u)|^q dz\Big)^{1/q}\\ &\leq C_k\Big(\frac{|h|}{R}\Big)^{1-n/q}\frac{1}{R^{n-1}}\Big(1+\frac{R}{\rho(x)}\Big)^{-k_1}\\ &= \frac{C|h|^\delta}{|x-u|^{n-1+\delta}}\Big(1+\frac{|x-u|}{\rho(x)}\Big)^{-k}.\end{aligned}$$

This completes the our proof. □

We denote by $\mathcal{K}^*$ and $\mathbf{K}^*$ the associated kernels of $\mathcal{R}^* = \nabla^2(-\Delta)$ and R^*, respectively. The following result will play an important role in the sequel.

Lemma 3.9. *Let $V \in RH_q$ with $q > n$ such that $|\nabla V| \in RH_{q_1}$ with $q_1 > n/2$ and $0 < \delta < \min\{1-\frac{n}{q}, 2-\frac{n}{q_1}\}$. Suppose that $\rho \lesssim \rho_1 \lesssim 1$. Then, there exists a positive constant C such that*

$$\begin{aligned}&\Big|[\mathbf{K}^*(x,z) - \mathcal{K}^*(x,z)] - [\mathbf{K}^*(y,z) - \mathcal{K}^*(y,z)]\Big|\\ &\lesssim \frac{|x-y|^\delta}{|x-z|^{n+\delta}}\Big(\frac{|x-z|}{\rho(x)}\Big)^{2-n/q} + \frac{|x-y|^\delta}{|x-z|^{n-1+\delta}}\Big(\frac{|x-z|}{\rho_1(x)}\Big)^{2-n/q_1}\\ &+ \frac{|x-y|^\delta}{|x-z|^{n-2+\delta}}\int\limits_{B(z,3|x-z|)}\frac{|\nabla V(u)|}{|z-u|^{n-1}}du,\end{aligned} \tag{3.8}$$

whenever $|x-z| \geq 2|x-y|$ and $|x-y| \leq A\rho(x)$ for some positive constant A.

Proof. We consider two cases.

Case 1. $|x-z| \geq \rho(x)/2$.

By Lemma 3.5 the lemma is proved .

Case 2. $|x-z| < \rho(x)/2$.

Note that $(-\Delta + V)\Gamma(x,y) = 0$. By standard arguments, we come up to

$$-\Delta(\nabla\Gamma) = -\nabla(V\Gamma).$$

Moreover, using the fact that $\Delta(\nabla\Gamma_0) = 0$, we have

$$\begin{aligned}(\nabla\Gamma - \nabla\Gamma_0)(x,z) &= -(-\Delta)^{-1}\nabla(V\Gamma)(x,z)\\ &= -\int\limits_{\mathbb{R}}^{n}\Gamma_0(x,u)(\nabla V(u))\Gamma(u,z)du\\ &\quad -\int\limits_{\mathbb{R}}^{n}\Gamma_0(x,u)V(u)(\nabla\Gamma(u,z))du\end{aligned}$$

Since $\Gamma(u, x) = \Gamma(x, u)$, we obtain

$$\begin{aligned}
&[\mathbf{K}^*(x,z) - \mathcal{K}^*(x,z)] - [\mathbf{K}^*(y,z) - \mathcal{K}^*(y,z)] \\
&= -\int\limits_{\mathbb{R}}^{n} \nabla\Gamma_0(z,u)(\nabla V(u))[\Gamma(x,u) - \Gamma(y,u)]du \\
&\quad - \int\limits_{\mathbb{R}}^{n} \nabla\Gamma_0(z,u)V(u)[\nabla\Gamma(x,u) - \nabla\Gamma(y,u)]du \\
&= -(A_1 + A_2).
\end{aligned}$$

We claim that

$$|A_2| \leq \frac{C|x-y|^{\delta}}{|x-z|^{n+\delta}}\Big(\frac{|x-z|}{\rho(x)}\Big)^{2-n/q}. \tag{3.9}$$

To show this, we split $\mathbb{R}^n$ into 4 regions:

$$E_1 = \left\{u : |x-u| < \frac{3}{2}|x-y|\right\},$$

$$E_2 = \left\{u : \frac{3}{2}|x-y| \leq |x-u| < \frac{1}{2}|x-z|\right\},$$

$$E_3 = \left\{u : \frac{1}{2}|x-z| \leq |x-u| < 2|x-z|\right\},$$

$$E_4 = \{u : |x-u| \geq 2|x-z|\}.$$

Write $I_j = \int\limits_{E_j} |\nabla\Gamma_0(z,u)|V(u)\,|\nabla\Gamma(x,u) - \nabla\Gamma(y,u)|\,du$ for $j = 1, 2, 3, 4$.

We have

$$\begin{aligned}
I_1 &\leq \int\limits_{E_1} |\nabla\Gamma_0(z,u)\nabla\Gamma(x,u)|\,V(u)du + \int\limits_{E_1} |\nabla\Gamma_0(z,u)\nabla\Gamma(y,u)|\,V(u)du \\
&:= J_1 + J_2.
\end{aligned}$$

We only estimate J_1 here. The same proof works for J_2.

Since $|x-z| \geq 2|x-y|$ implies $|z-u| \geq \frac{1}{4}|x-z|$ for all $u \in E_1$, and since $\delta < 1 - n/q$, by using Lemma 2.6 with $\epsilon = 1 - \delta$ and $r = |x-z| < \frac{\rho(x)}{2}$, we have

$$\begin{aligned}
\int\limits_{E_1} |\nabla\Gamma_0(z,u)\nabla\Gamma(x,u)|V(u)du &\leq \frac{C}{|x-z|^{n-1}} \int\limits_{B(x,2|x-y|)} \frac{V(u)}{|x-u|^{n-1}}du \\
&\leq \frac{C|x-y|^{\delta}}{|x-z|^{n-1}} \int\limits_{B(x,|x-z|)} \frac{V(u)}{|x-u|^{n-1+\delta}}du \\
&\leq \frac{C|x-y|^{\delta}}{|x-z|^{n+\delta}}\Big(\frac{|x-z|}{\rho(x)}\Big)^{2-n/q}.
\end{aligned}$$

Using Lemma 3.8 yields $I_j \lesssim \int\limits_{E_j} \frac{V(u)}{|u-z|^{n-1}} \frac{|x-y|^{\delta}}{|x-u|^{n-1+\delta}}\Big(1 + \frac{|x-u|}{\rho(x)}\Big)^{-k} du$ for $j = 2, 3, 4$.

Note that $|x-z| \geq 2|x-y|$ implies $|u-z| \geq \frac{1}{2}|x-z|$ for $u \in E_2$. It follows from Lemma 2.6 with $\epsilon = 1-\delta$ and $r = \frac{1}{2}|x-z|$ that

$$I_2 \leq \frac{C|x-y|^{\delta}}{|x-z|^{n-1}} \int_{B(x,\frac{|x-z|}{2})} \frac{V(u)}{|x-u|^{n-1+\delta}} du \leq \frac{C|x-y|^{\delta}}{|x-z|^{n+\delta}} \Big(\frac{|x-z|}{\rho(x)}\Big)^{2-n/q}.$$

Noticing that $E_3 \subset B(z, 3|x-z|)$, we observe that

$$I_3 \leq \frac{C|x-y|^{\delta}}{|x-z|^{n-1+\delta}} \int_{B(z,3|x-z|)} \frac{V(u)}{|u-z|^{n-1}} du \leq \frac{C|x-y|^{\delta}}{|x-z|^{n+\delta}} \Big(\frac{|x-z|}{\rho(x)}\Big)^{2-n/q},$$

where we have used Lemma 2.6 in the last inequality with $\epsilon = 1$, $r = |x-z|$ and the fact that $\rho(z) \sim \rho(x)$.

Finally, noticing $|x-u| \sim |u-z|$ if $u \in E_4$, we ensures that

$$I_4 \leq C|x-y|^{\delta} \int_{E_4} \frac{V(u)}{|x-u|^{2n-2+\delta}} \left[1 + \frac{|x-u|}{\rho(x)}\right]^{-k} du.$$

Set $F_1 = \{u \in \mathbb{R}^n : 2|x-z| \leq |x-u| < \rho(x)\}$ and $F_2 = \{u \in \mathbb{R}^n : |x-u| \geq \rho(x)\}$.

Then $E_4 = F_1 \cup F_2$. Hence, we can get estimate

$$\begin{aligned}
&\int_{F_1} \frac{V(u)}{|x-u|^{2n-2+\delta}} \left[1 + \frac{|x-u|}{\rho(x)}\right]^{-k} du \\
&\leq \Big(\int_{B(x,\rho(x))} V^q(u) du\Big)^{1/q} \Big(\int_{|x-u| \geq 2|x-z|} \frac{1}{|x-u|^{(2n-2+\delta)q'}} du\Big)^{1/q'} \\
&\leq \rho(x)^{n/q-n} \int_{B(x,\rho(x))} V(u) du \\
&\times \Big(\sum_{i=1}^{\infty} \int_{2^i|x-z| \leq |x-u| \leq 2^{i+1}|x-z|} \frac{1}{|x-u|^{(2n-2+\delta)q'}} du\Big)^{1/q'} \\
&\leq \frac{C}{|x-z|^{n+\delta}} \Big(\frac{|x-z|}{\rho(x)}\Big)^{2-n/q},
\end{aligned}$$

where we have used the definition of $\rho(x)$ in the last inequality.

Taking into account of $n + \delta \geq 2 - n/q$, for sufficiently large k leads us to

$$\int_{F_2} \frac{V(u)}{|x-u|^{2n-2+\delta}} \left[1 + \frac{|x-u|}{\rho(x)}\right]^{-k} du$$

$$\leq \int_{|x-u|\geq\rho(x)} \frac{V(u)}{|x-u|^{2n-2+\delta}} \Big(\frac{\rho(x)}{|x-u|}\Big)^k du$$

$$\leq \rho(x)^{-2n+2-\delta} \sum_{j=1}^{\infty} 2^{-j(2n-2+\delta+k)} \int_{|x-u|<2^j\rho(x)} V(u)du$$

$$\lesssim \rho(x)^{-2n+2-\delta} \sum_{j=1}^{\infty} 2^{-j(2n-2+\delta+k-Q\mu)} \int_{|x-u|<\rho(x)} V(u)du$$

$$\lesssim \rho(x)^{-n-\delta} \Big(\frac{\rho(x)}{|x-z|}\Big)^{n+\delta-2+n/q}$$

$$\lesssim \frac{1}{|x-z|^{n+\delta}} \Big(\frac{|x-z|}{\rho(x)}\Big)^{2-n/q},$$

where we have used that V belongs to D_μ for some $\mu \geq 1$ in the third inequality.

Therefore, we imply that

$$I_4 \leq \frac{C|x-y|^\delta}{|x-z|^{n+\delta}} \Big(\frac{|x-z|}{\rho(x)}\Big)^{2-n/q}.$$

Write $J_i = \int_{E_i} |\nabla\Gamma_0(z,u)|(\nabla V(u))[\Gamma(x,u) - \Gamma(y,u)]du$ for $i = 1, 2, 3, 4$. For J_1, we majorize the difference related to Γ by the sum of the absolute values of each term and estimate each integral separately. Since both are similar we work out one of them. First we notice that $|x-z| > 2|x-y|$ implies $|z-u| > \frac{1}{4}|x-z|$ for $u \in E_1$. Then, using Lemma 3.1, we deduce that

$$\int_{E_1} |\nabla\Gamma_0(z,u)\Gamma(x,u)||\nabla V(u)|du \lesssim \frac{1}{|x-z|^{n-1}} \int_{B(x,2|x-y|)} \frac{|\nabla V(u)|}{|x-u|^{n-2}} du$$

$$\lesssim \frac{|x-y|^\delta}{|x-z|^{n-1+\delta}} \Big(\frac{|x-z|}{\rho_1(x)}\Big)^{2-n/q_1},$$

where in the last inequality we have used Lemma 2.6 with $\epsilon = 2 - \delta$ and $r = |x-y| < 2\rho(x)$.

For the remaining regions we will use Lemma 3.6. To estimate J_2 we use Lemma 3.6 to get

$$J_2 \lesssim \frac{|x-y|^\delta}{|x-z|^{n-1}} \int_{B(x,\frac{1}{2}|x-z|)} \frac{|\nabla V(u)|}{|x-u|^{n-2+\delta}} du \lesssim \frac{|x-y|^\delta}{|x-z|^{n-1+\delta}} \Big(\frac{|x-z|}{\rho_1(x)}\Big)^{2-n/q_1},$$

where in the last inequality we have used Lemma 2.6 with $\epsilon = 2 - \delta$ and $r = \frac{1}{2}|x-z|$.

To deal with J_3 we notice $E_3 \subset B(z, 3|x-z|)$. Using again Lemma 3.6 we arrive to

$$J_3 \lesssim \frac{|x-y|^\delta}{|x-z|^{n-2+\delta}} \int_{B(z,3|x-z|)} \frac{|\nabla V(u)|}{|u-z|^{n-1}} du.$$

Finally, for $u \in E_4$ we have $|x-u| \sim |u-z|$ and hence, using Lemma 3.6

$$J_4 \leq |x-y|^{\delta} \int_{E_4} \frac{|\nabla V(u)|}{|x-u|^{2n-3+\delta}} \Big[1 + \Big(\frac{|x-u|}{\rho(u)}\Big)\Big]^{-k} du.$$

We set $E_4 = E_4^1 \cup E_4^2$, where $E_4^1 = \{u : 2|x-z| \leq |x-u| \leq \rho_1(x)\}$. Applying Hölder's inequality the above integral over E_4^1 is bounded by

$$\begin{aligned} &\Big(\int_{B(x,\rho_1(x))} |\nabla V|^{q_1}\Big)^{1/q_1}\Big(\int_{|x-u|>2|x-z|} \frac{1}{|x-u|^{(2n-3+\delta)q_1'}} du\Big)^{1/q_1'} \\ &\leq \frac{C}{|x-z|^{n-1+\delta}}\Big(\frac{|x-z|}{\rho_1(z)}\Big)^{2-n/q_1}, \end{aligned}$$

where in the last inequality we have used the reverse Hölder condition on $|\nabla V|$ and the definition of ρ_1.

To estimate the integral on E_4^2, we have

$$\rho(u) \leq C\rho(x)^{1-\sigma}|x-u|^{\sigma},$$

with $0 < \sigma < 1$. Therefore, we set $N = k(1-\sigma)$ to get

$$\begin{aligned} &\int_{|x-u|>\rho_1(x)} \frac{|\nabla V(u)|}{|x-u|^{2n-3+\delta}}\Big(\frac{\rho_1(x)}{|x-u|}\Big)^{N} du \\ &\leq \rho_1(x)^{-n+3-\delta} \sum_{j=1}^{\infty} 2^{-j(2n-3+\delta+N)} \int_{|x-u|<2^j\rho_1(x)} |\nabla V|. \end{aligned}$$

Since $|\nabla V|$ satisfies a doubling condition and we can choose k large enough, we have

$$\begin{aligned} &\int_{|x-u|>\rho_1(x)} \frac{|\nabla V(u)|}{|x-u|^{2n-3+\delta}}\Big(\frac{\rho_1(x)}{|x-u|}\Big)^{N} du \\ &\leq \rho_1(x)^{-n+3-\delta} \sum_{j=1}^{\infty} 2^{-j(2n-3+\delta+N)} \int_{|x-u|<2^j\rho_1(x)} |\nabla V| \\ &\lesssim \rho_1(x)^{-2n+3-\delta} \sum_{j=1}^{\infty} 2^{-j(2n-3+\delta+N-n\mu)} \int_{|x-u|<\rho_1(x)} |\nabla V(u)|du \\ &\lesssim \rho_1(x)^{-n+1-\delta} \lesssim \frac{1}{|x-z|^{n-1+\delta}}\Big(\frac{|x-z|}{\rho_1(z)}\Big)^{2-n/q_1}, \end{aligned}$$

where we have use that $|\nabla V|$ belongs to D_μ for some $\mu \geq 1$ and $n-1+\delta \geq 2-n/q_1$.

Now using the estimates in E_4^1 and E_4^2 reminding that $|x-z| \leq \rho_1(x)$, we obtain

$$J_4 \leq \frac{|x-y|^{\delta}}{|x-z|^{n-1+\delta}}\Big(\frac{|x-z|}{\rho_1(x)}\Big)^{2-n/q_1}$$

The proof is completed. □

4 Proofs of the main results

First, we prove the following theorem.

Theorem 4.1. *For any $0 \le \beta < 1, 1 \le \alpha < 1 + \frac{1-\beta}{n}$, and $\omega \in A_\infty \cap D_\alpha$, the operator $\mathcal{R}^* = (-\Delta)^{-1}\nabla^2$ is bounded on $BMO^\beta(\omega)$.*

Proof. For $B := B(x_0, r)$, set $g := f - f_{5B}$. We write $g = g_1 + g_2$, with $g_1 = g\chi_{5B}$.

Since $\mathcal{R}^*(f_{5B}) = 0$, we have $\mathcal{R}^*(f) = \mathcal{R}^*(g)$. Since $\omega \in A_\infty$, we have $\omega \in A_p$ for some $1 < p < \infty$ and hence $\omega^{1-p'} \in A_{p'}$. Since $\mathcal{R}^*$ is a Calderón-Zygmund operator, we have

$$\int_B |\mathcal{R}^* g(x) - (\mathcal{R}^* g)_B| dx$$
$$\le \int_B |\mathcal{R}^* g_1(x) - (\mathcal{R}^* g_1)_B| dx + \int_B |\mathcal{R}^* g_2(x) - (\mathcal{R}^* g_2)_B| dx = I_1 + I_2.$$

By the boundedness of $\mathcal{R}^*$ and the John-Nirenberg's inequality, we have

$$I_1 \le 2\int_B |\mathcal{R}^* g_1| \le \omega(B)^{1/p}\Big(\int_B |\mathcal{R}^* g_1|^{p'}\omega^{1-p'}\Big)^{1/p'}$$
$$\le C\omega(B)^{1/p}\Big(\int_{5B} |f - f_{5B}|^{p'}\omega^{1-p'}\Big)^{1/p'} \lesssim \|f\|_{BMO^\beta(\omega)}\omega(B)|B|^{\beta/n}.$$

On the other hand, observe that

$$I_2 \le \frac{1}{|B|}\int_B\int_B\int_{(5B)^c} |\mathcal{K}^*(x-z) - \mathcal{K}^*(y-z)||f(z) - f_{5B}| dz dx dy.$$

Applying Proposition 1.7 in [9], we come up to

$$|\mathcal{K}^*(x-y) - \mathcal{K}^*(x)| \le \frac{C|y|}{|x|^{n+1}},$$

for $|y| \le \frac{|x|}{2}$.

Note that $|x - z| \ge 2|x - y|$ for $x, y \in B$ and $z \in (5B)^c$. Therefore, we have

$$I_2 \le \frac{C}{|B|}\int_B\int_B\int_{(5B)^c} \frac{|x-y|}{|x-z|^{n+1}}|f(z) - f_{5B}| dz dx dy$$
$$\le Cr^{n+1}\int_{(5B)^c} \frac{|f(z) - f_{5B}|}{|x_0 - z|^{n+1}} dz$$
$$\le C\sum_{j=2}^{\infty} \frac{1}{2^{j(n+1)}}\int_{2^{j+1}B} |f(z) - f_{5B}| dz$$
$$\le Cr^\beta \|f\|_{BMO^\beta(\omega)}\omega(B)\sum_{j=2}^{\infty}(j+1)2^{-j(n+1-\beta-n\alpha)}.$$

The last series converges provided that $\alpha < 1 + \dfrac{1-\beta}{n}$. Combining these, we obtain the desired result. □

Proof of Theorem 1.2. Let $s > p'$ such that $\omega \in A_{p/s'} \cap RH_s$. We choose q_1 satisfying

$$n/2 < q_1 < q_1^* \le n, 0 \le \beta < \delta < \min\{1 - \frac{n}{q}, 2 - \frac{n}{q_1}\}, 1 \le \alpha < 1 + \frac{\delta - \beta}{n}$$

and such that $\omega \in A_{p_1/s'}$ for $\frac{1}{p_1} = \frac{1}{q_1} - \frac{1}{n}$.

Let $f \in BMO^\beta(\omega)$. By Proposition 2.1, we need to prove that

$$\int_B |R^* f| \lesssim \|f\|_{BMO^\beta(\omega)} \omega(B)|B|^{\beta/n}, \tag{4.10}$$

for all $B = B(x_0, \rho(x_0))$, $x_0 \in \mathbb{R}^n$, and

$$\int_B |R^* f - (R^* f)_B| \lesssim \|f\|_{BMO^\beta(\omega)} \omega(B)|B|^{\beta/n}, \tag{4.11}$$

with $B = B(x_0, r), r < \rho(x_0)$.

First, we prove (4.10). Let $B = B(x_0, \rho(x_0))$, set $g = f - f_B$. We write $g = g_1 + g_2$, with $g_1 = g\chi_{2B}$. Since $R^*(f_B) = 0, R^*(f) = R^*(g)$. Since R^* is bounded in $L^{p'}$, (2.7) yields that

$$\begin{aligned}\int_B |R^* g_1| &\le |B|^{1/p}\Big(\int_B |R^* g_1|^{p'}\Big)^{1/p'} \le C|B|^{1/p}\Big(\int_{2B} |f - f_B|^{p'}\Big)^{1/p'} \\ &\lesssim \|f\|_{BMO^\beta(\omega)} |B|^{\beta/n} \omega(B).\end{aligned}$$

For g_2 we estimate the size of $\mathbf{K}^*$ using Lemma 3.3. Now, note that for $x \in B$ and $z \in \mathbb{R}^n \setminus 2B, \rho_1(x) \sim \rho_1(x_0), |x_0 - z| \sim |x - z|$, we have

$$\begin{aligned}\int_B |R^* g_2| &\le \int_B \int_{(2B)^c} |\mathbf{K}^*(x,z)||f(z) - f_B| dz dx \\ &\le C_k \int_B \int_{(2B)^c} \Big(\frac{\rho(x)}{|x-z|}\Big)^k \frac{1}{|x-z|^n} |f(z) - f_B| dz dx \\ &+ C_k \int_B \int_{(2B)^c} \rho(x)^k \Big(\int_{B(z,2|x-z|)} \frac{|\nabla V(u)|}{|u-z|^{n-1}} du\Big) \frac{|f(z) - f_B|}{|x-z|^{k+n-2}} dz dx \\ &= I_1 + I_2.\end{aligned}$$

We have

$$I_1 \le C_k \rho(x_0)^{k+n} \int_{(2B)^c} \frac{|f(z) - f_B|}{|x_0 - z|^{k+n}} dz.$$

Note that

$$\begin{aligned}\int_{2^jB} |f - f_B| &\leq \int_{2^jB} |f - f_{2^jB}| + |2^jB| \sum_{l=0}^{j-1} |f_{2^{l+1}B-f_{2^lB}}| \\ &\leq \omega(2^jB)|2^jB|^{\beta/n}\|f\|_{BMO^\beta(\omega)} \\ &+ |2^jB| \sum_{l=0}^{j-1} |2^lB|^{\beta/n-1}\omega(2^lB)\|f\|_{BMO^\beta(\omega)} \\ &\leq 2^{j(n\alpha+\beta)}\omega(B)|B|^{\beta/n}\|f\|_{BMO^\beta(\omega)} \\ &+ 2^{jn}|B|^{\beta/n}\omega(B) \sum_{l=0}^{j-1} 2^{l(\beta-n+n\alpha)}\|f\|_{BMO^\beta(\omega)} \\ &\leq 2^{j(n\alpha+\beta)}\omega(B)|B|^{\beta/n}\|f\|_{BMO^\beta(\omega)}.\end{aligned}$$

Splitting the integral into dyadic annuli and using the doubling property, I_1 is bounded by

$$C_k \sum_{j=2}^{\infty} \frac{1}{2^{j(k+n)}} \int_{2^jB} |f(z) - f_B|dz \leq C_k\|f\|_{BMO^\beta(\omega)}\omega(B)|B|^{\beta/n} \sum_{j=2}^{\infty} 2^{-j(k+n-n\alpha-\beta)},$$

and the last sum is finite choosing k big enough.

Now, using that for $x \in B$ and $z \in \mathbb{R}^n \setminus 2B$,

$$\rho_1(x) \sim \rho_1(x_0), |x - z| \sim |x_0 - z|, B(z, 2|x - z|) \subset B(x_0, 4|x_0 - z|),$$

we have that

$$I_2 \lesssim \rho^2(x_0) \sum_{j=1}^{\infty} \frac{1}{2^{j(k+n-2)}} \int_{2^{j+1}B\setminus 2^jB} \Big(\int_{2^{j+3}} \frac{|\nabla V(u)|}{|x - z|^{n-1}}du\Big)|f(z) - f_B|dz.$$

Noticing that

$$\int_{2^{j+3}} \frac{|\nabla V(u)|}{|x - z|^{n-1}}du = \mathcal{I}_1(|\nabla V|\chi_{2^{j+3}B}).$$

We may use Lemma 2.8 and $\omega \in D_\alpha, |\nabla V| \in D_\mu$ to obtain the bound

$$C\|f\|_{BMO^\beta(\omega)}\omega(B)\rho(x_0)^{\beta+1} \sum_{j=1}^{\infty} \frac{j+2}{2^{j(k+2n-n\alpha-n\mu-3)}}.$$

Choosing k large enough to make the series convergent and note $\rho \lesssim \rho_1 \lesssim 1$ we arrive to the desired estimate. Now we take care of the oscillation of R^* on a ball $B = B(x_0, r)$ with $r < \rho(x_0)$. Then, we deduce that

$$\begin{aligned}\int_B |R^*f(x) - (R^*f)_B|dx &\leq \int_B |(R^* - \mathcal{R}^*)f(x) - [(R^* - \mathcal{R}^*)f]_B|dx \\ &+ \int_B |\mathcal{R}^*f(x) - (\mathcal{R}^*f)_B|dx = I + J,\end{aligned}$$

where $\mathcal{R}^* = (-\Delta)^{-1}\nabla^2$.

Note that $BMO^{\beta}_{\mathcal{L}}(\omega) \subset BMO^{\beta}(\omega)$. Using Theorem 4.1, we arrive as

$$J \leq C\|f\|_{BMO^{\beta}(\omega)}\omega(B)|B|^{\beta/n}.$$

To take care of I, we set $g = f - f_B$. We write $g = g_1 + g_2 + g_3$, with $g_1 = g\chi_{5B}$, $g_2 = g\chi_{(B_0\setminus 5B)}$ and $g_3 = g\chi_{B_0^c}$, where $B_0 = B(x_0, 5\rho(x_0))$.

Since $(R^* - \mathcal{R}^*)(f_B) = 0$, $(R^* - \mathcal{R}^*)(f) = (R^* - \mathcal{R}^*)(g)$. Then, we get that $I \leq I_1 + I_2 + I_3$, where

$$I_j = \int_B \Big|(R^* - \mathcal{R}^*)g_j(x) - [(R^* - \mathcal{R}^*)g_j]_B\Big| dx$$

for $j = 1, 2, 3$.

Now we proceed to estimate I_2. We apply Lemma 3.9 with $\beta < \delta < \min\{1 - n/q, 2 - n/q_1\}$

$$\begin{aligned}
I_2 &\leq \frac{1}{|B|}\int_B\int_B\int_{B_0\setminus 5B} |[\mathbf{K}^*(x,z) - \mathcal{K}^*(x,z)] - [\mathbf{K}^*(y,z) - \mathcal{K}^*(y,z)]| \times \\
&\times |f(z) - f_B| dz dx dy \\
&\leq \frac{1}{|B|}\int_B\int_B\int_{B_0\setminus 5B} \frac{|x-y|^{\delta}}{|x-z|^{n+\delta}}\Big(\frac{|x-z|}{\rho(x)}\Big)^{2-n/q}|f(z) - f_B| dz dx dy \\
&+ \frac{1}{|B|}\int_B\int_B\int_{B_0\setminus 5B} \Big(\frac{|x-y|^{\delta}}{|x-z|^{n-2+\delta}}\int_{B(z,3|x-z|)} \frac{|\nabla V(u)|}{|u-z|^{n-1}} du\Big)|f(z) - f_B| dz dx dy \\
&+ \frac{1}{|B|}\int_B\int_B\int_{B_0\setminus 5B} \frac{|x-y|^{\delta}}{|x-z|^{n-1+\delta}}\Big(\frac{|x-z|}{\rho_1(x)}\Big)^{2-n/q_1}|f(z) - f_B| dz dx dy \\
&:= I_{2,1} + I_{2,2} + I_{2,3}.
\end{aligned}$$

For every $x, y \in B$ and for every $z \in (5B)^c$, noting that $|x-z| \geq 2|x-y|$, $\rho(x_0) \sim \rho(x)$, $|x-z| \sim |x_0 - z|$ and $|x_0 - z| < 5\rho(x_0) \lesssim 1$,

$$\begin{aligned}
I_{2,1} &\leq \frac{C}{|B|}\int_B\int_B\int_{B_0\setminus 5B} \frac{|x-y|^{\delta}}{|x-z|^{n+\delta}}\Big(\frac{|x-z|}{\rho(x)}\Big)^{2-n/q}|f(z) - f_B| dz dx dy \\
&\lesssim \frac{r^{n+\delta}}{\rho(x_0)^{2-n/q}}\int_{B_0\setminus 5B} \frac{|f(z) - f_B|}{|x_0 - z|^{n+\delta-2+n/q}} dz.
\end{aligned}$$

Let j_0 be the integer part of $\log_2(5\rho(x_0)/r)$. Then, we have

$$\begin{aligned}
I_{2,1} &\lesssim \Big(\frac{r}{\rho(x_0)}\Big)^{2-n/q}\sum_{j=2}^{j_0}\frac{1}{2^{j(n+\delta-2+n/q)}}\int_{2^{j+1}B\setminus 2^jB} |f(z) - f_B| dz \\
&\lesssim \Big(\frac{r}{\rho(x_0)}\Big)^{2-n/q}\omega(B)|B|^{\beta/n}\|f\|_{BMO^{\beta}(\omega)}\sum_{j=2}^{j_0} j2^{j(2+n\alpha+\beta-n-\delta-n/q)} \\
&\lesssim \Big(\frac{r}{\rho(x_0)}\Big)^{n+\delta-n\alpha-\beta}\log_2(5\rho(x_0)/r)|B|^{\beta/n}\omega(B)\|f\|_{BMO^{\beta}(\omega)}.
\end{aligned}$$

Using the inequality $1 + \log_2(t) \le Ct^{\varepsilon/2}$ for $t > 1/8$, with ε small enough, we get

$$I_{2,1} \lesssim |B|^{\beta/n}\omega(B)\|f\|_{BMO^{\beta}(\omega)}.$$

We only have to take care of $I_{2,2}$ by $I_{2,3}$ is the same as $I_{2,1}$. Now, using that for $x \in B$ and $z \in \mathbb{R}^n \setminus 5B, \rho(x) \sim \rho(x_0), |x - z| \sim |x_0 - z|, B(z, 3|x - z|) \subset B(x_0, 5|x_0 - z|)$, we imply that

$$I_{2,2} \lesssim r^{\delta+n} \int_{B_0\setminus 5B} \frac{|f(z) - f_B|}{|x_0 - z|^{n+\delta-2}} \int_{B(x_0,5|x_0-z|)} \frac{|\nabla V(u)|}{|u - z|^{n-1}} du dz.$$

Breaking the integral in z dyadically and setting j_0 such that $2^{j_0-1}r \le \rho(x_0) \le 2^{j_0}r$

$$I_{2,2} \lesssim r^2 \sum_{j=3}^{j_0} \frac{1}{2^{j(n+\delta-2)}} \int_{2^jB} |f(z) - f_B| \mathcal{I}_1(\chi_{2^{j+3}B}|\nabla V|)(z) dz.$$

We deduce from Lemma 2.8 that

$$\begin{aligned}
I_{2,2} &\lesssim r^{\beta+1}\|f\|_{BMO^{\beta}(\omega)}\omega(B)\Big(\frac{r}{\rho(x_0)}\Big)^{n-n\alpha-\beta+2-n/q_1} \sum_{j=3}^{j_0} j2^{j(3-\delta-n/q_1)} \\
&\lesssim r^{\beta}\rho(x_0)\|f\|_{BMO^{\beta}(\omega)}\omega(B)\Big(\frac{r}{\rho(x_0)}\Big)^{n-n\alpha-\beta+3-n/q_1} \sum_{j=3}^{j_0} j2^{j(3-\delta-n/q_1)} \\
&\lesssim \|f\|_{BMO^{\beta}(\omega)}\omega(B)|B|^{\beta/n} \log_2(5\rho(x_0)/r)\Big(\frac{r}{\rho(x_0)}\Big)^{n+\delta-n\alpha-\beta}.
\end{aligned}$$

Using the inequality $1 + \log_2(t) \le Ct^{\varepsilon/2}$ for $t > 1/8$, with ε small enough, we obtain

$$I_{2,2} \lesssim \|f\|_{BMO^{\beta}(\omega)}\omega(B)|B|^{\beta/n}.$$

Now we take care of I_3. We use the smoothness of each kernel separately. For $\mathcal{K}^*$ we use Calderón-Zygmund condition and for $\mathbf{K}^*$ we use Lemma 3.5. We have

$$\begin{aligned}
I_3 &\lesssim r^{n+\delta} \int_{B_0^c} \frac{|f(z) - f_B|}{|x_0 - z|^{n+\delta}} dz \\
&+ \rho(x_0)^k r^{\delta+n} \int_{\mathbb{R}^n\setminus B(x_0,\rho(x_0))} \frac{|f(z) - f_B|}{|x_0 - z|^{k+n+\delta-2}} \int_{B(x_0,4|x_0-z|)} \frac{|\nabla V(u)|}{|u - z|^{n-1}} du dz \\
&:= B_1 + B_2.
\end{aligned}$$

We estimate B_1

$$\begin{aligned}
B_1 &\lesssim r^{n+\delta} \int_{B_0^c} \frac{|f(z) - f_B|}{|x_0 - z|^{n+\delta}} dz \lesssim \sum_{j=j_0}^{\infty} \frac{1}{2^{j(n+\delta)}} \int_{2^{j+1}B} |f(z) - f_B| dz \\
&\lesssim \|f\|_{BMO^{\beta}(\omega)}|B|^{\beta/n}\omega(B) \sum_{j=j_0}^{\infty} j \frac{1}{2^{j(n+\delta-n\alpha-\beta)}} \lesssim \|f\|_{BMO^{\beta}(\omega)}|B|^{\beta/n}\omega(B).
\end{aligned}$$

Moreover

$$B_2 \le C_k\rho(x_0)^k r^{2-k}\sum_{j=j_0}^{\infty}\frac{1}{2^{j(k+n+\delta-2)}}\int_{2^jB}|f(z)-f_B|\mathcal{I}_1(\chi_{2^{j+3}B}|\nabla V|)(z)dz$$
$$\le C_k\|f\|_{BMO^\beta(\omega)}r^\beta\omega(B)\rho_1(x_0)\Big(\frac{r}{\rho_1(x_0)}\Big)^{3+(\mu-1)n-k}\sum_{j=j_0}^{\infty}j\frac{1}{2^{j(k+2n+\delta-3-n\mu-n\alpha)}}.$$

Using the inequality $j\le\frac{2^{j\epsilon}}{\epsilon ln2}$ with ϵ small enough, we arrive at

$$B_2 \le C_k\|f\|_{BMO^\beta(\omega)}r^\beta\omega(B)\rho_1(x_0)\Big(\frac{r}{\rho_1(x_0)}\Big)^{3+(\mu-1)n-k}\sum_{j=j_0}^{\infty}\frac{1}{2^{j(k+2n+\delta-3-n\mu-n\alpha-\epsilon)}}.$$

Choosing k large enough to make the series convergent we get

$$C_k\|f\|_{BMO^\beta(\omega)}r^\beta\Big(\frac{r}{\rho_1(x_0)}\Big)^{n+\delta-n\alpha-\epsilon},$$

and the last factor is bounded since $r<\rho_1(x_0)$.

Finally,

$$I_1\le 2\int_B|(R^*-\mathcal{R}^*)g_1(x)|dx\lesssim\int_B\int_{5B}|\mathbf{K}^*(x,y)-\mathcal{K}^*(x,y)||f(y)-f_B|dydx.$$

Then, Lemma 3.4 now implies

$$I_1\lesssim\int_B\int_{5B}\Big[\frac{1}{|x-y|^{n-2}}\int_{B(y,|x-y|)}\frac{|\nabla V(z)|dz}{|y-z|^{n-1}}+\frac{1}{|x-y|^{n-1}}\Big(\frac{|x-y|}{\rho_1(x)}\Big)^{\delta_1}$$
$$+\frac{1}{|x-y|^n}\Big(\frac{|x-y|}{\rho(x)}\Big)^{\delta}\Big]|f(y)-f_B|dydx$$
$$:=I_{1,1}+I_{1,2}+I_{1,3}.$$

We have

$$\begin{aligned}
I_{1,3} &\lesssim \int_B\int_{5B}\frac{1}{|x-y|^n}\Big(\frac{|x-y|}{\rho(x)}\Big)^{\delta}|f(y)-f_B|dydx\\
&\lesssim \rho(x_0)^{-\delta}\int_{5B}\int_B\frac{1}{|x-y|^{n-\delta}}dx|f(y)-f_B|dy\\
&\lesssim \Big(\frac{r}{\rho(x_0)}\Big)^{\delta}\int_{5B}|f(y)-f_B|dy\lesssim\|f\|_{BMO^\beta(\omega)}|B|^{\beta/n}\omega(B).
\end{aligned}$$

Since $r < \rho(x_0) \lesssim \rho_1(x_0) \lesssim 1$, we also have

$$\begin{aligned} I_{1,2} &\lesssim \int_B \int_{5B} \frac{1}{|x-y|^{n-1}} \Big(\frac{|x-y|}{\rho_1(x)}\Big)^{\delta_1} |f(y) - f_B| dy dx \\ &\lesssim \rho(x_0)^{-\delta_1} \int_{5B} \int_B \frac{1}{|x-y|^{n-1-\delta_1}} dx |f(y) - f_B| dy \\ &\lesssim r\Big(\frac{r}{\rho(x_0)}\Big)^{\delta_1} \int_{5B} |f(y) - f_B| dy \lesssim \|f\|_{BMO^\beta(\omega)} |B|^{\beta/n} \omega(B). \end{aligned}$$

To estimate $I_{1,1}$ we notice that $B(y, |x-y|) \subset 6B$ for $x \in B$ and $y \in 5B$. Applying Lemma 2.8, we imply that

$$\begin{aligned} I_{1,1} &\lesssim \int_{5B} \int_B \frac{1}{|x-y|^{n-2}} dx |f(y) - f_B| \mathcal{I}_1(|\nabla V| \chi_{6B})(y) dy \\ &\lesssim r^2 \int_{2^3 B} |f(y) - f_B| \mathcal{I}_1(|\nabla V| \chi_{2^5 B})(y) dy \\ &\lesssim r^{\beta+1} \omega(B) \|f\|_{BMO^\beta(\omega)} \Big(\frac{r}{\rho_1(x)}\Big)^{n-n\alpha+2-n/q_1} \lesssim r^\beta \omega(B) \|f\|_{BMO^\beta(\omega)}, \end{aligned}$$

where we have used $r < \rho(x_0) \lesssim \rho_1(x_0) \lesssim 1$ in the last inequality. The proof is completed. □

References

[1] B. T. Anh, *Weighted estimates for commutators of some singular integrals related to Schrödinger operators*, Bulletin des Sciences Mathematiques. **138**, 270–292 (2014).

[2] B. T. Anh, *The weighted norm inequalities for Riesz transforms of magnetic Schrödinger operators*, Differential Integral Equations. **23**, 811–826 (2010).

[3] B. T. Anh and D. X. Thinh, *Boundedness of singular integrals and their commutators with BMO functions on Hardy spaces*, Adv. Diff. Eq. **18**, 459–494 (2013).

[4] B. Bongioanni, E. Harboure, and O. Salinas, *Riesz transform related to Schrödinger operators acting on BMO type spaces*, J. Math. Anal. Appl. **357**, 115–131 (2009).

[5] J. Cao, D. C. Chang, D. Yang, and S. Yang, textitBoundedness of second order Riesz transforms associated to Schrödinger operators on Musielak-Orlicz-Hardy spaces, Commun. Pure Appl. Anal. **13**, 1435–1463 (2014).

[6] J. Cao, D. C. Chang, D. Yang, and S. Yang, *Estimates for second-order Riesz transforms associated with magnetic Schrödinger operators on Musielak-Orlicz- Hardy spaces*, Appl. Anal. **93**, 2519–2545 (2014).

[7] T. Coulhon and X. T. Duong, *Riesz transforms for* $1 \le p \le 2$, Trans. Amer. Math. Soc. **351**, 1151–1169 (1999).

[8] J. Dong and Y. Liu, *The higher order Riesz transform and BMO space type associated to Schrödinger operator*, Math.Nachr. **285**, 1–11 (2011).

[9] G. B. Folland and E. M. Stein, *Hardy Spaces on Homogeneous Groups*, Math. Notes, 28, Princeton Univ. Press, Princeton, 1982.

[10] D. Gilberg and N. Trudinger, *Elliptic partial differential equations of second order*, Second Ed. Springer Verlag, 1983.

[11] L. Hömander, *Hypoelliptic second-order differential equations*, Acta Math. **119**, 147–171 (1967).

[12] Y. Liu and J. Dong, *The Higher Order Riesz Transform and BMO Type Space Associated with Schrödinger Operators on Stratified Lie Groups*, Hindawi Publishing Corporation, Journal of Function Spaces and Applications. **Vol 2013**, Article ID 483951, 13 pages.

[13] Y. Liu, L. Wang, and J. Dong, *Commutators of Higher Order Riesz Transform Associated with Schrödinger Operators*, Hindawi Publishing Corporation, Journal of Function Spaces and Applications, Article ID 842375 (2013).

[14] Y. Liu, *The Weighted Estimates of the Schrödinger Operators on the Nilpotent Lie Group*, Journal of Mathematical Research & Exposition. **30**, No.6, 1023–1031 (2010).

[15] G. Lu and R. L. Wheeden, *High order representation formulas and embedding theorems on stratified groups and generalizations*, Studia Mathematica, **142**, no.2, 101–133 (2000).

[16] H. Q. Li, *Estimations L^p des opérateurs de Schrödinger sur les groupes nilpotents*, J. Funct. Anal. **161**, 152–218 (1999).

[17] F. Ly, *Second order Riesz transforms associated to the Schrödinger operator for $p \leq 1$*, J. Math. Anal. Appl. **410**, 391–402 (2014).

[18] M. Morvidone, *Weighted BMO_ϕ spaces and the Hilbert transform*, Rev. Un. Mat. Argentina. **44**, 1–16 (2003).

[19] Z. Shen, L^p estimates for Schrödinger operators with certain potentials, *Ann. Inst. Fourier* **45**, 513-546 (1995).

[20] E. M. Stein, *Harmonic analysis: Real variable methods, orthogonality and oscillatory integrals*, Princeton Univ. Press, Princeton, NJ, 1993.

[21] D. Yang and S. Yang, *Second-order Riesz transforms and maximal inequalities associated with magnetic Schrödinger operators*, Canad. Math. Bull. **58**, 432–448 (2015).

[22] D. Yang and Y. Zhou, *Localized Hardy spaces H^1 related to admissible functions on RD-spaces and applications to Schrodinger operators*, Trans. Amer. Math. Soc. **363**, 1197–1239 (2011).

[23] D. Yang, D. Yang, and Y. Zhou, *Localized BMO and BLO spaces on RD-spaces and applications to Schrödinger operators*, Commun. Pure Appl. Anal. **9**, 779–812 (2010).

[24] D. Yang, D. Yang, and Y. Zhou, *Localized Morrey-Campanato Spaces on Metric Measure Spaces and Applications to Schrödinger Operators*, Nagoya Math. J.. **198**, 77–119 (2010).

[25] J. Zhong, *Harmonic analysis for some Schrödinger type operators*, Ph.D.Thesis, Princeton University, 1993.

On inverse problem for a class of fourth order strongly damped wave equations

Nam Danh Hua Quoc[1,*], *Can* Nguyen Huu[2,**], *Au* Vo Van[3,***], and *Binh* Tran Thanh[4,****]

[1]Department of Scientific Research Management, Thu Dau Mot University, Thu Dau Mot City, Binh Duong Province, Viet Nam
[2]Applied Analysis Research Group, Faculty of Mathematics and Statistics, Ton Duc Thang University, Ho Chi Minh City, Viet Nam
[3]Faculty of General Sciences, Can Tho University of Technology, Can Tho City, Viet Nam
[4]Faculty of Applied Mathematics, Sai Gon University, Ho Chi Minh City, Viet Nam

Abstract. In this paper, we study the initial inverse problem for a class of fourth order strongly damped linear wave equations. In the beginning, we show that the problem is ill-posed in the sense of Hadamard. Next, we propose the method called: the Fourier truncation method for stabilizing the problem. Convergence estimates are established under a priori regularity assumptions on the problem data.
Mathematics subject Classification 2000: 35K05, 35K99, 47J06, 47H10

1 Introduction

Throughout this paper, $\mathcal{H}$ will denote a Hilbert space endowed with the inner product $\langle \cdot, \cdot \rangle$ and the norm $\|\cdot\|$. Let T be a positive number and Ω be an open, bounded and connected domain in $\Omega \in \mathbb{R}^d$ $(d \geq 1)$ with a smooth boundary $\partial\Omega$. We consider the problem of finding a function $u\colon [0,T] \to \mathcal{H}$ satisfes the initial boundary value problem for a class of fourth order strongly damped linear wave equations

$$u_{tt} - \Delta u + \Delta^2 u - \gamma \Delta u_t = \mathcal{R}(x,t), \quad (x,t) \in \Omega \times [0,T], \tag{1.1}$$

subject to the conditions

$$\begin{cases} u = \varphi, \ \ u_t = \chi, & (x,t) \in \Omega \times \{T\}, \\ u = \Delta u = 0, & (x,t) \in \partial\Omega \times [0,T]. \end{cases} \tag{1.2}$$

There exists a sequence of positive real eigenvalues of the operator $-\Delta$, which denoted by $\{\lambda_k\}_{k\geq 1}$ where

$$\begin{cases} 0 < \lambda_1 \leq \lambda_2 \leq \lambda_3 \leq \ldots \leq \lambda_k \leq \ldots, \\ \lambda_k \to \infty \ \ as \ \ k \to \infty. \end{cases} \tag{1.3}$$

*e-mail: namdhq@tdmu.edu.vn
**e-mail: nguyenhuucan@tdtu.edu.vn (corresponding author)
***e-mail: vvau@ctuet.edu.vn
****e-mail: tranthanhbinh@sgu.edu.vn

Moreover, there exists an orthonormal basis $\{\xi_k\}_{k\geq 1}$, where ξ_k is an eigenfunction corresponding to λ_k

$$\begin{cases} -\Delta\xi_k(x) = \lambda_k\xi_k(x), & x \in \Omega, \\ \xi_k(x) = 0, & x \in \partial\Omega. \end{cases} \tag{1.4}$$

And $\gamma > 2$ is a positive constant, satisfies $(\gamma^2 - 4)\lambda_1 - 4 > 0$.

In practice, the data $(\varphi, \chi) \in \mathcal{H}\times\mathcal{H}$ is obtained by measurement contaminated with noise. Hence, instead of (φ, χ), we have the observation data $(\varphi^\alpha, \chi^\alpha)$ satisfying

$$\|\varphi - \varphi^\alpha\| + \|\chi - \chi^\alpha\| \leq \alpha, \tag{1.5}$$

where the constant $\alpha > 0$ represents a bound on the measurement error.

In the last years, the class of fourth order strongly damped wave equations has been investigated by many researchers, for example, existence, global classical solution, attractor, well-posedness, decay estimates, blowup, controllability, bootstrapping (e.g., [3]-[10]). Another important aspect of the qualitative study for the solutions of strongly damped wave equations is ill-posedness (in the sense of Hadamard).

In 2013, Xu Runzhang and Yanbing Yang [11] considered the global existence and asymptotic behaviour of solutions for a class of fourth order strongly damped nonlinear wave equations but they did not regularize it. Our work provide a regularization method for the solution of the ill-posed problem (1.1) - (1.2).

There are three sections in this paper. In Section 1, we have introduced our problem. The exact solution and ill-posedness of the inverse problem (1.1) - (1.2) are given in Section 2. Moreover, in Subsection 2.2, we give an example to show that the problem (1.1) - (1.2) is not stable. In Section 3, we develop a regularization method based on the truncated Fourier method and the stability estimate is established.

2 Exact solution and ill-posedness of the inverse problem

2.1 Exact solution

Assume that the problem (1.1) - (1.2) has a unique solution in the series form

$$u(x,t) = \sum_{k=1}^{\infty} \langle u, \xi_k\rangle\xi_k. \tag{2.6}$$

Then $u(x,t)$ satisfies the following system

$$\begin{cases} \dfrac{d^2}{dt^2}u_k(t) + \gamma\lambda_k\dfrac{d}{dt}u_k(t) + (\lambda_k + \lambda_k^2)u_k(t) = \mathcal{R}_k(t), & t \in (0,T), \\ u_k(T) = \varphi_k, \quad \dfrac{d}{dt}u_k(T) = \chi_k, \end{cases} \tag{2.7}$$

where $u_k = \langle u, \xi_k\rangle$, $\varphi_k = \langle \varphi, \xi_k\rangle$, $\chi_k = \langle \chi, \xi_k\rangle$, $\mathcal{R}_k = \langle \mathcal{R}, \xi_k\rangle$.

For any $\gamma > 2$ satisfies $(\gamma^2 - 4)\lambda_1 - 4 > 0$, the solution of (2.7) is given by:

$$u_k(t) = \frac{\mu_k^+ \exp(\mu_k^-(T-t)) - \mu_k^- \exp(\mu_k^+(T-t))}{\mu_k^+ - \mu_k^-}\varphi_k - \frac{\exp(\mu_k^+(T-t)) - \exp(\mu_k^-(T-t))}{\mu_k^+ - \mu_k^-}\chi_k$$
$$+ \int_t^T \frac{\exp(\mu_k^+(s-t)) - \exp(\mu_k^-(s-t))}{\mu_k^+ - \mu_k^-}\mathcal{R}_k(s)ds, \tag{2.8}$$

where we define

$$\mu_k^+ = \frac{1}{2}\left(\gamma\lambda_k + \delta_k^{\frac{1}{2}}\right), \quad \mu_k^- = \frac{1}{2}\left(\gamma\lambda_k - \delta_k^{\frac{1}{2}}\right), \quad \delta_k = (\gamma^2 - 4)\lambda_k^2 - 4\lambda_k. \tag{2.9}$$

Let us define the operators $\mathbb{G}(t)$ and $\mathbb{K}(t)$ as follow

$$\mathbb{G}(t)w = \sum_{k=1}^{\infty} \frac{\mu_k^+ \exp(\mu_k^- t) - \mu_k^- \exp(\mu_k^+ t)}{\mu_k^+ - \mu_k^-} \langle w, \xi_k \rangle \xi_k, \tag{2.10}$$

and

$$\mathbb{K}(t)w = \sum_{k=1}^{\infty} \frac{\exp(\mu_k^- t) - \exp(\mu_k^+ t)}{\mu_k^+ - \mu_k^-} \langle w, \xi_k \rangle \xi_k, \tag{2.11}$$

for $w \in \mathcal{H}$. From above observations, we can rewrite (2.8) as

$$u(x,t) = \mathbb{G}(T-t)\varphi - \mathbb{K}(T-t)\chi + \int_t^T \mathbb{K}(s-t)\mathcal{R}(s)ds. \tag{2.12}$$

2.2 Ill-posedness of the inverse problem

In this subsection, we give an example which shows that the solution $u^*(x,t)$ of problem (1.1) – (1.2) (if it exists) is not stable.

- Let φ^*, χ^*, and $\mathcal{R}^*$ be defined as follows

$$u^*(x,T) = \varphi^*(x) = 0, \quad \partial_t u^*(x,T) = \chi^*(x) = \frac{\xi_m(x)}{\sqrt{\lambda_m}}, \quad \mathcal{R}^* = \sum_{m=1}^{\infty} \frac{2\exp(-\gamma T\lambda_m)}{T^3}. \tag{2.13}$$

Then u^* satisfy the integral equation

$$u^*(x,t) = -\mathbb{K}(T-t)\chi^* + \int_t^T \mathbb{K}(s-t)\mathcal{R}^*(s)ds. \tag{2.14}$$

We have

$$\begin{aligned}\|\mathbb{K}(T-t)\chi^*\|^2 &= \left(\frac{\exp(\mu_m^-(T-t)) - \exp\mu_m^+(T-t))}{\mu_m^+ - \mu_m^-}\right)^2 \frac{1}{\lambda_m} \\ &= \frac{\exp(2\mu_m^+(T-t))\left(1 - \exp\left(-(T-t)\sqrt{\delta_m}\right)\right)}{\lambda_m\delta_m},\end{aligned}$$

where $\delta_m = (\gamma^2-4)\lambda_m^2 - 4\lambda_m$, noting that $\mu_m^+ - \mu_m^- = \sqrt{\delta_m}$.
This leads to

$$\|\mathbb{K}(T-t)\chi^*\|^2 = \frac{\exp(2\mu_m^+(T-t))\left(1 - \exp\left(-(T-t)\sqrt{(\gamma^2-4)\lambda_m^2 - 4\lambda_m}\right)\right)^2}{\lambda_m\left[(\gamma^2-4)\lambda_m^2 - 4\lambda_m\right]}.$$

It is easy to see that the function $f(t) = \exp(\mu_m^+(T-t))\left(1 - \exp\left(-(T-t)\sqrt{(\gamma^2-4)\lambda_1^2 - 4\lambda_1}\right)\right)$ is a decreasing function for $t \in (0, T)$, we deduce that

$$\sup_{0\le t\le T} \left\|\mathbb{K}(T-t)\chi^*\right\| \ge \sup_{0\le t\le T} \frac{\exp(\mu_m^+(T-t))\left(1 - \exp\left(-(T-t)\sqrt{(\gamma^2-4)\lambda_1^2 - 4\lambda_1}\right)\right)}{\sqrt{\lambda_m\left[(\gamma^2-4)\lambda_m^2 - 4\lambda_m\right]}}$$
$$= \frac{\exp(\mu_m^+ T)\left(1 - \exp\left(-T\sqrt{(\gamma^2-4)\lambda_1^2 - 4\lambda_1}\right)\right)}{\sqrt{\lambda_m\left[(\gamma^2-4)\lambda_m^2 - 4\lambda_m\right]}}$$
$$\ge \frac{\exp\left(\frac{\gamma\lambda_m}{2}T\right)\left(1 - \exp\left(-T\sqrt{(\gamma^2-4)\lambda_1^2 - 4\lambda_1}\right)\right)}{\sqrt{\lambda_m\left[(\gamma^2-4)\lambda_m^2 - 4\lambda_m\right]}}. \tag{2.15}$$

On the other hand, we get

$$\left\|\int_t^T \mathbb{K}(s-t)\mathcal{R}^* ds\right\| \le \int_t^T \sqrt{\sum_{m=1}^{\infty}\left[\frac{\exp(\mu_m^-(s-t)) - \exp(\mu_m^+(s-t))}{\mu_m^+ - \mu_m^-}\right]^2 \frac{4\exp(-2\gamma T\lambda_m)}{T^6}} ds$$
$$= \int_t^T \sqrt{\sum_{m=1}^{\infty}\exp\left(2(s-t)(\mu_m^+ + \mu_m^-)\right)\left[\frac{\exp\left(-\mu_m^+(s-t)\right) - \exp\left(-\mu_m^-(s-t)\right)}{\mu_m^+ - \mu_m^-}\right]^2 \frac{4\exp(-2\gamma T\lambda_m)}{T^6}} ds.$$

Using the inequality $|\exp(-y) - \exp(-z)| \le |y-z|$, for $y, z > 0$ and noting that $\mu_m^+ + \mu_m^+ = \gamma\lambda_m$, we obtain

$$\left\|\int_t^T \mathbb{K}(s-t)\mathcal{R}^* ds\right\| \le \int_t^T \sqrt{\sum_{m=1}^{\infty}\exp\left(2\gamma\lambda_m(s-t)\right)(s-t)^2\frac{4\exp(-2\gamma T\lambda_m)}{T^6}} ds$$
$$\le \int_t^T \frac{2(s-t)}{T^3} ds \le \frac{1}{T}. \tag{2.16}$$

Combining (2.15) and (2.16), we have

$$\|u^*\| \ge \left\|\mathbb{K}(T-t)\chi^*\right\| - \left\|\int_t^T \mathbb{K}(s-t)\mathcal{R}^*(s)ds\right\|$$
$$\ge \frac{\exp\left(\frac{\gamma\lambda_m}{2}T\right)\left(1 - \exp\left(-T\sqrt{(\gamma^2-4)\lambda_1^2 - 4\lambda_1}\right)\right)}{\sqrt{\lambda_m\left[(\gamma^2-4)\lambda_m^2 - 4\lambda_m\right]}} - \frac{1}{T}.$$

As $m \to \infty$, we get

$$\lim_{m\to\infty}\left(\|\varphi_m\| + \|\chi_m\|\right) = \lim_{m\to\infty}\frac{1}{\sqrt{\lambda_m}} = 0. \tag{2.17}$$

$$\lim_{m\to\infty}\|u^*\|_{C([0,T];\mathcal{H})} = \lim_{m\to\infty}\left(\frac{\exp\left(\frac{\gamma\lambda_m}{2}T\right)\left(1 - \exp\left(-T\sqrt{(\gamma^2-4)\lambda_1^2 - 4\lambda_1}\right)\right)}{\sqrt{\lambda_m\left[(\gamma^2-4)\lambda_m^2 - 4\lambda_m\right]}} - \frac{1}{T}\right) = \infty. \tag{2.18}$$

- Let $\varphi^{**} = \chi^{**} = \mathcal{R}^{**} = 0$, we deduce $u^{**} = 0$. This shows that a small perturbation of a physical measurement makes the corresponding solution have a large error. Thus the problem (1.1)-(1.2) is ill-posed in the sense of Hadamard.

Next section, we propose a regularization method for stabilizing the problem.

3 Fourier's truncation method

Let $N = \{k \in \mathbb{N}$ such that $k \geq 1$ and $\lambda_k \leq \Lambda(\alpha)\}$. Without loss of generality, we assume that $\Lambda(\alpha)$ be a regularization parameter satisfying

$$\lim_{\alpha \to 0} \Lambda(\alpha) = \infty.$$

For $w \in \mathcal{H}$, let us define the truncated version of (2.10) and (2.11) as

$$\mathbb{G}_{\Lambda(\alpha)}(t)w = \sum_{k \in N} \frac{\mu_k^+ \exp(\mu_k^- t) - \mu_k^- \exp(\mu_k^+ t)}{\mu_k^+ - \mu_k^-} \langle w, \xi_k \rangle \xi_k, \tag{3.19}$$

and

$$\mathbb{K}_{\Lambda(\alpha)}(t)w = \sum_{k \in N} \frac{\exp(\mu_k^- t) - \exp(\mu_k^+ t)}{\mu_k^+ - \mu_k^-} \langle w, \xi_k \rangle \xi_k. \tag{3.20}$$

Note that we pick $\Lambda(\alpha)$ so large that $\mathbb{G}, \mathbb{K}$ are retained.
Let us define the regularized solution by Fourier's truncation method as follows

$$u^{\Lambda(\alpha)} = \mathbb{G}_{\Lambda(\alpha)}(T-t)\varphi^\alpha - \mathbb{K}_{\Lambda(\alpha)}(T-t)\chi^\alpha + \int_t^T \mathbb{K}_{\Lambda(\alpha)}(\omega - t)\mathcal{R}(x, \omega)d\omega. \tag{3.21}$$

We have the following lemma which shall be useful in next results.

Lemma 3.1 *For $t \in [0, T]$, we have*

$$\left\|\mathbb{G}_{\Lambda(\alpha)}(t)\right\|_{\mathcal{L}(\mathcal{H})} \leq \exp\left(t\gamma\Lambda(\alpha)\right)\left(2 + \frac{4T}{\gamma^2}\right),$$
$$\left\|\mathbb{K}_{\Lambda(\alpha)}(t)\right\|_{\mathcal{L}(\mathcal{H})} \leq \exp\left(t\gamma\Lambda(\alpha)\right)T.$$

Proof.
• From (2.9) and (3.19), let $w \in \mathcal{H}$, we have

$$\begin{aligned}
\left\|\mathbb{G}_{\Lambda(\alpha)}(t)w\right\|^2 &= \sum_{k \in N} \left[\frac{\mu_k^+ \exp(\mu_k^- t) - \mu_k^- \exp(\mu_k^+ t)}{\mu_k^+ - \mu_k^-}\right]^2 \langle w, \xi_k \rangle^2 \\
&= \sum_{k \in N} \exp\left(2t(\mu_k^+ + \mu_k^-)\right)\left[\frac{\mu_k^+ \exp(-t\mu_k^+) - \mu_k^- \exp(-t\mu_k^-)}{\mu_k^+ - \mu_k^-}\right]^2 \langle w, \xi_k \rangle^2 \\
&\leq \sum_{k \in N} \exp\left(2t\gamma\Lambda(\alpha)\right)\left[\frac{\mu_k^+ \exp(-t\mu_k^+) - \mu_k^- \exp(-t\mu_k^-)}{\sqrt{\delta_k}}\right]^2 \langle w, \xi_k \rangle^2,
\end{aligned} \tag{3.22}$$

where $\mu_k^- + \mu_k^- = \gamma\lambda_k$ and $\mu_k^+ - \mu_k^- = \sqrt{\delta_k}$.
On the other hand, using the inequality $(a+b)^2 \leq 2(a^2+b^2)$ and $\left|\exp(-a) - \exp(-b)\right| \leq |a-b|$ for $a, b > 0$, we have

$$\begin{aligned}
\left(\frac{\mu_k^+ \exp(-t\mu_k^+) - \mu_k^- \exp(-t\mu_k^-)}{\sqrt{\delta_k}}\right)^2 &= \left(\exp(-t\mu_k^+) + \mu_k^- \frac{\exp(-t\mu_k^+) - \exp(-t\mu_k^-)}{\sqrt{\delta_k}}\right)^2 \\
&\leq 2\exp(-2t\mu_k^+) + 2\left|\mu_k^-\right|^2 t^2 \\
&\leq 2 + \left(\frac{2\lambda_k t}{\gamma\lambda_k + \sqrt{\delta_k}}\right)^2 \leq 2 + \frac{4T}{\gamma^2}.
\end{aligned} \tag{3.23}$$

Combining (3.22) and (3.23), we get

$$\left\|\mathbb{G}_{\Lambda(\alpha)}(t)w\right\|^2 \le \exp\Big(2t\gamma\Lambda(\alpha)\Big)\left(2 + \frac{4T}{\gamma^2}\right)^2 \|w\|^2. \tag{3.24}$$

• From (2.9) and (3.20), let $w \in \mathcal{H}$, we have

$$\begin{aligned}\left\|\mathbb{K}_{\Lambda(\alpha)}(t)w\right\|^2 &= \sum_{k\in N}\left(\frac{\exp(\mu_k^- t) - \exp(\mu_k^+ t)}{\mu_k^+ - \mu_k^-}\right)^2 \langle w, \xi_k\rangle^2 \\ &= \sum_{k\in N} \exp\Big(2t(\mu_k^+ + \mu_k^-)\Big)\left(\frac{\exp\big(-t\mu_k^+\big) - \exp\big(-t\mu_k^-\big)}{\mu_k^+ - \mu_k^-}\right)^2 \langle w, \xi_k\rangle^2 \\ &\le \sum_{k\in N} \exp\Big(2t\gamma\Lambda(\alpha)\Big)T^2 \langle w, \xi_k\rangle^2 = \exp\Big(2t\gamma\Lambda(\alpha)\Big)T^2\|w\|^2. \end{aligned} \tag{3.25}$$

This completes the proof of Lemma 3.1. □

Before we formulate the main theorem, we also introduce the Gevrey class of functions of order $p > 0$ and index $r > 0$, see e.g. [1], defined by the spectrum of the Laplacian is denoted by

$$\mathbf{B}_{p,r} := \left\{ f \in L^2(\Omega) : \sum_{k=1}^{\infty} \lambda_k^{2p} \exp(2r\lambda_k)\langle f, \xi_k(x)\rangle^2 < \infty \right\}, \tag{3.26}$$

and their norms given by

$$\|f\|_{\mathbf{B}_{p,r}} = \sqrt{\sum_{k=1}^{\infty} \lambda_k^{2p} \exp(2r\lambda_k)\langle f, \xi_k(x)\rangle^2}. \tag{3.27}$$

Theorem 3.1 *Let $\Lambda(\alpha)$ be a regularization parameter such that*

$$\Lambda(\alpha) > \frac{4}{\gamma^2 - 4}, \quad \textit{where } \lim_{\alpha\to 0} \Lambda(\alpha) = \infty \textit{ and } \lim_{\alpha\to 0} \exp\Big(T\gamma\Lambda(\alpha)\Big)\alpha = 0.$$

Suppose that the problem (2.8) has a unique solution $u \in L^\infty(0, T; \mathbf{B}_{\beta,\gamma T})$ and the integral equation (3.21) has a unique solution $u^{\Lambda(\alpha)} \in C([0, T]; \mathcal{H})$. Then the estimate between the exact solution and the regularized solution is given by

$$\left\|u - u^{\Lambda(\alpha)}\right\| \le \exp\Big(t\gamma\Lambda(\alpha)\Big)T_{max}\alpha + \lambda_{N^\alpha}^{-\beta} \exp\left(-\gamma t\lambda_{N^\alpha}\right) \|u\|_{L^\infty(0,T;\mathbf{B}_{\beta,\gamma T})}, \tag{3.28}$$

where $T_{max} = \max\left(2 + \frac{4T}{\gamma^2}, T\right)$, N^α is the integer part of $\Lambda(\alpha)$ and $\gamma > 2$ is a positive constant, satisfies $(\gamma^2 - 4)\lambda_1 - 4 > 0$.

Proof.

First of all, we denote

$$U^{\Lambda(\alpha)}(t) = \mathbb{G}_{\Lambda(\alpha)}(T - t)\varphi - \mathbb{K}_{\Lambda(\alpha)}(T - t)\chi + \int_t^T \mathbb{K}_{\Lambda(\alpha)}(\omega - t)\mathcal{R}(x, \omega)d\omega. \tag{3.29}$$

Estimating (3.28) is divided into 2 steps.

Step 1. Estimate $\left\|u^{\Lambda(\alpha)}(\cdot, t) - U^{\Lambda(\alpha)}(\cdot, t)\right\|$.

Using (1.5), (3.21), (3.29) and Lemma 3.1, we have

$$\begin{aligned}\left\|u^{\Lambda(\alpha)}(\cdot,t)-U^{\Lambda(\alpha)}(\cdot,t)\right\| &\le \left\|\mathbb{G}_{\Lambda(\alpha)}(T-t)(\varphi-\varphi^{\alpha})\right\|+\left\|\mathbb{K}_{\Lambda(\alpha)}(T-t)(\chi-\chi^{\alpha})\right\|\\ &\le \exp\Big(t\gamma\Lambda(\alpha)\Big)\left(2+\frac{4T}{\gamma^2}\right)(\varphi-\varphi^{\alpha})+\exp\Big(t\gamma\Lambda(\alpha)\Big)T(\chi-\chi^{\alpha})\\ &\le \exp\Big(t\gamma\Lambda(\alpha)\Big)T_{max}\alpha, \end{aligned}\tag{3.30}$$

where $T_{max}=\max\left(2+\dfrac{4T}{\gamma^2},T\right)$.

Step 2. Estimate $\left\|u(\cdot,t)-U^{\Lambda(\alpha)}(\cdot,t)\right\|$.
It is easy to see that

$$\sum_{k=1}^{N^{\alpha}} u_k(t)\xi_k=\mathbb{G}_{\Lambda(\alpha)}(T-t)\varphi+\mathbb{K}_{\Lambda(\alpha)}(T-t)\chi-\int_t^T \mathbb{K}_{\Lambda(\alpha)}(\omega-t)\mathcal{R}(x,\omega)\mathrm{d}\omega.$$

Therefore, using Lemma 3.1, we obtain

$$\begin{aligned}\left\|u(\cdot,t)-U^{\Lambda(\alpha)}(\cdot,t)\right\| &\le \left\|u(\cdot,t)-\sum_{k=1}^{N^{\alpha}}u_k(t)\xi_k\right\|+\left\|\sum_{k=1}^{N^{\alpha}}u_k(t)\xi_k-U^{\Lambda(\alpha)}(\cdot,t)\right\|\\ &\le \sqrt{\sum_{k=N^{\alpha}+1}^{\infty}\lambda_{N^{\alpha}}^{-2\beta}\exp\left(-2\gamma t\lambda_{N^{\alpha}}\right)\lambda_k^{2\beta}\exp\left(2\gamma t\lambda_k\right)|u_k(t)|^2}\\ &\le \lambda_{N^{\alpha}}^{-\beta}\exp\left(-\gamma t\lambda_{N^{\alpha}}\right)\|u\|_{L^{\infty}(0,T;\mathbf{B}_{\beta,\gamma T})}. \end{aligned}\tag{3.31}$$

Thanks to triangle inequality $\left\|u-u^{\Lambda(\alpha)}\right\|\le\left\|u-U^{\Lambda(\alpha)}\right\|+\left\|u^{\Lambda(\alpha)}-U^{\Lambda(\alpha)}\right\|$, combining (3.30) and (3.31) we have completed the proof of Theorem 3.1. □

References

[1] C. Cao, M.A. Rammaha, and E.S. Titi, *The Navier-Stokes equations on the rotating 2-D sphere: Gevrey regularity and asymptotic degrees of freedom*, Z. Angew. Math. Phys., 341-360 (1999).

[2] L. C. Evans, *Partial Differential Equation*, American Mathematical Society,Providence, Rhode Island, Volume 19 (1997).

[3] W. Fitzgibbon. *Strongly damped quasilinear evolution equations, Journal of Mathematical Analysis and Applications*, 79(2):536–550 (1981).

[4] V. Kalantarov and S. Zelik, *Finite-dimensional attractors for the quasi-linear strongly-damped wave equation*, Journal of Differential Equations, 247(4):1120–1155 (2009).

[5] S. Larsson, V. Thomée, and L. B. Wahlbin, *Finite-element methods for a strongly damped wave equation*, IMA journal of numerical analysis, 11(1):115–142 (1991).

[6] P. Massatt, *Limiting behavior for strongly damped nonlinear wave equations*, Journal of Differential Equations, 48(3):334–349 (1983).

[7] Nguyen Huy Tuan, Doan Vuong Nguyen, Vo Van Au, Daniel Lesnic, *Recovering the initial distribution for strongly damped wave equation*, Applied Mathematics Letters, 73: 69-77 (2017).

[8] V. Pata and M. Squassina, *On the strongly damped wave equation*, Communications in mathematical physics, 253(3):511–533 (2005).

[9] V. Pata and S.V. Zelik, *Smooth attractors for strongly damped wave equations*, Nonlinearity, 1495-1506 (2006).

[10] V. Thomée and L. Wahlbin, *Maximum-norm estimates for finite-element methods for a strongly damped wave equation*, BIT Numerical Mathematics, 44(1):165–179 (2004).

[11] Xu Runzhang and Yanbing Yang, *Global existence and asymptotic behaviour of solutions for a class of fourth order strongly damped nonlinear wave equations*, Quarterly of Applied Mathematics , 71(3): 401–415 (2013).

A truncation regularization method for a time fractional diffusion equation with an in-homogeneous source

Luu Vu Cam Hoan[1,2,*], Ho Duy Binh[3,4,**], and Tran Bao Ngoc[5,***]

[1]Faculty of Basic Science, Posts and Telecommunications, Institute of Technology, Ho Chi Minh City, Vietnam
[2]Faculty of Mathematics and Computer Science, University of Science, Vietnam National University (VNU-HCMC), Ho Chi Minh City, Vietnam
[3]Faculty of Fundamental Science, Nguyen Hue University, Dong Nai, Viet Nam
[4]Department of Applied Mathematics, Faculty of Applied Science, University of Technology (VNU-HCM), Ho Chi Minh City, Viet Nam
[5]Department of Mathematics, Faculty of Science, Nong Lam University, Ho Chi Minh City, Viet Nam

Abstract. In the present paper, we consider a time-fractional inverse diffusion problem with an in-homogeneous source, where data is given at $x = 1$ and the solution is required in the interval $0 < x < 1$. This problem is ill-posed, i.e. the solution (if it exists) does not depend continuously on the data. We propose a regularization method to solve it based on the solution given by the Fourier method.

Keywords: Ill-posed problem, time fractional diffusion equation, regularization, regularized truncation method.

1 Introduction

In recent years, fractional calculus and fractional differential equations have gained much special attention by many authors. They appear naturally in wide sciences area of physics, chemical engineering, biology, signal processing, electrical, control theory, finance, population dynamics, etc; refer to [1–3].

Time-fractional diffusion equation arises by replacing the standard time partial derivative in the diffusion equation with a time-fractional partial derivative. They have arisen from a continuous-time random walk model with temporal memory and source where anomalous diffusion such as occurs in spatial environments, or in slow heat flux diffusion in thermal conductions. In this paper, we study a more general ill-posed problem for the time-fractional diffusion equation with the in-homogeneous source in a one-dimensional semi-infinite domain as follows

$$\frac{\partial^{\beta} u(x,t)}{\partial t^{\beta}} + \frac{\partial u(x,t)}{\partial x} = g(x,t), \qquad x > 0, \quad t > 0, \tag{1}$$

*e-mail: lvcamhoan@gmail.com
**e-mail: binhlucquan@gmail.com
***e-mail: tranbaongoc@hcmuaf.edu.vn

with the Cauchy condition and initial condition

$$u(1,t) = f(t), \quad t \geq 0, \tag{2}$$

$$\lim_{x \to +\infty} u(x,t) = u(x,0) = 0, \tag{3}$$

where b is a constant diffusivity coefficient. The fractional derivative $\frac{\partial^\beta}{\partial t^\beta}$ is the Caputo fractional derivative of order β defined by

$$\frac{\partial^\beta u(x,t)}{\partial t^\beta} = \frac{1}{\beta(1-\beta)} \int_0^t \frac{\partial u(x,s)}{\partial s} \frac{ds}{(t-s)^\beta},$$

for $0 < \beta < 1$ whenever the right-hand-side exists and

$$\frac{\partial^\beta u(x,t)}{\partial t^\beta} = \frac{\partial u(x,t)}{\partial t}$$

for $\beta = 1$ where Γ is the Gamma function.

In 2007, D.A. Murio considered the fractional inverse heat conduction problem in a quarter plane which includes a classical derivative of first-order respectively the spatial variable as follow

$$\begin{aligned} u_x(x,t) &= -a^{-\frac{1}{2}} \left({}_0^{RL}D_t^{\frac{1}{2}} u(x,t) \right) + u_\infty (\pi a t)^{-\frac{1}{2}}, x > 0, t > 0 \\ u(1,t) &= f(t) \end{aligned}$$

with the initial condition

$$\begin{aligned} u(0,t) &= g(t) \\ -u_x(0,t) &= h(t), \end{aligned}$$

for all $t \geq 0$ in which the time fractional derivative be understood in the sense of Riemann-Liouville of order $1/2$, see [7]. In 2011, G.H. Zheng and T. Wei had considered a homogeneous time fractional diffusion problem, where the time fractional derivative be understood in sense of Dzerbayshan-Caputo also in a quarter plane, in form of

$$\begin{aligned} -a u_x(x,t) &= {}_0D_t^\beta u(x,t), \quad x > 0, t > 0, 0 < \beta < 1, \\ u(1,t) &= f(t) \end{aligned}$$

corresponding to the initial condition

$$u(x,0) = \lim_{x \to +\infty} u(x,t) = 0.$$

However, in many phenomena, diffusion progresses occur in spatially inhomogeneous environments or the slow diffusion progress require problem models which contains the inhomogeneous or nonlinear source term. The authors couldn't find any results for TFIDE in that cases. To solve that problem, it is required difficult techniques and new ideas to deal with the fractional terms and an in-homogeneous term which can not be treated by using known ideas.

That is the reason why we need a regularization method to approximate the problem (4). In the present paper, we choose the truncation method. There are two new points that are devised when we overcame the difficult points as we had discussed in the head of this section. Firstly, we note that solution of the classical time diffusion equation, where the time derivative order is an integer number, has just formulated by a form of real computations while there arise complex computations in case of the fractional derivatives, see [5, 8, 9, 11]. Secondly, techniques of using Gronwall inequality are used very sophisticated providing that we apply the Gronwall inequality for a product of the error and some appropriate exponential functions but not only for the error. This technique plays a key role to solve TFIDEs. Moreover, a hypothesis is considered utmost flexible which will be assumed to holds on the exact data, so that our result, by the way, had improved strongly. This helps us in grasping the role of solving in-homogeneous TFIDEs, also recognize the difficult points as we have discussed and the worthy validity of our results.

We divide the paper into four sections. The ill-posedness and the method of regularization for the problem (1) - (3) will be analyze in section 2. In section 3, the main results are speech. Some discussion and conclusion are given to explain the results more clearly. The proof part of our results will be given in section 4.

2 Regularized problem

2.1 Representation for solution

By extending all functions to the whole line $-\infty < t < \infty$ which are replaced by zero value for $t < 0$ and taking a Fourier transformation to the problem (1) with respect to t, we obtain

$$(i\xi)^\beta \mathcal{F}(u(x,.))(\xi) + \mathcal{F}(u_x(x,.))(\xi) = \mathcal{F}(g(x,.))(\xi),$$
$$\mathcal{F}(u(1,.))(\xi) = \mathcal{F}(f(.))(\xi),$$

for $x > 0, \xi \in \mathbb{R}$ where we denote by $\mathcal{F}(v(.))$ the Fourier transforms of v defined by the following equation

$$\mathcal{F}(v(.))(\xi) = \frac{1}{\sqrt{2\pi}}\int_{-\infty}^{\infty} v(t)e^{-i\xi t}dt.$$

Solving the above differential equation, the solution of problem (1) - (3) can be obtained in form of

$$\mathcal{F}(u(x,.))(\xi) = \exp\Big((i\xi)^\beta(1-x)\Big)\mathcal{F}(f(.))(\xi)$$
$$+\int_x^1 \exp\Big((i\xi)^\beta(\eta-x)\Big)\mathcal{F}(g(\eta,.))(\xi)d\eta,$$

or

$$u(x,t) = \frac{1}{\sqrt{2\pi}}\int_{-\infty}^{+\infty} \exp\Big((i\xi)^\beta(1-x)\Big)\mathcal{F}(f(.))(\xi)e^{i\xi t}d\xi$$
$$\frac{1}{\sqrt{2\pi}}\int_{-\infty}^{+\infty}\int_x^1 \exp\Big((i\xi)^\beta(\eta-x)\Big)\mathcal{F}(g(\eta,.))(\xi)e^{i\xi t}d\eta d\xi, \tag{4}$$

where the real part and im-real part of $(i\xi)^\beta$ be given by

$$(i\xi)^\beta = \Re((i\xi)^\beta) + i\Im((i\xi)^\beta), \tag{5}$$
$$\Re((i\xi)^\beta) = |\xi|^\beta \cos\frac{\beta\pi}{2}, \tag{6}$$
$$\Im((i\xi)^\beta) = |\xi|^\beta \text{sign}(\xi)\sin\frac{\beta\pi}{2}. \tag{7}$$

2.2 Ill-posedness and truncation regularization method

We emphasize that $\exp\big((i\xi)^\beta(1-x)\big)$ and $\exp\big((i\xi)^\beta(\eta-x)\big)$ can be go to infinity module. So by applying the Parseval theorem, the exact solution u may be go to infinity $L^2(\mathbb{R})-$ norm value. For finding the solution $u(x,.)$ in $L^2(\mathbb{R})$ for every $0 \le x \le 1$, the exact data functions

$$\mathcal{F}(f(.))(\xi), \quad \mathcal{F}(g(x,.))(\xi)$$

must rapidly decay when $|\xi| \to \infty$. But we couldn't have the exact data, we only have the noisy data from the progress of real measurement which instead of the exact data.

Therefore, we can not expect the noisy data functions f_ϵ, g_ϵ have the same decay in high frequency f, g although

$$\|f_\epsilon - f\|_{L^2(\mathbb{R})} + \|g_\epsilon - g\|_{L^2(0,1;L^2(\mathbb{R}))} \le \epsilon, \tag{8}$$

for small ϵ, i.e., small errors in high frequency domain can blow up and destroy the solution for $0 < x < 1$. Thus the time fractional inverse diffusion problem (1)-(3) is sererely ill-posed. That is the reason why it's necessary to propose a regularization approximation for the exact solution $u(x,t)$. In this section, we considere the following approximative problem of finding $u(0,t)$ for all time $t > 0$ as follows

$$\begin{aligned} u_\epsilon^\alpha(x,t) &= \frac{1}{\sqrt{2\pi}} \int_{-\infty}^{+\infty} \exp\big((i\xi)^\beta(1-x)\big)\mathcal{F}\big(f^\epsilon(.)\big)(\xi)\chi_{[-\alpha;\alpha]}(\xi)e^{i\xi t}d\xi \\ &+ \frac{1}{\sqrt{2\pi}} \int_{-\infty}^{+\infty}\int_x^1 \exp\big((i\xi)^\beta(\eta-x)\big)\mathcal{F}\big(g^\epsilon(\eta,.)\big)(\xi)d\eta\chi_{[-\alpha;\alpha]}(\xi)e^{i\xi t}d\xi, \end{aligned} \tag{9}$$

where the noisy data f^ϵ, g^ϵ satisfying the noisy assumption (8) and $\alpha := \alpha(\epsilon)$ is a positive constant which depends on ϵ and will be specified later as a parameter of the regularization method.

In order to obtain the convergence estimate of regularization method, we give the following a prior bound assumption

$$\|u(0,.)\|_{L^2(\mathbb{R})} \le E, \tag{10}$$

where E is a positive constant.

3 Main result

In this section, we present our results. For establishing the main theorem, some lemmas will be given. In addition, it is useful to introduce u^α defined by the following function

$$\begin{aligned} u^\alpha(x,t) &= \frac{1}{\sqrt{2\pi}} \int_{-\infty}^{+\infty} \exp\big((i\xi)^\beta(1-x)\big)\mathcal{F}\big(f(.)\big)(\xi)\chi_{[-\alpha;\alpha]}(\xi)e^{i\xi t}d\xi \\ &+ \frac{1}{\sqrt{2\pi}} \int_{-\infty}^{+\infty}\int_x^1 \exp\big((i\xi)^\beta(\eta-x)\big)\mathcal{F}\big(g(\eta,.)\big)(\xi)d\eta\chi_{[-\alpha;\alpha]}(\xi)e^{i\xi t}d\xi, \end{aligned} \tag{11}$$

where f, g are the exact data functions and α is the parameter of the regularization method.

Lemma 1 *Let u and u^{α} be the exact solution and regularized solution respectively defined by (4) and (11) which are corresponding to the exact data (f,g) and the noisy data $(f^{\epsilon},g^{\epsilon})$. Assume the prior bound (10) hold. If there exists a positive constant M such that*

$$\int_{-\infty}^{+\infty} \exp\left(2\eta\Re((i\xi)^{\beta})\right)|\mathcal{F}\left(g(\eta,.)\right)(\xi)|^2 d\xi \le M^2 \tag{12}$$

then

$$\|u(x,.)-u^{\alpha}(x,.)\|_{L^2(\mathbb{R})} \le (E+M)\exp\left(-x\alpha^{\beta}\cos\frac{\beta\pi}{2}\right), \tag{13}$$

for all $0<x<1$.

Proof. It follows from (4) and (11) that

$$u(x,t)-u^{\alpha}(x,t) = \frac{1}{\sqrt{2\pi}}\int_{|\xi|>\alpha} \exp\left((i\xi)^{\beta}(1-x)\right)\mathcal{F}\left(f(.)\right)(\xi)e^{i\xi t}d\xi$$
$$+\frac{1}{\sqrt{2\pi}}\int_{|\xi|>\alpha}\int_{x}^{1} \exp\left((i\xi)^{\beta}(\eta-x)\right)\mathcal{F}\left(g(\eta,.)\right)(\xi)d\eta e^{i\xi t}d\xi. \tag{14}$$

If we represent $u(x,t)$ as follows

$$u(x,t) = \frac{1}{\sqrt{2\pi}}\int_{-\infty}^{+\infty} \exp\left((i\xi)^{\beta}(1-x)\right)Q(\xi)e^{i\xi t}d\xi$$
$$+\frac{1}{\sqrt{2\pi}}\int_{-\infty}^{+\infty}\int_{x}^{1} \exp\left((i\xi)^{\beta}(\eta-x)\right)\mathcal{F}\left(g(\eta,.)\right)(\xi)e^{i\xi t}d\eta d\xi,$$

then

$$Q(\xi) = \exp\left(-(i\xi)^{\beta}\right)\mathcal{F}(u(0,.))(\xi)$$
$$-\exp\left(-(i\xi)^{\beta}\right)\int_{0}^{1}\exp\left((i\xi)^{\beta}\eta\right)\mathcal{F}\left(g(\eta,.)\right)(\xi)d\eta$$

by substituting 0 into x. This implies

$$u(x,t) = \frac{1}{\sqrt{2\pi}}\int_{-\infty}^{+\infty} \exp\left(-(i\xi)^{\beta}x\right)\mathcal{F}(u(0,.))(\xi)e^{i\xi t}d\xi$$
$$-\frac{1}{\sqrt{2\pi}}\int_{-\infty}^{+\infty}\int_{0}^{x} \exp\left((i\xi)^{\beta}(\eta-x)\right)\mathcal{F}\left(g(\eta,.)\right)(\xi)e^{i\xi t}d\eta d\xi. \tag{15}$$

We deduce from (15) that

$$u(x,t)-u^{\alpha}(x,t) = \frac{1}{\sqrt{2\pi}}\int_{|\xi|>\alpha} \exp\left(-(i\xi)^{\beta}x\right)\mathcal{F}(u(0,.))(\xi)e^{i\xi t}d\xi$$
$$-\frac{1}{\sqrt{2\pi}}\int_{|\xi|>\alpha}\int_{0}^{x} \exp\left((i\xi)^{\beta}(\eta-x)\right)\mathcal{F}\left(g(\eta,.)\right)(\xi)e^{i\xi t}d\eta d\xi.$$

By applying the Parseval identity, we have

$$\|u(x,.)-u^{\alpha}(x,.)\|^2_{L^2(\mathbb{R}} = \int_{|\xi|>\alpha} \exp\left(-2\Re((i\xi)^{\beta})x\right)|\mathcal{F}(u(0,.))(\xi)|^2 d\xi$$
$$+\int_{|\xi|>\alpha}\int_0^x \exp\left(2\Re((i\xi)^{\beta})(\eta-x)\right)|\mathcal{F}\left(g(\eta,.)\right)(\xi)|^2 d\eta d\xi$$
$$\le \exp\left(-2x\alpha^{\beta}\cos\frac{\beta\pi}{2}\right)\|\mathcal{F}(u(0,.))\|^2_{L^2(\mathbb{R})}$$
$$+\exp\left(-2x\alpha^{\beta}\cos\frac{\beta\pi}{2}\right)\int_0^x\left(\int_{|\xi|>\alpha}\exp\left(2\eta\Re((i\xi)^{\beta})\right)|\mathcal{F}\left(g(\eta,.)\right)(\xi)|^2 d\xi\right)d\eta.$$

We note that

$$\|\mathcal{F}(u(0,.))\|_{L^2(\mathbb{R}} = \|u(0,.)\|_{L^2(\mathbb{R}},$$
$$\|\mathcal{F}\left(g(\eta,.)\right)\|_{L^2(\mathbb{R}} = \|g(\eta,.)\|_{L^2(\mathbb{R})},$$

by using Parserval identity. Hence we imply

$$\|u(x,.)-u^{\alpha}(x,.)\|^2_{L^2(\mathbb{R}} \le \exp\left(-2x\alpha^{\beta}\cos\frac{\beta\pi}{2}\right)\|u(0,.)\|^2_{L^2(\mathbb{R})}$$
$$+\exp\left(-2x\alpha^{\beta}\cos\frac{\beta\pi}{2}\right)\int_0^x\int_{|\xi|>\alpha}\exp\left(2\eta\Re((i\xi)^{\beta})\right)|\mathcal{F}\left(g(\eta,.)\right)(\xi)|^2 d\xi d\eta$$
$$\le (E+M)^2\exp\left(-2x\alpha^{\beta}\cos\frac{\beta\pi}{2}\right),$$

where the prior estimate (10) and the assumption (20) on exact solution have been used.

Lemma 2 *Let u_ϵ^α and u^α be respectively defined by (9) and (11) which are corresponding to the noisy data (f^ϵ, g^ϵ) and the exact data. Assume that the noisy assumption (8) holds then we have*

$$\left\|u^{\alpha}(x,.)-u^{\alpha}_{\epsilon}(x,.)\right\|_{L^2(\mathbb{R})} \le 2\epsilon\exp\left((1-x)\alpha^{\beta}\cos\frac{\beta\pi}{2}\right), \tag{16}$$

for all $0<x<1$.

Proof. By denoting

$$I_1(x,t) := \frac{1}{\sqrt{2\pi}}\int_{|\xi|\le\alpha}\exp\left((i\xi)^{\beta}(1-x)\right)\mathcal{F}\left(f(.)-f^{\epsilon}(.)\right)(\xi)\chi_{[-\alpha;\alpha]}(\xi)e^{i\xi t}d\xi,$$
$$I_2(x,t) := \frac{1}{\sqrt{2\pi}}\int_{|\xi|\le\alpha}\int_x^1\exp\left((i\xi)^{\beta}(\eta-x)\right)\mathcal{F}\left(g(\eta,.)-g^{\epsilon}(\eta,.)\right)(\xi)d\eta\chi_{[-\alpha;\alpha]}(\xi)e^{i\xi t}d\xi.$$

Then it deduces from (9) and (11) that

$$u^{\alpha}(x,t)-u^{\alpha}_{\epsilon}(x,t) = I_1(x,t)+I_2(x,t). \tag{17}$$

We have

$$\|I_1(x,.)\|^2_{L^2(\mathbb{R})} = \int_{|\xi|\le\alpha}\exp\left(2\Re((i\xi)^{\beta})(1-x)\right)|\mathcal{F}(f^{\epsilon}-f)(\xi)|^2 d\xi$$
$$\le \exp\left(2(1-x)\alpha^{\beta}\cos\frac{\beta\pi}{2}\right)\|\mathcal{F}(f_{\epsilon}-f)\|^2_{L^2(\mathbb{R})}$$
$$\le \exp\left(2(1-x)\alpha^{\beta}\cos\frac{\beta\pi}{2}\right)\|f_{\epsilon}-f\|^2_{L^2(\mathbb{R})},$$

where we have used Parseval identity. Consequently,

$$\|I_1(x,.)\|^2_{L^2(\mathbb{R})} \le \epsilon^2 \exp\left(2(1-x)\alpha^\beta \cos\frac{\beta\pi}{2}\right), \tag{18}$$

where we have used the noisy assumption (8). In addition, we have

$$\begin{aligned}\|I_2(x,.)\|^2_{L^2(\mathbb{R})} &= \int_{|\xi|\le\alpha}\left|\int_x^1 \exp\left((i\xi)^\beta(\eta-x)\right)\left[\mathcal{F}(g^\epsilon(\eta,.)-g(\eta,.))(\xi)\right]d\eta\right|^2 d\xi\\ &\le \int_x^1\int_{|\xi|\le\alpha} \exp\left(2\Re((i\xi)^\beta)(\eta-x)\right)|\mathcal{F}(g^\epsilon(\eta,.)-g(\eta,.))(\xi)|^2 d\xi d\eta.\end{aligned}$$

Thus

$$\begin{aligned}\|I_2(x,.)\|^2_{L^2(\mathbb{R})} &\le \int_x^1 \exp\left(2(\eta-x)\alpha^\beta\cos\frac{\beta\pi}{2}\right)\|\mathcal{F}(g^\epsilon(\eta,.)-g(\eta,.))\|^2_{L^2(\mathbb{R})}d\eta\\ &\le \exp\left(2(1-x)\alpha^\beta\cos\frac{\beta\pi}{2}\right)\int_x^1 \|g^\epsilon(\eta,.)-g(\eta,.)\|^2_{L^2(\mathbb{R})}d\eta\\ &\le \exp\left(2(1-x)\alpha^\beta\cos\frac{\beta\pi}{2}\right)\|g^\epsilon-g\|^2_{L^2(0,1;L^2(\mathbb{R}))},\end{aligned}$$

where we have used the Parseval identity

$$\|\mathcal{F}(g^\epsilon(\eta,.)-g(\eta,.))\|_{L^2(\mathbb{R})} = \|g^\epsilon(\eta,.)-g(\eta,.)\|_{L^2(\mathbb{R})}.$$

By using the noisy assumption (8), this follows

$$\|I_2(x,.)\|^2_{L^2(\mathbb{R})} \le \epsilon^2 \exp\left(2(1-x)\alpha^\beta\cos\frac{\beta\pi}{2}\right). \tag{19}$$

By combining (18) with (19), we derive

$$\begin{aligned}\left\|u^\alpha(x,.)-u^\alpha_\epsilon(x,.)\right\|_{L^2(\mathbb{R})} &\le \|I_1(x,.)\|_{L^2(\mathbb{R})} + \|I_2(x,.)\|_{L^2(\mathbb{R})}\\ &\le 2\epsilon\exp\left((1-x)\alpha^\beta\cos\frac{\beta\pi}{2}\right).\end{aligned}$$

This completes the proof of Lemma.

Theorem 1 *Let u^ϵ_α and u be the regularized solution and exact solution defined by (9) and (4) which are corresponding to the noisy data (f^ϵ, g^ϵ) and the exact data (f,g). Assume that the noisy assumption (8) and (10) hold. If*

$$\int_{-\infty}^{+\infty} \exp\left(2\eta\Re((i\xi)^\beta)\right)|\mathcal{F}(g(\eta,.))(\xi)|^2 d\xi \le M^2, \tag{20}$$

then we have

$$\left\|u^\alpha_\epsilon(x,.)-u(x,.)\right\|_{L^2(\mathbb{R})} \le (E+M)\exp\left(-x\alpha^\beta\cos\frac{\beta\pi}{2}\right) + 2\epsilon\exp\left((1-x)\alpha^\beta\cos\frac{\beta\pi}{2}\right), \tag{21}$$

for all $0 < x < 1$.

Proof. It follows from

$$u(x,t) - u^\alpha_\epsilon(x,t) = u(x,t) - u^\alpha(x,t) + u^\alpha(x,t) - u^\alpha_\epsilon(x,t)$$

that

$$\left\|u^\alpha_\epsilon(x,.) - u(x,.)\right\|_{L^2(\mathbb{R})} \le \|u(x,.) - u^\alpha(x,.)\|_{L^2(\mathbb{R})} + \left\|u^\alpha(x,.) - u^\alpha_\epsilon(x,.)\right\|_{L^2(\mathbb{R})},$$

where u^α is defined by (11). We note that the assumptions of Lemma 1 and Lemma 2 hold. Therefore, by applying Lemma 1 and Lemma 2, we obtain

$$\left\|u^\alpha_\epsilon(x,.) - u(x,.)\right\|_{L^2(\mathbb{R})} \le (E+M)\exp\left(-x\alpha^\beta \cos\frac{\beta\pi}{2}\right) + 2\epsilon \exp\left((1-x)\alpha^\beta \cos\frac{\beta\pi}{2}\right),$$

This implies (21).

Remark 2 (convergence) In the above theorem, the error

$$\left\|u^\alpha_\epsilon(x,.) - u(x,.)\right\|_{L^2(\mathbb{R})}$$

converges to 0 if the parameter of regularization method satisfies the following condition

$$\alpha \to \infty \quad \text{as} \quad \epsilon \to 0, \text{ and} \tag{22}$$

$$\epsilon \exp\left((1-x)\alpha^\beta \cos\frac{\beta\pi}{2}\right) \to 0 \quad \text{as} \quad \epsilon \to 0. \tag{23}$$

Remark 3 (convergent rate) Let us choose

$$\alpha = \left(\frac{\ln\left(\ln\left(\frac{1}{\epsilon}\right)\right)}{\cos\frac{\beta\pi}{2}}\right)^{1/\beta}$$

then (22) and (23) hold. Moreover, we derive

$$\left\|u^\epsilon_\alpha(x,.) - u(x,.)\right\|_{L^2(\mathbb{R})} \le (E+M)\left(\ln\left(\frac{1}{\epsilon}\right)\right)^{-x} + 2\epsilon\left(\ln\left(\frac{1}{\epsilon}\right)\right)^{1-x}, \tag{24}$$

for $0 < x < 1$.

References

[1] S.G. Samko, A.A. Kilbas, O.I. Marichev, *Fractional integrals and derivatives* (Theory and Applications, Gordon and Breach Science, Naukai Tekhnika, Minsk, 1987).
[2] I. Podlubny, *Fractional differential equations* (Academic Press, London, 1999).
[3] K. Diethelm, *The analysis of fractional differential equation* (Springer, Berlin, 2010).
[4] J.A.T. Machado, V. Kiryakova, F. Mainardi, A poster about the recent history of fractional calculus, Fract. Calc. Appl. Anal., **Vol. 13**, 329-334 (2010).
[5] D.A. Murio, Time fractional IHCP with Caputo fractional derivatives, Comput. Math. with Appl., **Vol. 56**, 2371- 2381 (2008).
[6] K.B. Oldham, J. Spanier, A general solution of the diffusion equation for semifinite geometries, J. Math. Anal. Appl., **Vol. 39**, 655 - 669 (1972).
[7] D.A. Murio, Stable numerical solution of a fractional-diffusion inverse heat conduction problem, Comput. Math. Appl., **Vol. 53**, 1492 - 1501 (2007).

[8] W. Cheng, C.L. Fu, Z. Qian, Two regularization methods for a spherically symmetric inverse heat conduction problem, Comput. Math. Appl., **Vol. 56**, 1138 - 1145 (2008).
[9] G.H. Zheng, T. Wei, Spectral regularization method for a Cauchy problem of the time fractional advection-dispersion equation, J. Comput. Appl. Math., **Vol. 233**, 2631 - 2640 (2010).
[10] G.H. Zheng, T. Wei, Spectral regularization method for the time fractional inverse advection-dispersion equation, Math. Comput. Simul., **Vol 81**, 37 - 51 (2010).
[11] X. Xiong, H. Guo, X. Liu, An inverse problem for a fractional diffusion equation, J. Comput. Appl. Math., **Vol. 236**, 4474 - 4484 (2012).
[12] X. Xiong, Q. Zhou, Y.C. Hon, An inverse problem for fractional diffusion equation in 2-dimensional case: Stability analysis and regularization, J. Math. Anal. Appl., **Vol. 393**, 185 - 199 (2012).

Application of Contractive-like Mapping Principles to Fuzzy Functional Differential Equation

Vu Ho[1,*]

[1]Faculty of Mathematical Economics, Banking University of Ho Chi Minh City, Vietnam

Abstract. In this paper, we prove the existence and uniqueness of solution for the fuzzy functional differential equation under generalized Hukuhara derivative via contractive-like mapping principles.

Keywords: generalized Hukuhara derivative; contractive mapping; fuzzy differential equation.
2010 Mathematics Subject Classification: 34F05, 34G20, 34A12, 26E50.

1 Introduction

Fuzzy differential equations (FDEs) forms a suitable setting for the mathematical modeling of real-world problems in which uncertainty or vagueness pervades. A rich collection of results from the theory of FDEs is contained in the monograph of Lakshmikantham and Mohapatra [1] and references therein.

The method of fuzzy mapping was initially introduced by Chang and Zadeh [2]. Later, Dubois and Prade [3, 4] presented a form of elementary fuzzy calculus based on the extension principle [5]. Puri and Ralescu [6] suggested two definitions for the fuzzy derivative of fuzzy functions. The first method was based on H-difference notation and was further investigated by Kaleva [7]. Several approaches were later proposed for FDEs and the existence of their solutions (e.g. [8, 9]). There are several approaches to the study of fuzzy differential equations. One popular approach is based on H-differentiability. The approach based on H-derivative has the disadvantage that it leads to solutions which have an increasing length of their support.

Bede and Gal [10] solved the above mentioned approach under strongly generalized differentiability of fuzzy-number-valued functions. In this case the derivative exists and the solution of FDEs may have decreased the length of the support, but the uniqueness is lost. Other researchers have proposed several approaches to the solutions of FDEs (e.g. [11–14],[15]). Almost authors applied the fixed point theorems like the Darbo's theorem and the classical Banach fixed point theorem [16], Schauder's fixed point theorem [17] and the method of successive approximations [8],... as a tool to prove the existence and uniqueness of the solution of FDEs. In [13], Lupulescu proved the local existence and uniqueness via the method of successive approximations and for global existence and uniqueness via the Banach fixed-point theorem. Allahviranloo et al. [17] also investigated existence and uniqueness of the solution of nonlinear fuzzy Volterra integral equations. Using Arzela-Ascoli's theorem and Schauder's fixed point theorem, authors proved thr existence and uniqueness of the solution for this kind

*e-mail: vuh@buh.edu.vn

of equations. In [14], using the contractive-like mapping principles, authors discussed the existence and uniqueness of solution of fuzzy differential equation under generalized Hukuhara derivative.

In this paper, we prove the existence and uniqueness of solution for fuzzy functional differential equation under generalized Hukuhara derivative via the some fixed point theorems, established in [18].

This paper is organized as follows: In Section 2 we shall review some results the existence and uniqueness of fixed points for mappings defined in partially ordered sets. Also, we recall some fundamental results of fuzzy sets, ordering relations over fuzzy sets and fuzzy distance, which will be necessary to study the existence and uniqueness of solution for the fuzzy functional differential equation. In Section 3 we study on the existence and uniqueness of solution for the fuzzy functional differential equation.

2 Preliminaries

In this section, we shall review some results the existence and uniqueness of fixed points for mappings defined in partially ordered sets. Also, we recall some fundamental results of fuzzy sets, ordering relations over fuzzy sets and fuzzy distance.

Definition 2.1 *(see [18]) An altering distance function is a function $\Lambda : \mathbb{R}_+ \to \mathbb{R}_+$ which satisfies*

(i) Λ is continuous and non-decreasing.

(ii) $\Lambda(t) = 0$ if and only if $t = 0$.

Theorem 2.1 *(see [18]) Let $(X, \leq)$ be a be a partially ordered set and suppose that there exists a metric d in X such that (X, d) is a complete metric space. Let $F : X \to X$ be a monotone non-decreasing mapping such that*

$$\Lambda(d(F(x), F(y))) \leq \Lambda(d(x, y)) - \Upsilon(d(x, y)), \text{ for all } x \geq y,$$

for some altering distance functions Λ and Υ. Suppose that either F is continuous or X is such that

$$\text{if a non-decreasing sequence } (x_n)_{n\in\mathbb{N}} \to 0 \text{ in } X, \text{ then } x_n \leq x \text{ for all } n \in \mathbb{N}.$$

If there exists $x_0 \in X$ with $x_0 \leq F(x_0)$, then f has a fixed point.

Theorem 2.2 *(see [18]) Let $(X, \leq)$ be a be a partially ordered set and suppose that there exists a metric d in X such that (X, d) is a complete metric space. Let $F : X \to X$ be a monotone non-decreasing mapping such that*

$$\Lambda(d(F(x), F(y))) \leq \Lambda(d(x, y)) - \Upsilon(d(x, y)), \text{ for all } x \geq y,$$

for some altering distance functions Λ and Υ. Suppose that either f is continuous or X is such that

$$\text{if a non-increasing sequence } (x_n)_{n\in\mathbb{N}} \to 0 \text{ in } X, \text{ then } x_n \geq x \text{ for all } n \in \mathbb{N}.$$

If there exists $x_0 \in X$ with $x_0 \geq F(x_0)$, then F has a fixed point.

Let $\mathscr{K}_c(\mathbb{R}^d)$ denote the collection of all nonempty compact and convex subsets of $\mathbb{R}^d$. The addition and scalar multiplication in $\mathscr{K}_c(\mathbb{R}^d)$, we define as usual, i.e. $A, B \in \mathscr{K}_c(\mathbb{R}^d)$ and $\lambda \in \mathbb{R}$, then we have $A + B = \{a + b \mid a \in A, b \in B\}$, $\lambda A = \{\lambda a \mid a \in A\}$. The Hausdorff metric d_H in $\mathscr{K}_c(\mathbb{R}^d)$ is defined as follows

$$d_H(A, B) = \max\{\sup_{a\in A} \inf_{b\in B} \|a - b\|_{\mathbb{R}^d}, \sup_{b\in B} \inf_{a\in A} \|a - b\|_{\mathbb{R}^d}\},$$

where $A, B \in \mathscr{K}_c(\mathbb{R}^d)$. It is known that $(\mathscr{K}_c(\mathbb{R}^d), d_H)$ is a complete metric space.

Denote $E^d = \{u : \mathbb{R}^d \longrightarrow [0, 1]$ such that $x(z)$ satisfies (i)-(iv) stated below$\}$

(i) x is normal, i.e, there exists an $z_0 \in \mathbb{R}^d$ such that $x(z_0) = 1$;

(ii) x is fuzzy convex, that is, for $0 \le \lambda \le 1$, $x(\lambda z_1 + (1 - \lambda)z_2) \ge \min\{x(z_1), x(z_2)\}$, for any $z_1, z_2 \in \mathbb{R}^d$;

(iii) x is upper semicontinuous;

(iv) $[x]^0 = cl\{z \in \mathbb{R}^d : x(z) > 0\}$ is compact.

Then E^d is called the space of fuzzy sets.

For $\alpha \in (0, 1]$, denote $[x]^\alpha = \{z \in \mathbb{R}^d \mid x(z) \ge \alpha\}$. We will call this set an α - cut (α - level set) of the fuzzy set x. For $x \in E^d$ one has that $[x]^\alpha \in \mathscr{K}_c(\mathbb{R}^d)$ for every $\alpha \in [0, 1]$.

The supremum on E^d is defined by

$$d_\infty(x_1, x_2) = \sup_{\alpha\in[0,1]} d_H([x_1]^\alpha, [x_2]^\alpha)$$

for every $x_1, x_2 \in E^d$, and (E^d, d_∞) is a complete metric space.

Define $D : E^d \times E^d \longrightarrow \mathbb{R}_+$ by the expression

$$D(x, y) = \sup_{t\in[a,b]} d_\infty(x_1(t), x_2(t)).$$

It is easy to see that D is a metric in E^d. In fact, (E^d, D) is a complete metric space.

For every $x, y, z \in E^d$ and $\lambda \in \mathbb{R}_+$, we have

$$\begin{aligned} &D(x + z, y + z) = D(x, y),\\ &D(\lambda x, \lambda y) = \lambda D(x, y),\\ &D(x, y) \le D(x, z) + D(y, z). \end{aligned}$$

Definition 2.2 *([10])* Let $x : (a, b) \longrightarrow E^d$, $t \in [a, b]$. We say that x is differentiable at t, if there exists $x'(t) \in E^d$ such that

(i)for all $h > 0$ sufficiently small, there exist $x\,(t + h) \ominus x(t)$, $x(t) \ominus x(t - h)$ and the limits

$$\lim_{h\to 0^+} \frac{x\,(t + h) \ominus x(t)}{h} = \lim_{h\to 0^+} \frac{x\,(t) \ominus x(t - h)}{h} = x'(t)$$

or

(ii)for all $h > 0$ sufficiently small, there exist $x\,(t) \ominus x(t + h)$, $x(t - h) \ominus x(t)$ and the limits

$$\lim_{h\to 0^+} \frac{x\,(t) \ominus x(t + h)}{-h} = \lim_{h\to 0^+} \frac{x\,(t - h) \ominus x(t)}{-h} = x'(t).$$

In the sequel, we define the following partial orderings $\preceq$ and $\preceq$ in the space E^d and E^1:

$$x, y \in E^d, \quad x \preceq y \Longleftrightarrow [x]^\alpha \subseteq [y]^\alpha, \text{ for all } \alpha \in [0,1],$$

and

$$x \preceq y \Longrightarrow x + z \preceq y + z, \quad \text{for } x, y, z \in E^1,$$
$$x \preceq y \Longrightarrow x + z \preceq y + z, \quad \text{for } x, y, z \in E^d,$$

and for fuzzy functions,

$$f \preceq g \Longrightarrow \int_0^t f(s)ds \preceq \int_0^t g(s)ds, \quad \text{for } t \in [0,a], \ f, g \in C([0,a], E^1),$$
$$f \preceq g \Longrightarrow \int_0^t f(s)ds \preceq \int_0^t g(s)ds, \quad \text{for } t \in [0,a], \ f, g \in C([0,a], E^d).$$

3 Main result

For a given $\sigma > 0$, we denote by C_σ the space $C([-\sigma, 0], E^d)$. Also, we denote by

$$D_\sigma(x, y) = \sup_{t \in [-\sigma, 0]} d_\infty(x(t), y(t))$$

the metric on the space C_σ. Let $x(\cdot) \in C([-\sigma, \infty), E^d)$. Then, for each $t \in [0, \infty)$ we denote by x_t the element of C_σ defined by $x_t(s) = x(t + s)$, $s \in [-\sigma, 0]$.

In this section, we consider the fuzzy functional differential equation under generalized Hukuhara derivative as follows:

$$\begin{cases} u'(t) = f(t, u_t), & t \in [0,a], \\ u(t) = \varphi(t), & t \in [-\sigma, 0]. \end{cases} \tag{1}$$

Theorem 3.1 *(see [12]) A fuzzy mapping $u : [-\sigma, a] \to E^d$ is a solution to the problem (1) on $[-\sigma, a]$ if and only if u is a continuous fuzzy mapping and it satisfies to one of the following fuzzy integral equations:*

(i) If u is (i)-differentiable on $[0,a]$, then

$$u(t) = \begin{cases} \varphi(t), & \text{for } t \in [-\sigma, 0], \\ \varphi(0) + \int_0^t f(s, u_s)ds, & \text{for } t \in [0,a], \end{cases} \tag{2}$$

(ii) If u is (ii)-differentiable on $[0,a]$, then

$$u(t) = \begin{cases} \varphi(t), & \text{for } t \in [-\sigma, 0], \\ \varphi(0) \ominus (-1)\int_0^t f(s, u_s)ds, & \text{for } t \in [0,a]. \end{cases} \tag{3}$$

Definition 3.1 *(see [?])* A lower solution for the problem (1) is a function $\mu \in C([0,a], E^d) \cap C^1([-\sigma, a], E^d)$ such that

$$\mu'(t) \preceq f(t, u_t), \quad t \in [0,a], \quad \mu_0 \preceq \varphi.$$

A upper solution for the problem (1) is a function $\mu \in C([0,a], E^d) \cap C^1([-\sigma, a], E^d)$ such that

$$\mu'(t) \succeq f(t, u_t), \quad t \in [0,a], \quad \mu_0 \succeq \varphi.$$

Remark 3.1 If μ is (i)-differentiable (or (ii)-differentiable), then μ is said a lower (i)-solution (or a lower (ii)-solution), respectively. If μ is (i)-differentiable (or (ii)-differentiable), then μ is said an upper (i)-solution (or (ii)-differentiable), respectively.

Theorem 3.2 *Suppose that there exists $\mu \in C([0,a],E^d) \cap C^1([-\sigma,a],E^d)$ a lower (i)-solution of the problem (1). Let $f : [0,a] \times C_\sigma \to E^d$ be continuous such that*

(i) f is non-decreasing in the second variable, i.e., if $\varphi \geq \phi$ then $f(t,\varphi) \geq f(t,\phi)$;

(ii) f is weakly contractive for comparable elements, i.e., for some altering distance functions Λ and Υ, it holds

$$\Lambda\big(d_\infty(f(t,\varphi), f(t,\phi))\big) \leq \Lambda\big(D_\sigma(\varphi,\phi)\big) - \Upsilon\big(D_\sigma(\varphi,\phi)\big), \text{ if } \varphi \geq \phi,$$

for every $t \in [0,a]$ and $\varphi, \phi \in C_\sigma$.

Then, the problem (1) has a unique (i)-solution on $[-\sigma,a]$.

Proof. We consider the space $C([-\sigma,a],E^d)$ equipped with the complete metric

$$D_\rho(u,v) = \sup_{t\in[-\sigma,a]} \{d_\infty(u(t),v(t))\,e^{-\rho t}\}, \quad u,v \in C([-\sigma,a],E^d), \quad \text{for } \rho > 0.$$

It is easy to see that this metric is equivalent to the metric $D(u,v)$, because

$$D(u,v)e^{-\rho a} \leq D_\rho(u,v) \leq D(u,v) \tag{4}$$

for all $u,v \in C([-\sigma,a],E^d)$. Moreover, $(C([-\sigma,a],E^d), D_\sigma)$ is a complete metric space.

Now, let the operator $\mathscr{T} : C([-\sigma,a],E^d) \longrightarrow C([-\sigma,a],E^d)$ be defined by

$$[\mathscr{T}u](t) = \begin{cases} \varphi(t), & \text{for } \ t \in [-\sigma,0], \\ \varphi(0) + \int_0^t f(s,u_s)ds, & \text{for } \ t \in [0,a]. \end{cases} \tag{5}$$

If there exists $u \in C([-\sigma,a],E^d)$ is a fixed point of $\mathscr{T}$, then $u \in C^1([0,a],E^d)$ is a solution of the problem (1) and conversely.

In order to prove the theorem, we shall show the conditions of Theorem (2.1) are satisfied. First, we prove that the operator $\mathscr{T}$ is a non-decreasing and continuous. Indeed, the continuity is trivial and take $u \geq v$ on intervals $[0,a]$ and for every $t \in [-\sigma,0]$, we have

$$[\mathscr{T}u](t) = [\mathscr{T}v](t) = \varphi(t).$$

For every $t \in [0,a]$, we obtain

$$[\mathscr{T}u](t) = \varphi(0) + \int_0^t f(s,u_s)ds \geq \varphi(0) + \int_0^t f(s,v_s)ds = [\mathscr{T}v](t).$$

From the assumption (ii), we have

$$\Lambda\big(d_\infty(f(t,u), f(t,v))\big) \leq \Lambda\big(D_\sigma(u,v)\big), \quad \text{if } u,v \in C_\sigma \text{ and } u \geq v. \tag{6}$$

Next, we will prove that the inequality (6) holds, for all $u \geq v$. Suppose that

$$D_\sigma(u,v) < d_\infty\big(f(t,u), f(t,v)\big), \quad \text{for all } u \geq v$$

and Λ is an altering distance function. Then we have

$$\Lambda\big(D_\sigma(u,v)\big) \leq \Lambda\big(d_\infty\big(f(t,u), f(t,v)\big)\big). \tag{7}$$

By the inequality (6) and (7), we derive that

$$\Lambda\big(D_\sigma(u,v)\big) = \Lambda\big(d_\infty\big(f(t,u), f(t,v)\big)\big), \quad \text{for all } u \geq v. \tag{8}$$

Combine assumption (ii) and (8), it follows that $0 \leq -\Upsilon\big(D_\sigma(u,v)\big)$, for all $u \geq v$, and therefore, $\Upsilon\big(D_\sigma(u,v)\big) = 0$. This implies $D_\sigma(u,v) = 0$. Thus, we have $\Lambda\big(d_\infty\big(f(t,u), f(t,v)\big)\big) = \Upsilon\big(D_\sigma(u,v)\big) = 0$. Since Υ is an altering distance function, we derive that $d_\infty\big(f(t,u), f(t,v)\big) = 0$. This contradict the assumption above, that is, for all $u \geq v$ we have

$$d_\infty\big(f(t,u), f(t,v)\big) \leq D_\sigma(u,v). \tag{9}$$

For all $u \geq v$, if $t \in [-\sigma, 0]$, then $d_\infty([\mathscr{T}u](t), [\mathscr{T}v](t)) = d_\infty(\varphi(t), \varphi(t)) = 0$, and if $t \in [0,a]$, we get

$$\begin{aligned} d_\infty([\mathscr{T}u](t), [\mathscr{T}v](t)) &= d_\infty\left(\varphi(0) + \int_0^t f(s,u_s)ds, \varphi(0) + \int_0^t f(s,v_s)ds\right) \\ &\leq \int_0^t d_\infty\big(f(s,u_s), f(s,v_s)\big)ds \leq \int_0^t D_\sigma(u,v)ds \\ &\leq \int_0^t \sup_{r\in[-\sigma,0]} d_\infty\big(u(r+s), v(r+s)\big)ds, \end{aligned}$$

then

$$\begin{aligned} D_\rho\big(\mathscr{T}u, \mathscr{T}v\big) &= \sup_{t\in[-\sigma,a]} \left\{d_\infty([\mathscr{T}u](t), [\mathscr{T}v](t))\, e^{-\rho t}\right\} \\ &\leq \sup_{t\in[-\sigma,a]} \left\{\int_0^t \sup_{r\in[-\sigma,0]} d_\infty\big(u(r+s), v(r+s)\big)ds\, e^{-\rho t}\right\} \\ &= \sup_{t\in[-\sigma,a]} \left\{\int_0^t \sup_{r\in[-\sigma,0]} d_\infty\big(u(r+s), v(r+s)\big)\, e^{-\rho(r+s)}\, e^{\rho(r+s)}ds\, e^{-\rho t}\right\} \\ &\leq \sup_{t\in[-\sigma,a]} \left\{\int_0^t \sup_{r\in[-\sigma,0]} e^{\rho(r+s)}ds\, e^{-\rho t}\right\} D_\rho(u,v) \\ &= \sup_{t\in[-\sigma,a]} \left\{\int_0^t e^{\rho s}ds\, e^{-\rho t}\right\} D_\rho(u,v) \\ &= \sup_{t\in[-\sigma,a]} \left\{\frac{1-e^{-\rho t}}{\rho}\right\} D_\rho(u,v) = \frac{1-e^{-\rho a}}{\rho} D_\rho(u,v). \end{aligned} \tag{10}$$

Since Λ is an altering distance function and by the inequality (10), we infer that

$$\begin{aligned} \Lambda\big(D_\rho\big(\mathscr{T}u, \mathscr{T}v\big)\big) &\leq \Lambda\left(\frac{1-e^{-\rho a}}{\rho} D_\rho(u,v)\right) \\ &= \Lambda\big(D_\rho(u,v)\big) - \left(\Lambda\big(D_\rho(u,v)\big) - \Lambda\left(\frac{1-e^{-\rho a}}{\rho} D_\rho(u,v)\right)\right). \end{aligned}$$

Therefore, if $\Upsilon(t) = \Lambda(t) - \Lambda\left(\frac{1-e^{-\rho a}}{\rho} t\right)$, it follows that

$$\Lambda\big(D_\rho\big(\mathscr{T}u, \mathscr{T}v\big)\big) \leq \Lambda\big(D_\rho(\varphi,\phi)\big) - \Upsilon\big(D_\rho(\varphi,\phi)\big), \quad \text{for all } u \geq v.$$

From 4 and (8), we obtain

$$\Lambda\big(d_\infty(f(t,u),f(t,v))\big) \le \Lambda\big(D_\sigma(\varphi,\phi)\big) - \Upsilon\big(D_\sigma(\varphi,\phi)\big), \quad \text{for all } u \ge v.$$

Finally, using the existence of the lower (i)-solution and Theorem 2.1, we shall prove that μ is such that $\mu \le \mathscr{T}\mu$. Indeed, if $t \in [-\sigma, 0]$, $\mu_0 \le \varphi$ then $\mu(t) \le \varphi(t) = [\mathscr{T}\mu](t)$ and for $t \in [0,a]$,

$$\mu(t) = \mu(0) + \int_0^t \mu'(s)ds \le \varphi(0) + \int_0^t f(s,\mu_s)ds = [\mathscr{T}\mu](t).$$

Since every pair of functions in $C([-\sigma,a],E^d)$ has an upper bound, the operator $\mathscr{T}$ has a unique fixed point $u \in C([-\sigma,a],E^d)$ and u is the unique solution of the problem (1) on $[-\sigma,a]$.

Theorem 3.3 *Replacing the existence of a lower (i)-solution of the problem (1) by the existence of a upper (i)-solution of the problem (1), the conclusion of Theorem 3.2 is still valid.*

Proof. The proof of Theorem 3.3 is similarly the proof Theorem 3.2.

Theorem 3.4 *Suppose that there exists $\mu \in C([0,a],E^d) \cap C^1([-\sigma,a],E^d)$ a lower (ii)-solution of the problem (1). Let $f : [0,a] \times C_\sigma \to E^d$ be continuous such that*

*(**i**) for all $\alpha \in [0,1]$ and $\varphi \in C_\sigma$,*

$$diam([\varphi(0)]^\alpha) \ge diam\left(\left[\int_{t_0}^t f(s,u_s)ds\right]^\alpha\right);$$

*(**ii**) f is non-decreasing in the second variable, i.e., if $\varphi \ge \phi$ then $f(t,\varphi) \ge f(t,\phi)$;*

*(**iii**) f is weakly contractive for comparable elements, i.e., for some altering distance functions Λ and Υ, it holds*

$$\Lambda\big(d_\infty(f(t,\varphi),f(t,\phi))\big) \le \Lambda\big(D_\sigma(\varphi,\phi)\big) - \Upsilon\big(D_\sigma(\varphi,\phi)\big), \ \textit{if } \varphi \ge \phi,$$

for every $t \in [0,a]$ and $\varphi,\phi \in C_\sigma$.

Then, the problem (1) has a unique (ii)-solution on $[-\sigma,a]$.

Proof. By the condition (i), we imply that the existence of Hukuhara differences $\varphi(0) + (-1)\int_0^t f(s,u_s)ds$ is guaranteed. Now, let the operator $\mathscr{A} : C([-\sigma,a],E^d) \to C([-\sigma,a],E^d)$ be defined by

$$[\mathscr{A}u](t) = \begin{cases} \varphi(t), & \text{for } \ t \in [-\sigma,0], \\ \varphi(0) \ominus (-1)\int_0^t f(s,u_s)ds, & \text{for } \ t \in [0,a]. \end{cases} \tag{11}$$

Notice that if there exists $u \in C([-\sigma,a],E^d)$ is a fixed point of $\mathscr{A}$, then $u \in C^1([0,a],E^d)$ is a solution of the problem (1) and conversely.

Similarly to Theorem 3.2, we shall show the conditions of Theorem (2.1) are satisfied. For all $u_s \ge v_s$ on $[-\sigma,a]$, we have

$$[\mathscr{A}u](t) = [\mathscr{A}v](t) = \varphi(t) \quad \text{for } t \in [-\sigma,0]$$

and

$$[\mathscr{A}u](t) = \varphi(0) \ominus (-1) \int_0^t f(s, u_s)ds \geq \varphi(0) \ominus (-1) \int_0^t f(s, v_s)ds = [\mathscr{A}v](t), \quad \text{for } t \in [0, a].$$

Therefore, $\mathscr{A}$ has the operator non-decreasing.

For all $u \geq v$, if $t \in [-\sigma, 0]$ then $d_\infty([\mathscr{A}u](t), [\mathscr{A}v](t)) = d_\infty(\varphi(t), \varphi(t)) = 0$ and if $t \in [0, a]$, then we get

$$\begin{aligned} d_\infty([\mathscr{A}u](t), [\mathscr{A}v](t)) &= d_\infty\left(\varphi(0) \ominus (-1) \int_0^t f(s, u_s)ds, \varphi(0) \ominus (-1) \int_0^t f(s, v_s)ds\right) \\ &\leq \int_0^t d_\infty\big(f(s, u_s), f(s, v_s)\big)ds \leq \int_0^t D_\sigma(u_s, v_s)ds \\ &\leq \int_0^t \sup_{r\in[-\sigma,0]} d_\infty\big(u(r+s), v(r+s)\big)ds. \end{aligned}$$

By similar calculations as (10), we have.

$$\begin{aligned} D_\rho\big(\mathscr{T}u, \mathscr{T}v\big) &= \sup_{t\in[-\sigma,a]} \left\{d_\infty([\mathscr{T}u](t), [\mathscr{T}v](t))\, e^{-\rho t}\right\} \\ &\leq \sup_{t\in[-\sigma,a]} \left\{\frac{1 - e^{-\rho t}}{\rho}\right\} D_\rho(u, v) = \frac{1 - e^{-\rho a}}{\rho} D_\rho(u, v). \end{aligned} \tag{12}$$

Therefore, if Λ is some increasing altering distance function, it holds

$$\begin{aligned} \Lambda\big(D_\rho(\mathscr{T}u, \mathscr{T}v)\big) &\leq \Lambda\left(\frac{1 - e^{-\rho a}}{\rho} D_\rho(u, v)\right) \\ &= \Lambda\big(D_\rho(u, v)\big) - \left(\Lambda\big(D_\rho(u, v)\big) - \Lambda\left(\frac{1 - e^{-\rho a}}{\rho} D_\rho(u, v)\right)\right). \end{aligned}$$

Then, from 4 and (8), we derive that

$$\Lambda\big(d_\infty\big(f(t, u), f(t, v)\big)\big) \leq \Lambda\big(D_\sigma(\varphi, \phi)\big) - \Upsilon\big(D_\sigma(\varphi, \phi)\big), \quad \text{for all } u \geq v.$$

where $\Upsilon(t) = \Lambda(t) - \Lambda\left(\frac{1 - e^{-\rho a}}{\rho} t\right)$.

Finally, using the existence of the lower (ii)-solution and Theorem 3.1, we shall prove that μ is such that $\mu \leq \mathscr{A}\mu$. Indeed, if $t \in [-\sigma, 0]$, $\mu_0 \leq \varphi$ then $\mu(t) \leq \varphi(t) = [\mathscr{A}\mu](t)$ and for $t \in [0, a]$,

$$\mu(t) = \mu(0) \ominus (-1) \int_0^t \mu'(s)ds \leq \varphi(0) \ominus (-1) \int_0^t f(s, \mu_s)ds = [\mathscr{A}\mu](t).$$

Thus $\mu(t) \leq [\mathscr{A}\mu](t)$. We see that the operator $\mathscr{A}$ verifies all conditions of Theorem 2.1, that is, $\mathscr{A}$ has a fixed point in $C([-\sigma, E^d])$. Given that $C([-\sigma, E^d])$ verifies that every pair of elements of $C([-\sigma, E^d])$ has an upper bound, the operator $\mathscr{A}$ has a unique fixed point.

Theorem 3.5 *Replacing the existence of a lower (ii)-solution of the problem (11) by the existence of a upper (ii)-solution of the problem (11), the conclusion of Theorem 3.5 is still valid.*

Proof. The proof of Theorem 3.5 is similarly the proof Theorem 3.4.

Example 3.6 *Let us consider a fuzzy time-delay Malthusian model:*

$$\begin{cases} u'(t) = \lambda u(t-1), & t \geq 0, \\ u(t) = u_0, & -1 \leq t \leq 0, \end{cases} \tag{13}$$

where $u(t)$ is the population at time t, $u_0 = (-1, 0, 1)$ and $\lambda > 0$.

In [12, 13], it is shown that the problem (13) has a unique solution on $[-1, \infty)$.

But this example, we want to prove that the conditions of Theorem 3.2 (or Theorem 3.2) are satisfied.

Now, we check the existence of a lower (i)-solution for the problem (13). Let $f : [0, \infty) \times C([-1, 0], E^1) \to E^1$ be defined by $f(t, u) = \lambda u(t-1)$ is a continuous and non-increasing in the second variable, that is,

$$f(t, u) = \lambda u(t-1) \geq f(t, v) = \lambda v(t-1), \quad \text{if } u \geq v, \quad \text{for } t \in [0, \infty). \tag{14}$$

Let the function $\Lambda, \Upsilon : \mathbb{R}_+ \to \mathbb{R}_+$ be defined by $\Lambda(t) = t$ and $\Upsilon(t) = (1-\lambda)t$. It is easy to see that Λ, Υ satisfy the conditions of Definition 2.1, that is, Λ and Υ are some altering distance function.

By a direct calculation, we get

$$\begin{aligned} d_\infty(f(t, u), f(t, v)) &= \lambda d_\infty(u(t-1), v(t-1)) \\ &= \lambda \sup_{\alpha \in [0,1]} \max \left\{ \left| u^l_\alpha(t-1) - v^l_\alpha(t-1) \right|, \left| u^r_\alpha(t-1) - v^r_\alpha(t-1) \right| \right\} \\ &= \lambda \sup_{\alpha \in [0,1]} \max \left\{ \sup_{s \in [-1,0]} \left| u^l_\alpha(s) - v^l_\alpha(s) \right|, \sup_{s \in [-1,0]} \left| u^r_\alpha(s) - v^r_\alpha(s) \right| \right\} \\ &= \lambda \sup_{\alpha \in [0,1]} \left\{ \sup_{s \in [-1,0]} d_H\big([u(s)]^\alpha, [v(s)]^\alpha\big) \right\} \\ &\leq \lambda \sup_{s \in [-1,0]} d_\infty(u(s), v(s)) = \lambda D_\sigma(u, v). \end{aligned} \tag{15}$$

Since Λ is an altering distance function and from (15), we infer that

$$\begin{aligned} \Lambda\big(d_\infty(f(t, u), f(t, v))\big) &\leq \Lambda\big(\lambda D_\sigma(u, v)\big) \\ &\leq \Lambda\big(D_\sigma(u, v)\big) - \Big(\Lambda\big(D_\sigma(u, v)\big) - \Lambda\big(\lambda D_\sigma(u, v)\big)\Big) \\ &\leq \Lambda\big(D_\sigma(u, v)\big) - \Upsilon\big(D_\sigma(u, v)\big), \end{aligned} \tag{16}$$

where $\Upsilon(t) = (1-\lambda)t$.

By using Theorem 3.2, the existence of a lower (i)-solution for the problem provides the existence of a unique solution on $[-1, \infty)$.

It is easy to see that the conditions of Theorem (3.3) are satisfied, that is, the problem (13) has a unique (ii)-solution on $[-1, \infty)$.

Example 3.7 *Consider the fuzzy functional differential equation that describe the fuzzy Ehrlich ascities tumor model*

$$\begin{cases} u'(t) = ru(t-1)\big(1 - u(t-1)\big), & t \geq 0, \\ u(t) = u_0, & t \in [-1, 0], \end{cases} \tag{17}$$

where $u_0 = (-1, 0, 1)$ and $r \in [0, 2)$.

Similar to example above, we can easily see that all the conditions of Theorem (3.2) are verified and we can conclude the existence and uniqueness of solution of problem (17) by means of Theorem (3.2), though the unique solution for problem (17).

4 Conclusions

In this paper, we studied the new existence and uniqueness of solution for fuzzy functional differential equation under generalized Hukuhara derivative and by using some recent results of fixed point of weakly contractive mappings on partially ordered sets, established in [18]. We obtain a generalization of these results, if we replace the condition (h1) of Theorem 4.1 in [13] by the following condition $\Lambda\big(d_\infty(f(t,\varphi), f(t,\phi))\big) \leq \Lambda\big(D_\sigma(\varphi,\phi)\big) - \Upsilon\big(D_\sigma(\varphi,\phi)\big)$, where Λ, Υ are some altering distance functions.

References

[1] V. Lakshmikantham, R.N. Mohapatra, *Theory of Fuzzy Differential Equations and Inclusions* (Taylor and Francis Publishers, London, 2003), ISBN 9780415300735

[2] S.S.L. Chang, L.A. Zadeh (World Scientific Publishing Co., Inc., River Edge, NJ, USA, 1996), chap. On Fuzzy Mapping and Control, pp. 180–184, ISBN 981-02-2422-2, `http://dl.acm.org/citation.cfm?id=234347.234388`

[3] D. Dubois, H. Prade, Fuzzy Sets and Systems **8**, 1 (1982)

[4] D. Dubois, H. Prade, Fuzzy Sets and Systems **8**, 105 (1982)

[5] L. Zadeh, Information and Control **8**, 338 (1965)

[6] M.L. Puri, D.A. Ralescu, Journal of Mathematical Analysis and Applications **91**, 552 (1983)

[7] O. Kaleva, Fuzzy Sets and Systems **24**, 301 (1987), fuzzy Numbers

[8] S. Song, C. Wu, Fuzzy Sets and Systems **110**, 55 (2000)

[9] C. Wu, S. Song, E. Lee, Journal of Mathematical Analysis and Applications **202**, 629 (1996)

[10] B. Bede, S.G. Gal, Fuzzy Sets Syst. **151**, 581 (2005)

[11] S.S. Behzadi, T. Allahviranloo, Iranian Journal of Fuzzy Systems **13**, 71 (2016)

[12] A. Khastan, J. Nieto, R. Rodríguez-López, Information Sciences **275**, 145 (2014)

[13] V. Lupulescu, Fuzzy Sets and Systems **160**, 1547 (2009)

[14] E.J. Villamizar-Roa, V. Angulo-Castillo, Y. Chalco-Cano, Fuzzy Sets and Systems **265**, 24 (2015)

[15] O.S. Fard, D.F.M. Torres, M.R. Zadeh, Applicable Analysis and Discrete Mathematics pp. 152–167 (2016)

[16] P. Subrahmanyam, S. Sudarsanam, Fuzzy Sets and Systems **81**, 237 (1996)

[17] T. Allahviranloo, P. Salehi, M. Nejatiyan, Iranian Journal of Fuzzy Systems **12**, 75 (2015)

[18] J. Harjani, K. Sadarangani, Nonlinear Analysis: Theory, Methods & Applications **72**, 1188 (2010)

What is this fruit? Neural network application for Vietnamese fruit recognition

Mateusz Baryła[1,*]

[1]Faculty of Applied Mathematics, Wroclaw University of Science and Technology

Abstract. We use deep learning for problems in computer vision, image recognition and classification. Deep learning methods for fruit recognition are built with methods where features (in our case fruits key features) are processed and sent through multiple layers where transformations and computations are done sequentially to form a prediction model. Deep learning algorithms draws inspiration from many fields especially applied maths fundamentals like linear algebra, probability, information theory and numerical optimization. To the best of our knowledge this is the first web application for fruit recognition. Thanks to that users will be able to recognize most of the Vietnamese fruits without knowledge of Vietnamese language. What is more they can get a short description for each fruit and a video how to eat it. In the paper we compare different models of convolutional neural networks in order to find the best possible model of CNN. This system will is fine-tuned, what means that it learns on examples provided by users of application. Having that we propose algorithm which detects non-fruits pictures uploaded by user.

1 Introduction

Vietnam is located in tropical zone and famous with many kinds of fruits. There is a list of Top Tropical Fruits in Vietnam- they are really popular and both foreigners and Vietnamese should try them. Because of my studies which are Applied Mathematics and Computer Science I want to combine my two majors and create a web application with usage of complex math which are in Convolutional Neural Networks. Convolutional neural networks are primarily used to solve image recognition problems. Due to their simple architecture they offer a simple method of getting started with ANNs. AI is not new it actually started over 65 years ago. Today we finally have a cost efficient way to store transfer and compute a massive amount of data in a way that was never possible before.

Over the last several years machine learning techniques, especially when applied to neural networks, have played an increasingly important role in the design of image recognition systems.One of the main reasons for developing *www.whatisthisfruit.info* web application is to help people in Vietnam with fruits recognition by applying artificial intelligence solutions. Using this solution user is does not have to know Vietnamese language in order to ge to know fruits name, actually he or she might use my application, takes the picture and then review results.

*e-mail: 230038@student.pwr.edu.pl

The paper is structured as follows. Firstly I discuss related work obtained for fruits recognition. Next I introduce readers to the mathematical background of artificial neural networks. Later I discuss CNNs and their architecture. After that I present dataset and preprocessing used in preparing images following with results achieved with my different models. I also explain way of fine-tuning model, and architecture of web-application. Last but not least I propose future work and sum up my results.

2 Related work

Recent work in deep neural networks has led to the development of many interesting fruit detection systems.

Paper [1] presents a novel approach to fruit detection using deep convolutional neural networks. It's aim is to build detection system which is accurate, fast and reliable because it is a vital element of an autonomous agricultural robotic platform.In this work model is adapted, through transfer learning, for the task of fruit detection using imagery obtained from two modalities: colour (RGB) and Near-Infrared (NIR). Early and late fusion methods are explored for combining the multi-modal (RGB and NIR) information. In addition to improved accuracy, this approach is also much quicker to deploy for new fruits, as it requires bounding box annotation rather than pixel-level annotation (annotating bounding boxes is approximately an order of magnitude quicker to perform).

Research paper [3] propose a new, high-qality dataset of images containing most popular fruits. The dataset can be downloaded directly from [4] or [5]. It contains 49461 images of 74 fruits. The dataset is special because it doesn't contain noisy background. Authors use C++ programming language to remove background and leave only fruit on a given image.

In other papers researchers tried to detect fruits based on color shape or texture. There is also a method [2] which uses k-nearest neighbor to increase accuracy of fruit recognition.

In my work I am focused on CNNs architecture with the real-life use-case and then making people's lives easier. I've tried many different models of CNN architecture in order to find the best one and then deploy it to public use as a web application. In my work I describe 5 of them which has given the best results. What is more my system is fine-tuned what means that it is going to give better results with time.

3 Artificial neural neurons

Artificial Neural Networks are computational processing systems of which are heavily inspired by way biological nervous systems (such as the human brain operate). ANNs collect their knowledge by finding patterns and relationship in data and learn through experience, not from programming. Every ANNs is formed from hundreds of single units, artificial neurons or processing elements, connected with weights which make the whole neural structure and are organized in layers. In 1943, neurophysiologist Warren McCulloch and mathematician Walter Pitts wrote a paper on how neurons might work. In order to describe how neurons in the brain might work, they modeled a simple neural network using electrical circuits, the first model of a "McCulloch-Pitts neuron". The McCulloch-Pitts model was an extremely simple artificial neuron. The inputs could be either a zero or a one. And the output was a zero or a one.It is a neuron of a set of inputs $I_1, I_2, I_3, ..., I_m$ and one output y. The linear threshold gate simply classifies the set of inputs into two different classes. Thus the output y is binary. Such a function can be described mathematically using these equations:

$$Sum = \sum_{i=1}^{N} I_i W_i \tag{1}$$

$$y = f(Sum) \tag{2}$$

$W_1, W_2, W_3, \cdots, W_m$ are weight values normalized in the range of either $(0, 1)$ or $(-1, 1)$ and associated with each input line. Sum is the weighted sum, and T is threshold constant. The function is linear step function at threshold T.

The McCulloch-Pitts model of a neuron is simple yet has substantial computing potential. It also has a precise mathematical definition. However, this model is so simplistic that it only generates a binary output and also the weight and threshold values are fixed.

In the first stage, the linear combination of inputs is calculated. Each value of input array is associated with its weight value, which is normally between 0 and 1. Also, the summation function often takes an extra input value θ with weight value of 1 to represent threshold or bias of a neuron. The summation function will be then performed as

$$x = \sum_{i=1}^{N} A_i W_i + \Theta \tag{3}$$

The sum-of-product value is then passed into the second stage to perform the activation function which generates the output from the neuron. The activation function squashes the amplitude the output in the range of $(0, 1)$, or alternately $(-1, 1)$ [14]. The behavior of the activation function will describe the characteristics of an artificial neuron model.

One convenient form of such "semi-linear" function is the logistic sigmoid function, or in short, sigmoid function. As the input x tends to large positive value, the output value y approaches to 1. Similarly, the output gets close to 0 as x goes negative. However, the output value is neither close to 0 nor 1 near the threshold point. This function is expressed mathematically as follows:

$$y = \frac{1}{1 + exp(-x)} \tag{4}$$

Additionally, the sigmoid function describes the "closeness" to the threshold point by the slope. As x approaches to $-\infty$ or ∞, the slope is zero; the slope increases as x approaches to 0. This characteristic often plays an important role in learning of neural networks.

Perhaps, the most primary significance of a neural network is the ability to learn the incoming information and to improve the performance of processing information. The term learning refers to many concepts by various viewpoints, and it is difficult to agree on a precise definition of the term. In neural networks, we define learning as the following sequence of events[14]:

- Stimulation by an environment in which the network is embedded.
- Changes in free parameters of the network as the result of stimulation.
- Responses in a new way to the environment for improved performance.

A learning algorithm is a prescribed set of well-defined rules for learning of a neural network. There are many types of learning algorithms; the common goal of learning is the adjustment of connection weights.

There are two classes of learning: supervised and unsupervised learning. Supervised learning requires an external source of information in order to adjust the network. On the other hand, in unsupervised learning, there is no external agent that overlooks the process of learning. Instead, the network is adjusted through internal monitoring of performance. In this thesis, I deal with supervised learning.

Backpropagation neural networks employ one of the most popular neural network learning algorithms, the Backpropagation (BP) algorithm. Backpropagation is also the most suitable learning method for multilayer networks. Perhaps, the reason why the backpropagation made the major turning point is because the learning rule has a solid mathematical foundation and it is practical. The backpropagation processing unit should be in the form modified from a linear percepton so that the activation function is nonlinear and smoothed out at the threshold point. The suggested form of the activation function is the sigmoid function as mentioned previously. With sigmoid function, we can obtain not only output from the neuron but also information about how close we are to the threshold point using the slope of the sigmoid function. Mathematically, we can derive the slope from the equation as follows:

$$\frac{d}{dx}f(x) = \frac{exp(-x)}{(1+exp(-x))^{-2}} = \frac{1}{1+exp(-x)}\frac{exp(-x)}{1+exp(-x)} =$$
$$= \frac{1}{1+exp(-x)}\left[1-\frac{1}{1+exp(-x)}\right] = f(x)[1-f(x)] \tag{5}$$

3.0.1 Backpropagation Learning Algorithm

The backpropagation algorithm trains a given feed-forward multilayer neural network for a given set of input patterns with known classifications. When each entry of the sample set is presented to the network, the network examines its output response to the sample input pattern. The output response is then compared to the known and desired output and the error value is calculated. Based on the error, the connection weights are adjusted. The backpropagation algorithm is based on Widrow-Hoff delta learning rule in which the weight adjustment is done through mean square error of the output response to the sample input. The set of these sample patterns are repeatedly presented to the network until the error value is minimized.

Algorithm of backpropagation [15]:

1. Initialize connection weights into small random values
2. Present the pth sample input vector ofpattern $X_p = (X_{p1}, X_{p1}, \cdots, X_{pN_0})$ and the correspodning output target $T_p = (T_{p1}, T_{p2}, \cdots, T_{pN_M})$ to the network
3. Pass the input values to the first layer, layer 1. For every input node i in layer 0, perform: $Y_{0i} = X_{p_i}$
4. For every neuron i in every layer $j = 1, 2, \cdots, M$ from input to output layer find the output from the neuron

$$Y_{ji} = f\left(\sum_{k=1}^{N_{j-1}} Y_{(j-1)k}W_{jik}\right) \tag{6}$$

where

$$f(x) = \frac{1}{1+exp(-x)}$$

5. Obtain output values, for every output node i in layer M, perform:

$$O_{pi} = Y_{Mi}.$$

6. Calculate error value δ_{ji} for every neuron i in every layer in backward order $j = M, M-1, ..., 2, 1$, from output to input layer, followed by weight adjustments. For the output layer, the error value is:

$$\delta_{Mi} = Y_{Mi}(1 - Y_{Mi})(T_{pi} - Y_{Mi}), \tag{7}$$

and for hidden layers:

$$\delta_{ji} = Y_{ji}(1 - Y_{ji}) \sum_{k=1}^{N_{j+1}} \delta_{(j+1)k} W_{(j+1)ki} \tag{8}$$

The weight adjustment can be done for every connection from neuron k in layer $(i-1)$ to every neuron i in every layer i:

$$W^{+}_{jik} = W_{jik} + \beta\delta_{ji}Y_{ji} \tag{9}$$

where β represents weight adjustment factor normalized between 0 and 1.

The actions in steps 2 through 6 will be repeated for every training sample pattern p, and repeated for these sets until the root mean square (RMS) of output errors is minimized. Now attempt to derive the error and weight adjustment equations shown above. Let's begin with the Root Mean Square (RMS) of the errors in the output layer defined as:

$$E_p = \frac{1}{2} \sum_{j=1}^{N_M} (T_{pj} - O_{pj})^2, \tag{10}$$

for the pth sample pattern. In generalized delta rule [15] the error value δ_{ji} associated with the ith neuron in layer j is the rate of change in the RMS error E_p respect to the sum-of-product of the neuron:

$$\delta_{ji} = -\frac{\partial E_p}{\partial net_{ji}}, \tag{11}$$

where net_{ji} represents the sum-of-product value. With the chain rule, we can obtain the rate of change in the RMS error E_p in response to weight change:

$$\frac{\partial E_p}{\partial W_{jik}} = \frac{\partial E_p}{\partial net_{ji}} \frac{\partial net_{ji}}{\partial W_{jik}} = -\delta_{ji} \frac{\partial}{\partial W_{jik}} \left[Y_{(j-1)0} W_{(j-1)i0} + \cdots + Y_{(j-1)} k W_{(j-1)ik} + \cdots \right] =$$
$$= -\delta_{ji} \frac{\partial}{\partial W_{jik}} Y_{(j-1)ik} W_{(j-1)ik} = -\delta_{ji} Y_{(j-1)ik} \tag{12}$$

The weight change is proportional to value above [15].

$$\Delta W_{jik} = \beta\delta_{ji} Y_{(j-1)k} \tag{13}$$

where β is a constant. Because of that, weight change can be performed as:

$$W^{+}_{jik} = W_{jik} + \Delta W_{jik} \tag{14}$$

Now get back to the equation (14) to find an error value associate with the neuron. Again, using the chain rule, we get:

$$\delta_{ji} = -\frac{\partial E_p}{\partial Y_{ji}} \frac{\partial Y_{ji}}{\partial net_{ji}} \tag{15}$$

For output layer $j = M$ and $Y_{Mi} = O_{pi}$. Thus

$$\delta_{Mi} = -\frac{\partial E_p}{\partial O_{pi}} \frac{\partial Y_{Mi}}{\partial net_{Mi}} = -\frac{\partial}{\partial O_{pi}} \left[\frac{1}{2} \left[(T_{P1} - O_{p1})^2 + \cdots + (T_{Pi} - O_{pi})^2 + \cdots \right] \right] \frac{\partial}{\partial net_{M_i}} f(net_{Mi}) =$$
$$= -\frac{\partial}{\partial O_{pi}} \left[\frac{1}{2} \sum_{i=1}^{K} \left(T_{P1} - O_{p1} \right)^2 \right] f'(net_{Mi}) \tag{16}$$

Using equation

$$\delta_{Mi} = (T_{pi} - O_{pi}) \left[f(net_{mi}) \left[1 - f(net_{mi}) \right] \right] = (T_{pi} - O_{pi})(O_{pi})(1 - O_{pi}). \tag{17}$$

For error values associated with the hidden layer neurons, we cannot use target values. For this reason, the part $\frac{\partial E_p}{\partial Y_{ji}}$ in equation needs to be found using a different approach. Next the chain rule applied to the sum-of-product values of neurons in the layer $j + 1$ (front layer)

$$\frac{\partial E_p}{\partial Y_{ji}} = \frac{\partial E_p}{\partial net_{(j+1)1}} \frac{\partial net_{(j+1)1}}{\partial Y_{ji}} + \frac{\partial E_p}{\partial net_{(j+1)2}} \frac{\partial net_{(j+1)2}}{\partial Y_{ji}} + \cdots =$$
$$= \sum_{a=1}^{N_{j+1}} \left[\frac{\partial E_p}{\partial net_{(j+1)a}} \frac{\partial net_{(j+1)a}}{\partial Y_{ji}} \right] =$$
$$= \sum_{a=1}^{N_{j+1}} \left[-\delta_{(j+1)a} \frac{\partial}{\partial Y_{ji}} (W_{(j+1)a0} Y_{j0} + \cdots + W_{(j+1)ai} Y_{ji} + \cdots) \right]$$
$$= \sum_{a=1}^{N_{j+1}} \left[-\delta_{(j+1)a} \frac{\partial}{\partial Y_{ji}} Y_{ji} W_{(j+1)ai} \right] = \sum_{a=1}^{N_{j+1}} \left[-\delta_{(j+1)a} W_{(j+1)ai} \right] \tag{18}$$

Finally, combined with $\frac{\partial Y_{ji}}{\partial net_{ji}}$ the final form of error value associated with the ith neuron in layer j:

$$\delta_{ji} = \sum_{a=1}^{N_{j+1}} \left[-\delta_{(j+1)a} W_{(j+1)ai} \right] \frac{\partial Y_{ji}}{\partial net_{ji}} =$$
$$= Y_{ji}(1 - Y_{ji}) \sum_{a=1}^{N_{j+1}} \left[-\delta_{(j+1)a} W_{(j+1)ai} \right] \tag{19}$$

The backpropagation algorithm, as just described, employs gradient descent by following the slope of RMS error value E_p downward along with the change in all the weight values. The weight values are constantly adjusted until the value of E_p is no longer decreasing. Since the RMS error value is very complex function with many parameter values of weights, it is possible that the backpropagation network may converge into a local minima instead of the desired global minimum.

A trained backpropagation network is able to detect and classify an input pattern that has not been seen during learning. This feature is called generalization, borrowed from the psychology terms. Neural networks are known to be good at classifying noisy input patterns, but not at classifying a pattern that is intermediate between two solid patterns from the training samples. In other words, neural networks are good at interpolation but not extrapolation. Also, there may exist overfitted input data, the unseen input pattern such that it can be classified into one of the trained output response undesirably. Suggested solution includes modification of network architecture and more adequate training samples.

4 Convolutional neural networks

Multilayer Neural Networks trained with the backpropagation algorithm constitute the best example of a successful Gradient-Based Learning technique. Convolutional Neural Networks that are specifically designed to deal with the variability of 2D shapes are shown to outperform all other techniques. ConvNet architectures make the explicit assumption that the inputs are images which allows us to encode certain properties into the architecture. These then make the forward function more efficient to implement and vastly reduce the amount of parameters in the network. Convolutional Neural Networks are very similar to ordinary Neural Networks they are made up of neurons that have learnable weights and biases. Each neuron receives some inputs, performs a dot product and optionally follows it with a non-linearity. The whole network still expresses a single differentiable score function: from the raw image pixels on one end to class scores at the other. And they still have a loss function (e.g. SVM/Softmax) on the last (fully-connected) layer and all the tips/tricks we developed for learning regular Neural Networks still apply.

4.1 Convolution

The Conv layer is the core building block that does most of the convolutional heavy lifting. Convolution is a mathematical operation on two functions to produce a third function that is typically viewed as a modified version of one of the original functions, giving the integral of the pointwise multiplication of the two functions. Basically every small spatially (along width and height) filter is taken and slide this filter over the whole image. For example a typical filter of a ConvNet might have size 5x5x3(i.e. 5 pixels width and height, and 3 because images have depth 3, the color channels). Pixel values of the image are then multiplied with the pixel values in the filter. Intuitively, the network will learn filters that activate when they see some type of visual feature. Thanks to Stochastic Gradient Descent there is no need to specify them, because they can be learned themselves. Important features of classes can be found at any place because they are initialized randomly and are location-invariant.

4.1.1 Local Connectivity

When dealing with high-dimensional inputs such as images, it is impractical to connect neurons to all neurons in the previous volume. Instead each neuron connects only to a local region of the input volume.

4.1.2 Spatial arrangement

There are three hyperparameters control the size of the output volume: the depth, stride and zero-padding.

- **depth** of the output volume corresponds to the number of filters, each of them should be able to determine different features in image
- **stride** is the number of pixels with which thee filters move one step at a time
- **zero-padding** is a hyperparameter that allows to control the spatial size of the output volumes by padding the input volume with zeros around the border

It is able to compute the spatial size of the output volume as a function of the input volume size **W**, the receptive field size of the Conv Layer neurons (**F**), the stride with which they are applied (**S**), and the amount of zero padding used (**P**) on the border. The correct formula for calculating how many neurons will fit is given by $\frac{W-F+2P}{S+1}$.

4.1.3 Parameter Sharing

Parameter sharing scheme is used in Convolutional Layers to control the number of parameters. The number of parameters at the first ConvNet layer can be reduced by assumption that if the feature is useful to compute at some spatial position (x, y), then it should also be useful to compute at a different position (x_2, y_2). In other words, denoting a single 2-dimensional slice of depth as a depth slice

4.2 Pooling

It is common to periodically insert a Pooling layer in-between successive Conv layers in a ConvNet architecture in order to control overfitting. It is a way to take large images and shrink them down while preserving the most important information in them. Thanks to Max Pooling filters can explore bigger parts of the image, additionally because of the loss in pixels, this is a common practice to increase the number of filters. The Pooling Layer operates independently on every depth slice of the input and resizes it spatially, using the MAX operation.

4.3 ReLU

The ReLU function is $f(x) = max(0, x)$. Math behind this function is very simple wherever a negative number occurs swap it out for a 0. Purpose of it is usually to speed up training, because the computations are really simple.

4.4 Fully connected layers

Neurons in a fully connected layer take the high-level filtered images and translate them into votes. The output from the convolutional layers represents high-level features in the data. Of course that output can be easy connected to the output layer, but this is a common practice to add a fully connected layer which is a cheap way of learning non-linear combinations of these features and making the model end-to-end trainable. In this ConvNets transform the original image layer by layer from the original pixel values to the final class scores.

4.5 Loss function and training accuracy

The loss function is used to guide the training process of a neural network. For classification problem, mean squared error and cross entropy loss are widely used. Cross entropy loss is a loss function that commonly used in classification or regression problems. Cross entropy is more advanced than mean squared error. The introduction of cross entropy comes from maximum likelihood estimation in statistics. The formula of cross entropy is

$$H_{y'}(y) = -\sum_{i} y_i' log(y_i).$$

Where y_i' is the ground truth label of ith training instance and y_i is the prediction result of classifier. Cross entropy is probably the most important loss function in deep learning, it can be seen almost everywhere, but usage of cross entropy can be very different.

5 Architecture of the proposed CNN

Most of the architectures of Convolutional Neural Networks are made up of only three layer types: CONV, POOL and FC. Most ConvNet architecture follows the pattern:

$$INPUT -> [[CONV -> RELU] * n -> POOL] * M -> [FC -> RELU] * k -> FC$$

where * indicates repetition and the *POOL*? indicates an optional pooling layer. Assumption:

$$N >= 0, M >= 0, K >= 0$$

Model 1			Model 2		
layer type	kernel size/ pool size	filters/ strides	layer type	kernel size/ pool size	filters/ strides
Convolutional	(5,5)	16	Convolutional	(5,5)	32
MaxPooling	(2,2)	(2,2)	MaxPooling	(2,2)	(2,2)
Convolutional	(5,5)	32	Convolutional	(5,5)	64
MaxPooling	(2,2)	(2,2)	MaxPooling	(2,2)	(2,2)
Convolutional	(5,5)	64	Convolutional	(5,5)	128
MaxPooling	(2,2)	(2,2)	MaxPooling	(2,2)	(2,2)

Model 3			Model 4		
layer type	kernel size/ pool size	filters/ strides	layer type	kernel size/ pool size	filters/ strides
Convolutional	(5,5)	16	Convolutional	(5,5)	16
MaxPooling	(2,2)	(2,2)	MaxPooling	(2,2)	(2,2)
Convolutional	(5,5)	32	Convolutional	(5,5)	32
MaxPooling	(2,2)	(2,2)	MaxPooling	(2,2)	(2,2)
Convolutional	(5,5)	64	Convolutional	(5,5)	64
MaxPooling	(2,2)	(2,2)	MaxPooling	(2,2)	(2,2)
Convolutional	(5,5)	128			
MaxPooling	(2,2)	(2,2)			

Model 5		
layer type	kernel size/ pool size	filters/ strides
Convolutional	(5,5)	32
MaxPooling	(2,2)	(2,2)
Convolutional	(5,5)	64
MaxPooling	(2,2)	(2,2)

There are no hard guidelines how perfect architecture should look like for given problem. The basic principle which I follow is to keep the feature space wide and shallow in the initial stages of the network and then make it narrower and deeper towards the end. I am using small filters because I expect small and local feature in picture uploaded by user. I start by using small filters in order to collect as much local information as possible. Then I increase the filter width to reduce the generated feature space width to represent more global information. I don't want to shrink my image too much so the stride (2,2) is used. I keep adding new layers until my model start overfit to the data (model 3). I was trying to use dropout for this model but it hasn't improved results much. I follow advice of classic layers like Conv-Pool Conv-Pool and keeping trend in the number of filters 16-32-64 or 32-64-128.

For model 1 less preprocessing is done. I perform in it only shear ranging, zooming and horizontal flipping. For other model preprocessing described in section 6. The general algorithm of training looks as follows:

- pass an input image to the first convolutional layer. Apply filters in the convolutional layer and extract relevant features from the input image to pass them further
- pass convoluted output as an activation map
- if the image needs to be retained use zero padding
- add pooling layer
- the output layer in CNN is fully connected layer with the number of classes desired by the network
- generate output through the output layer and compare to the output layer for error calculation. Use loss function to compute MSE, then calculate gradient descent
- perform backpropagation

6 Dataset and preprocessing

The first step in process is collecting data for the model to be trained. Having a high-quality data set is essential for obtaining good results from datasets. There are some free great free and paid standard datasets. I used ImageNet unfortunately it contains not only fruits but also noisy background which often leads to misleading classifier. The dataset is completed with dataset from Kaggle [5] in order to collect images of Vietnamese fruits. I preprocess manually every picture from dataset by cropping it to only one fruit. What is more I use Keras library for data augmentation for example for shearing, zooming, flipping both horizontal and vertical, rotating but I don't save them, I only store them in memory during program execution, and as soon as program finishes garbage collector delete them all. Data augmented images are not included in the below table. The dataset was split in 2 parts: training set and testing set.

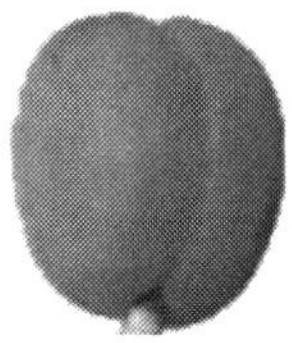

Figure 1. Apricot

Figure 2. Pineapple

Figure 3. Dragon fruit

Figure 4. Apricot

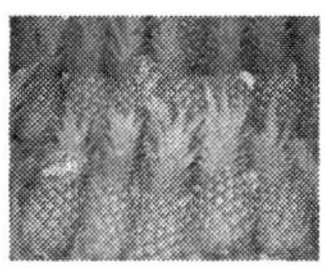

Figure 5. Pineapple

Figure 6. Dragon fruit

Fruit	Number of training images	Number of test images
Apricot	560	139
Avocado	606	159
Carambola	528	132
Clementine	555	138
Cocos	563	141
Dates	468	182
Dragon Fruit	331	154
Durian	820	205
Guava	567	141
Kaki	735	182
Kiwi	617	154
Longan	281	70
Melon	659	164
Papaya	589	129
Pineapple	727	182
Quince	575	143
Rambutan	167	40
Tamarillo	150	37

Table 1. Dataset

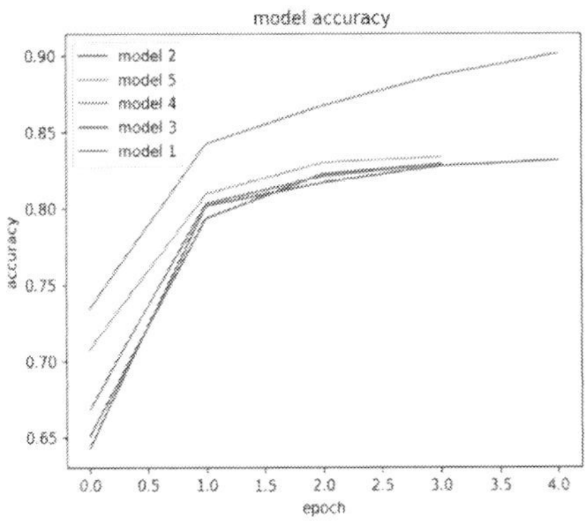

Figure 7. Model accuracy for dataset

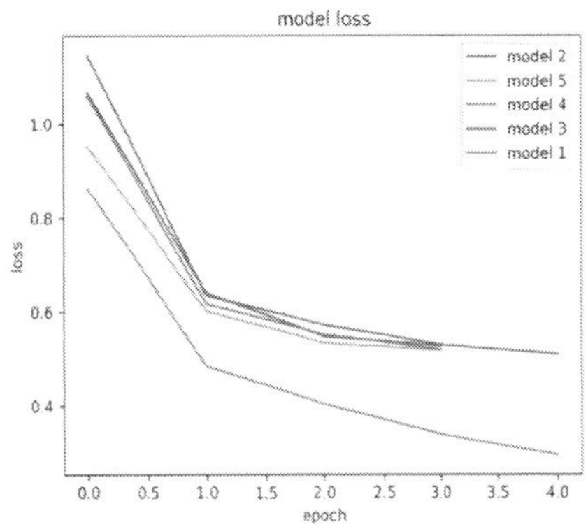

Figure 8. Model accuracy for dataset

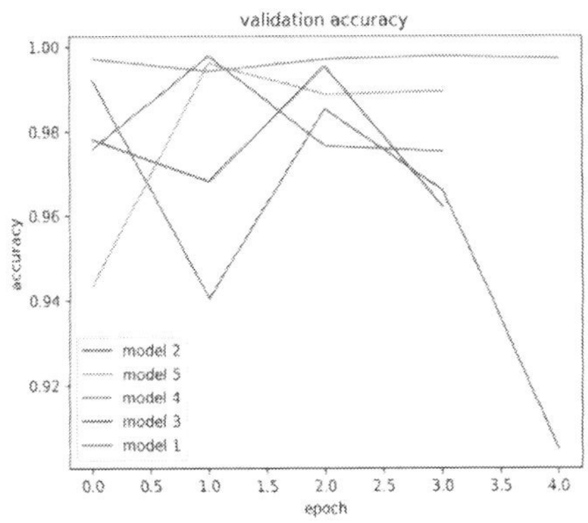

Figure 9. Validation accuracy for dataset

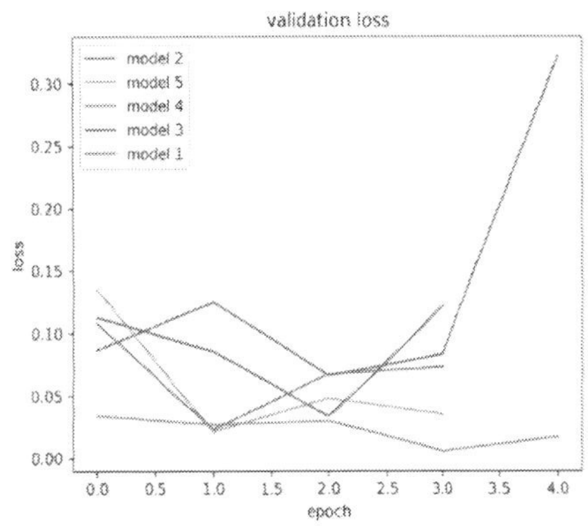

Figure 10. Validation loss for dataset

It is said that the lower the cost, the better the model. On the other hand the model shouldn't overfit to the training data (when model memorizes only the training examples and is not able to assign correct labels for the test set) which can be easily identify by when validation accuracy is much lower than training accuracy.

After each epoch the current model and validation accuracy and loss are saved. Results can be seen in Figures 7,8,9,10. It is clear that model 4 is the most reliable one with the best parameters. The comparison with other models shows that even though number of layers is similar more processing gives far better results than first model. On the other hand model perform best using pictures without noisy background and user of web application is encouraged to to take pictures in that manner.

7 Fine tuning model

When it comes to an online training we want the best quality models using as little time and computational power as possible. Instead of computing gradient on the whole dataset, I estimate it on a very small subset , with a batch size that allows me to stream the training samples. Online model consists in estimating the parameters of the model iteratively by means of stochastic gradient descent (SGD). Batch gradient descent performs redundant computations for large datasets as it recomputes gradients for similar examples before each parameter update. SGD does away with this redundancy by performing one update at a time. Is therefore much faster and can also be used to learn online. SGD enables it to jump to new and potentially better local minimum. It has been shown that with a slowly decreasing learning rate SGD almost certainly converges to a local or the global minimum for non-convex and convex optimization.

I fed my CNN with images of a new image uploaded by the user. After getting response from webserver user can help with updating model by answering question "Do you think that the answer is correct?" with possible answers "Yes" or "No".

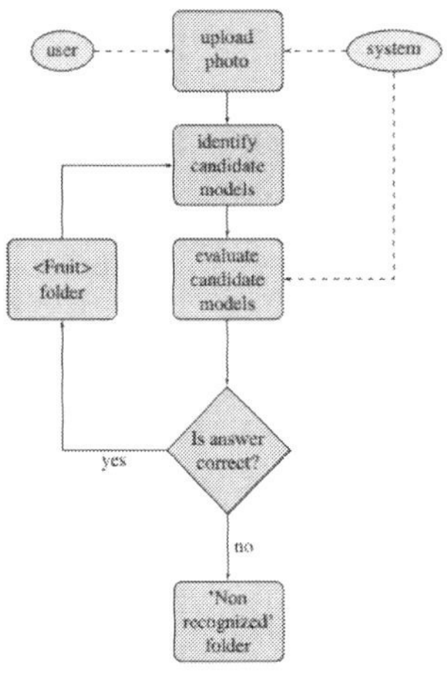

Then server administrator has to check Non recognized fruits in order to decide what were pictures unrecognized by the system. For fine tuning I change ADAM optimizer to a stochastic gradient descent with some specified batch size. Running SGD asynchronously is faster, additionally it is able to parallelize SGD on one machine without the need for a large computing cluster. Tensorflow is architecture which is used for optimizing parallelized and distributed SGD. The distributed version relies on a computation graph that is split into subgraph for every device while communication takes place using Send/Receive node pairs. Generally in order to avoid providing training examples in a meaningful order to my model as this may bias the optimization algorithm. It is often a good idea to shuffle the training data after every epoch. I also monitor error on a validation set during training and do not update model if my validation error does not improve enough. Model is fine-tuned every Sunday in the night. The new weights are assigned and then model is updated.

8 Keras, the Python Deep Learning library

Keras [7] is a high-level neural networks API written in Python and capable of running on top of Tensorflow[8] which I use to to train models. This library supports both convolutional networks and recurrent networks as well as combinations of the two and runs seamlessly on CPU and GPU. A single model is presented as a sequence object or a graph of standalone. It is possible to combine neural layers, cost functions, optimizers, initialization schemes, activation functions and regularization schemes.

9 Behind web application

In order to deploy my web application for users I used PythonAnywhere which is online integrated development environment (IDE) based on the Python programming language and is recommended as a way of hosting machine learning-based web applications. Web applications hosted by the service can be written using and WSGI-based application framework. The Web Server Gateway Interface is a simple calling convention for web servers to forward request to web framework written in Python.

I used Flask web framework to create web application which takes image as an input and returns three predicted pictures with labels and levels of prediction. Flask is a web framework that provides you with tools, libraries and technologies that allow you to build a web application in python. Flask is a part of the categories of the micro-framework. Micro-framework are normally framework with little to no dependencies to external libraries. This has pros and cons. Pros would be that the framework is light, there are little dependency to update and watch for security bugs, cons is that some time you will have to do more work by yourself or increase yourself the list of dependencies by adding plugins.

10 Future work

The system will be fed through **the YOLO(you only look once)** real time object detection system which produces a number of detections depending on the number of fruits in the image. The YOLO network is going to be modified to only find one class, fruits detected by model. The next thing is to work more on a project which can be able to detect fruit with noisy background. It is vital to spend more time on working on a model of CNN which will be able to predict better fruit when the picture with noisy background was taken.

11 Conclusions

I have created a web application which is going to help foreigners in Vietnam. It can be found highly useful especially in situations when seller doesn't know English name of fruit which is very usual in real situations during trip to Vietnam. During my Student Exchange Programme in Vietnam I faced many situations when I used my application for fruit recognition. There is also a promo video of this web application which shows how simply everyone can get to know what is this fruit.

Acknowledgements

I thank Dr Michał Przewoźniczek, from Wroclaw University of Science and Technology for comments that greatly improved the research paper.

References

[1] I. Sa, Z.Ge, F. Dayoub, B. Upcroft , T. Perez & C. McCool, DeepFruits: A Fruit Detection System Using Deep Neural Networks, **Vol. 16(8)**, pp. 1222-, (2016)

[2] P. Ninawe, S. Pandey, A Completion on Fruit Recognition System Using K-Nearest Neighbors Algorithm, International Journal of Advanced Research in Computer Engineering & Technology (IJARCET) **Vol. 3 (7)** (2014)

[3] Horea Muresan, Mihai Oltean, Fruit recognition from images using deep learning, Acta Univ. Sapientiae, Informatica **Vol. 10**, Issue 1, pp. 26-42, 2018.

[4] Fruit 360 Dataset on GitHub. https://github.com/Horea94/Fruit-Images-Dataset

[5] Fruit 360 Dataset on Kaggle. https://www.kaggle.com/moltean/fruits

[6] http://www.image-net.org/

[7] https://keras.io/

[8] https://www.tensorflow.org/

[9] https://en.wikipedia.org/wiki/Rectifier_(neural_networks)

[10] https://cs231n.github.io/convolutional-networks/

[11] Hassoun, Mohamad H. *Fundamentals of Artificial Neural Networks.* (The MIT Press, Bambridge, MA, 1995)

[12] Zurada, Jacek M. *Introduction to Artificial Neural System.* (West Publishing Company, St. Paul, MN, 1992)

[13] Bose, N. K. and Liang, P. *Neural Network Fundamentals with Graphs, Algorithms, and Applications* (McGraw-Hill, New York, NY, 1996)

[14] Haykin, Simon. *Neural Networks: A Comprehensive Foundation, second edition* (Prentice-Hall, Upper Saddle River, NJ, 1999)

[15] Beale, R. and Jackson, T. Hilger, *Neural Computing: An Introduction* (Philadelphia, PA, 1991)

Optimal Lipschitz extensions

Thanh Viet Phan[1,*]

[1]Applied Analysis Research Group, Faculty of Mathematics and Statistics, Ton Duc Thang University, Ho Chi Minh City, Vietnam

Abstract. The classical Lipschitz extension problem in concerned for conditions on a pair of metric spaces (X, d_X) and (Y, d_Y) such that for all $\Omega \subset X$ and for all Lipschitz function $f : \Omega \to Y$, then there is a function $g : X \to Y$ that extends f and has the same Lipschitz constant as f. In this paper we discuss some results and open questions related to that issue.

1 Introduction

In this paper, we discuss the Lipschitz extension problems.

Let (X, d_X) and (Y, d_Y) be metric spaces and let Ω be a subset of X. Let $f : \Omega \to Y$ be a Lipschitz function. We denote

$$\mathrm{Lip}(f, \Omega) := \sup_{\substack{x,y\in\Omega \\ x\neq y}} \frac{d_Y(f(x), f(y))}{d_X(x, y)}$$

to be the Lipschitz constant of f on Ω.

Question 1.1 *Whether there exists a function $g : X \to Y$ that extends f and has the same Lipschitz constant as f, i.e.* $\mathrm{Lip}(f, \Omega) = \mathrm{Lip}(g, X)$ *?*

The pair (X, Y) satisfies the property in Question 1.1 for all $\Omega \subset X$ and for all Lipschitz function $f : \Omega \to Y$ is said to have the *isometric extension property*. In Section 2 we present some famous examples for a pair of metric spaces (X, Y) to have this property.

We consider a pair of metric space (X, Y) that has isometric extension property. Let Ω be a subset of X and let $f : \Omega \to Y$ be a Lipschitz function. If g extends f and $\mathrm{Lip}(g, X) = \mathrm{Lip}(f, \Omega)$, then we say that g is a *minimal Lipschitz extension* (MLE) of f. Of course, the MLE may not be unique, and it is possible to search for extension with additional properties. In a series of papers in the 1960's [2–4], Aronsson proposed the notion of an *absolutely minimal Lipschitz extension* (AMLE):

Definition 1.2 *A function u defined on X is called an AMLE of f if u is a MLE of f and u satisfies*

$$\mathrm{Lip}(u, V) = \mathrm{Lip}(u, \partial V), \quad \textit{for any open } V \subset\subset X\backslash\Omega, \tag{1}$$

where ∂V denotes the boundary of V and $V \subset\subset X$ means that $\overline{V} \subset X$.

*e-mail: phanthanhviet@tdtu.edu.vn

Question 1.3 *Whether there exists an AMLE $u : X \to Y$ of f ? If this extension exists, is it unique ?*

After the works of Aronsson, there has been many studies devoted to the study of AMLEs. Although this problem was originated in real and functional analysis, it has been found relevant to many areas in applied mathematics (see [16, 23] for instance). In Section 3 we discuss the existence and uniqueness of AMLE, as well as problems related to them.

Now, given a function $f : \Omega \subset \mathbb{R}^n \to \mathbb{R}$. If f is differentiable in some sense, can we extend it to a function on the whole $\mathbb{R}^n$ with the same differentiability? This question goes back to Whitney [29] in 1934. The problem is particularly interesting when f is not smooth, and one looks for an extension satisfying certain constraints. Motivated by Taylor's expansion for smooth functions,

$$f(y) = f(x) + \langle \nabla f(x), y - x \rangle + o(\|y - x\|),$$

Whitney[29] and Glaeser[15] have introduced the notion of *1-fields* for non-smooth functions: for $f : \Omega \to \mathbb{R}^n$ and $Df : \Omega \to \mathbb{R}^n$, we define the 1-field $F : \Omega \to \mathcal{P}^1(\mathbb{R}^n, \mathbb{R})$, where

$$\mathcal{P}^1(\mathbb{R}^n, \mathbb{R}) := \{P : \mathbb{R}^n \to \mathbb{R} \mid \exists p \in \mathbb{R}, v \in \mathbb{R}^n : P(a) = p + \langle v, a \rangle, \forall a \in \mathbb{R}^n\}$$

(the set of first degree polynomials), by the identity

$$F(x)(y) := f(x) + \langle Df(x), y - x \rangle, \quad \forall x \in \Omega, y \in \mathbb{R}^n.$$

The 1-field F can be interpreted as a smooth approximation for f on Ω, and hence the extension problem for f can be related to the extension problem for F. Note that F is well-defined even if f and Df has no relation as a priori.

In 2009, Le Gruyer [20] introduced the Lipschitz constant of 1-field F:

$$\Gamma^1(F; \Omega) := 2 \sup_{\substack{x,y \in \Omega \\ x \neq y}} \sup_{a \in \mathbb{R}^n} \frac{|F(x)(a) - F(y)(a)|}{\|x - a\|^2 + \|y - a\|^2}, \tag{2}$$

Question 1.4 *Does the Lipschitz constant Γ^1 of 1-field play a role in the extension problem similarly to the Lipschitz constant* Lip *of continuous function ?*

Question 1.4 and related issues to the 1-field extension are discussed in Section 4.

2 Kirszbraun theorem and McShane-Whitney extremal extensions

It is rare for a pair of metric spaces (X, Y) to have the isometric extension property. In this section we present some famous examples for a pair of metric spaces (X, Y) to have the isometric extension property.

2.1 Kirszbraun theorem

Kirszbraun in 1934 [18] proved that if X and Y are Euclidean spaces both equipped with the Euclidean norm then (X, Y) have the isometric extension property. Later Frederick Valentine in 1943 [28] generalized it to pairs of Hilbert spaces of arbitrary dimension. This theorem is called Kirszbraun theorem, sometimes it is also called Kirszbraun-Valentine theorem.

Theorem 2.1 *(Kirszbraun theorem) Let H_1 and H_2 be Hilbert spaces. If Ω is a subset of H_1, and $f : \Omega \to H_2$ is a Lipschitz function, then there is a function $g : H_1 \to H_2$ satisfying*

$$g = f \text{ on } \Omega, \quad \text{and } \operatorname{Lip}(g, H_1) = \operatorname{Lip}(f, \Omega).$$

The proof of the theorem uses geometric features of Hilbert spaces, the corresponding statement for Banach spaces is not true in general, not even for finite-dimensional Banach spaces. We can construct counterexamples where the domain is a subset of $\mathbb{R}^n$ with the maximum norm and the map is valued in $\mathbb{R}^m$ with Euclidean norm. A simple counterexample is the following:

$$X = \mathbb{R}^2 \text{ with } \|x\|_1 = \sup\{|x_1|, |x_2|\},$$
$$Y = \mathbb{R}^2 \text{ with } \|x\|_2 = (x_1^2 + x_2^2)^{1/2},$$
$$\Omega = \{(1,-1), (-1,1), (1,1)\} \subset X, \quad f : \Omega \to Y,$$
$$f(1,-1) = (1,0), \quad f(-1,1) = (-1,0), \quad f(1,1) = (0, \sqrt{3}).$$

Then

$$\|x - y\|_1 = \|f(x) - f(y)\|_2 = 2, \quad \forall x, y \in \Omega.$$

Thus $\operatorname{Lip}(f, \Omega) = 1$.

On the other hand, we have $\|x\| = 1$ for all $x \in \Omega$, but there exists no $\xi \in \mathbb{R}^2$ such that

$$\|f(x) - \xi\| \leq 1, \quad \forall x \in \Omega.$$

Thus f has no extension u to $\Omega \cup \{(0,0)\}$ with $\operatorname{Lip}(u, \Omega \cup \{(0,0)\}) = 1$.

More generally, the theorem fails for $\mathbb{R}^m$ equipped with any ℓ_p norm ($p \neq 2$) (see Schwartz 1969 [26, p. 20]).

It is easy to see that the Kirszbraun theorem is equivalent to the following theorem.

Theorem 2.2 *[8, Proposition 1.13] Let H_1 and H_2 be Hilbert spaces, $\{x_\beta\}_{\beta\in I}$ a subset of H_1, $\{y_\beta\}_{\beta\in I}$ a subset of H_2, and $\{r_\beta\}_{\beta\in I}$ a collection of nonnegative real numbers. If*

$$\|y_\beta - y_\gamma\| \leq \|x_\beta - x_\gamma\| \tag{3}$$

for all $\beta, \gamma \in I$, then $\bigcap_{\beta\in I} B(y_\beta, r_\beta) \neq \emptyset$ whenever $\bigcap_{\beta\in I} B(x_\beta, r_\beta) \neq \emptyset$.

If $H_1 = H_2$ then the contraction hypothesis (3) in Theorem 2.2 implies that we can rearranging the ball $B(x_\beta, r_\beta)$ in a way that the centers of the balls are closer to each other. In addition, if $\dim H_1 = \dim H_2 < +\infty$ and the index I is finite, then it is natural to conjecture that the volume of the intersection should increase as we push the spheres together. In fact, we have

Problem 2.3 *[19, 24, Kneser-Poulsen conjecture] Let $\{r_1, ..., r_N\}$ be a collection of nonnegative numbers and $\{x_1, ..., x_N\}$ and $\{y_1, ..., y_N\}$ two subsets of $\mathbb{R}^n$. If $\{y_1, ..., y_N\}$ is a contraction of $\{x_1, ..., x_N\}$, i.e*

$$\|y_i - y_j\| \leq \|x_i - x_j\|,$$

for all $1 \leq i, j \leq N$. Then

$$Vol\left(\bigcap_{i=1}^{N} B(y_i, r_i)\right) \geq Vol\left(\bigcap_{i=1}^{N} B(x_i, r_i)\right)$$

The above was conjectured independently by Kneser [19] in 1955 and Poulsen [24] in 1954. It is quite surprising that the conjecture remains unproven for a long time. The conjecture is solved only when $n = 2$ by Bezdek and Connelly [9] in 2002 and it is still open when $n \geq 3$.

2.2 McShane-Whitney extremal extensions

McShane [22] and Whitney [29] in 1934 proved that if $Y = \mathbb{R}$ and for any arbitrary metric space X then (X, Y) have the isometric extension property. More precisely, they prove that for any metric space X and any Ω subset of X, every Lipschitz function $f : \Omega \to \mathbb{R}$ has two MLEs which can be written explicitly :

$$m^+(\xi) = \inf_{x\in\Omega}\{f(x) + Lip(f,\Omega)d_X(x,\xi)\} \text{ for } \xi \in X, \tag{4}$$

$$m^-(\xi) = \sup_{x\in\Omega}\{f(x) - Lip(f,\Omega)d_X(x,\xi)\} \text{ for } \xi \in X. \tag{5}$$

Moreover, $m^\pm$ are extremal: the first is maximal and the second is minimal, that is

$$m^-(x) \leq g(x) \leq m^+(x), \forall x \in X,$$

for any g other MLE of f.

Clearly MLEs are unique if and only if $m^+ = m^-$ on X. This rarely happens.

3 Absolutely minimal Lipschitz extension

We consider the case $X \subset \mathbb{R}^n$ and $Y = \mathbb{R}$ both equipped with the Euclidean norm. Let $\Omega = \partial X$ and $f : \Omega \to \mathbb{R}$. From McShane - Whitney formulas (4) and (5), we have two extremal MLEs m^+ and m^- of f. Thus, unless $m^+ \equiv m^-$, we have no uniqueness of MLE of f. In this section we discuss about the question of the existence and uniqueness of AMLE of f.

3.1 The existence and uniqueness of AMLE

We rewrite the original problem in (1) by calculus of variations problems in the sup-norm:

$$||\nabla u||_{L^\infty(V)} \leq ||\nabla v||_{L^\infty(V)}, \tag{6}$$

for all $V \subset X$, and for all "sufficiently regular" v satisfying $v = u$ on ∂V.

It is not clear at all that (6) is equivalent to (1). If V is convex then $\text{Lip}(v, V) = ||\nabla v||_{L^\infty(V)}$ for any locally Lipschitz continuous function v. But if V is not convex, then $\text{Lip}(v, V)$ may be infinite when $||\nabla v||_{L^\infty(V)} = 1$. The equivalence, which may not be generally appreciated, is established in [7].

A central question has been to understand minimization problems involving (6) and related functionals, and especially to develop appropriate PDE methods. The most popular line of research has arisen from the idea of interpreting the AMLE as a formal limit of u_p, as $p \to \infty$, where u_p is the minimizing of the functional

$$I_p[v] := \int_X ||\nabla v||^p dx, \tag{7}$$

where $X \subset \mathbb{R}^n$ is open, $v \in W^{1,p}(X;\mathbb{R})$ with boundary condition $v = f$ on ∂X, and $\nabla v = (u_{x_1}, ..., u_{x_n})$ is the gradient of v.

The problem (7) leads as usual to the Euler-Lagrange equation $\Delta_p u = 0$, but it was unclear what is the correct "Euler-Lagrange" equation for the problem (6). In 1967, Aronsson [4] discovered the "Euler-Lagrange" equation for the problem (6) by approximation procedure. Let us explain the idea at least formally. He considered for finite p the nonlinear p–Laplacian operator:

$$\Delta_p u = div(||\nabla u||^{p-2}\nabla u) = (p-2)||\nabla u||^{p-4}\Delta_\infty u + ||\nabla u||^{p-2}\Delta u,$$

where

$$\Delta_\infty u := \sum_{j,k=1}^{n} u_{x_j} u_{x_k} u_{x_j x_k}.$$

Dividing the equation $\Delta_p u = 0$ by $(p-2)\|\nabla u\|^{p-4}$ leads to

$$\Delta_\infty u + \frac{1}{p-2}\|\nabla u\|^2 \Delta u = 0,$$

Let p tend to infinity. This leads to the equation

$$\Delta_\infty u = 0, \tag{8}$$

where

$$\Delta_\infty u := \sum_{j,k=1}^{n} u_{x_j} u_{x_k} u_{x_j x_k}.$$

This nonlinear equation is a highly degenerate elliptic equation. The above "Euler-Lagrange" equation is nowadays known as the *infinity Laplace equation*.

In 1967, Aronsson [4] proved that if X is the closure of the bounded domain in $\mathbb{R}^n$ and f is a Lipschitz function given on ∂X then a C^2 function $u : X \to \mathbb{R}$ is an AMLE of f if and only if it is a solution of the infinity Laplace equation (8) with boundary condition $u = f$ on ∂X.

However, AMLEs are not C^2 in general, the classical solutions of the eikonal equation $|Du| = constant$ is an example of AMLEs which are not C^2. The existence and uniqueness of AMLEs was unknown in general, and it was also unclear what is correct notion for non-smooth solution of infinity Laplace equation (8). An important breakthrough in this direction was made by Jensen by using the *viscosity solution* concept:

Definition 3.1 *(i) An upper semi-continuous function $u : \overline{X} \to \mathbb{R}$ is a sub-solution of (8) if $\Delta_\infty \phi(x) \geq 0$, for every $(x, \phi) \in X \times C^2(X)$ such that $(u-\phi)(x) \geq (u-\phi)(y)$ for all $y \in X$.*

(ii) An lower semi-continuous function $u : \overline{X} \to \mathbb{R}$ is a super-solution of (8) if $\Delta_\infty \phi(x) \leq 0$, for every $(x, \phi) \in X \times C^2(\Omega)$ such that $(u-\phi)(x) \leq (u-\phi)(y)$ for all $y \in X$.

(iii) A continuous function $u : \overline{X} \to \mathbb{R}$ is a viscosity solution of (8) if it is both a sub-solution and super-solution.

This concept introduced by Crandall and Lions in a famous paper in 1983 [12] is a generalization of the classical concept to treat the problem for non-smooth function. The modern theory of viscosity solutions was developed by Crandall, Evans, Jensen, Ishii, Lions, and others. First, it was designed for first order equations. Later, it was extended to second order equations. The solutions must obey a Comparison Principle.

In 1993, Jensen [16] proved the existence and uniqueness of viscosity solution of infinity Laplace equation with boundary condition $u = f$ on ∂X . Jensen showed as well that this solution is also identified by Aronsson: for every open bounded subset V of X if u is the viscosity solution of infinity Laplace equation then

$$\|\nabla u\|_{L^\infty(V)} \leq \|\nabla v\|_{L^\infty(V)}, \tag{9}$$

for each $v \in C(\overline{V})$ and $u = v$ on ∂V.

After the foundational papers of Aronsson, the uniqueness of AMLE remained as the most important open problem. Jensen's work generated considerable interest in the long developments in the existence and uniqueness theory of AMLEs.

3.2 Comparison with cones and generalizing the AMLE in metric spaces

The existence and uniqueness theory of AMLEs was settled in Jensen's influential work. Jensen's proof was rather indirect. Moreover, the arguments relied on the theory of variational problems and the viscosity solution machinery developed for second-order elliptic equations. A more direct uniqueness proof was found by Armstrong and Smart in 2010 [1] by using the *comparison with cones* concept:

Definition 3.2 *(i) A function $u \in C(X)$ is said to enjoys comparison with cones from above in X if for every bounded open subset V of X and every $x_0 \in \mathbb{R}^n$, $a, b \in \mathbb{R}$ for which*

$$u(x) \leq C(x) = a + b\|x - x_0\|$$

holds on $\partial(V\backslash\{x_0\})$, we have $u \leq C$ in V as well.

(ii) A function $u \in C(X)$ is said to enjoys comparison with cones from below in X if for every bounded open subset V of X and every $x_0 \in \mathbb{R}^n$, $a, b \in \mathbb{R}$ for which

$$u(x) \geq C(x) = a + b\|x - x_0\|$$

holds on $\partial(V\backslash\{x_0\})$, we have $u \geq C$ in V as well.

(iii) A function $u \in C(X)$ is said to enjoys comparison with cones in X if it enjoin comparison with cones both above and below.

Notice that any solution of an eikonal equation $|Dv|$ = constant is a classical solution of $-\Delta_\infty v = 0$ wherever it is smooth. Thus the cone $C(x) = a + b\|x - x_0\|$ is a classical solution for $x \neq x_0$.

In 2001, Crandall, Evans, and Gariepy [11] proved that if u is a continuous function on a bounded open set $X \subset \mathbb{R}^n$ then the concepts of AMLE, the viscosity solution of infinity Laplace equation and the function enjoying comparison with cones are equivalent.

In 2010, Armstrong and Smart [1] presented an elegant and elementary proof of the uniqueness of the functions enjoying comparison with cones. This proof makes no use of partial differential equations and does not need the viscosity solution machinery developed for second-order elliptic equations. This implies a new and easy proof for the uniqueness of AMLE and the uniqueness of the viscosity solution of the infinity Laplace equation.

The definition of comparison with cones easily extends to other metric spaces. In 2007, Champion and De Pascale [10] adapted this definition to length spaces.

In 2002, Juutinen [17] used Perron's method to establish the existence of AMLE extensions for separable length space domains.

In 2009, Peres, Schramm, Sheffield and Wilson [23] used game-theoretic techniques to prove the existence and uniqueness of AMLE for general length spaces[1]. It relied on some complicated probabilistic arguments and a beautiful connection between the infinity Laplace equation and random-turn "tug of war" game.

Extending results on AMLEs to vector valued functions presents many difficulties, which in turn has limited the number of results in this direction, see e.g. [14, 27].

Problem 3.3 *For $X = \mathbb{R}^n$ and $Y = \mathbb{R}^m$ $(m \geq 2)$ both equipped with the Euclidean norm, is it true that for every $\Omega \subset X$, any Lipschitz function $f : \Omega \to Y$ admits an AMLE $g : X \to Y$?*

[1]A metric space (X, d_X) is a *length space* if for all $x, y \in X$, the distance $d_X(x, y)$ is the infimum of the lengths of curves in X that connect x to y.

3.3 Regularity

We study the differentiability properties of viscosity solutions of the PDE

$$\Delta_\infty u = 0, \quad \text{in } X, \tag{10}$$

where $X \subset \mathbb{R}^n$ is an open set.

A viscosity solution u of (10) is called an *infinity harmonic function.*

Example 3.4 *The best known infinity harmonic function was exhibited by Aronsson in 1984 [5]: $u(x,y) = x^{4/3} - y^{4/3}$. The first derivatives of u are Holder continuous with exponent $\frac{1}{3}$; the second derivatives do not exist on the lines $x = 0$ and $y = 0$. Therefore, $u \in C^1$ but $u \notin C^2$.*

Problem 3.5 *Are infinity harmonic functions necessarily continuously differentiable?*

In 2011, Evans and Smart[13] established that the infinity harmonic function is everywhere differentiable. In 2006, Savin [25] has shown that infinity harmonic functions in $n = 2$ variables are in fact continuously differentiable. Savin's arguments uses the topology of $\mathbb{R}^2$ very strongly, and it is difficult to general Savin's arguments for case $n > 2$. The question of whether infinity harmonic functions are necessarily C^1 in general remains the most conspicuous open problem in the area.

Problem 3.6 *Does an infinity harmonic function belong to $C^{1,\frac{1}{3}}_{loc}$?*

Savin [25] proved that in the plane all infinity harmonic functions have continuous gradients . Moreover from Example (3.4), the first derivatives of $u(x,y) = x^{4/3} - y^{4/3}$ are Holder continuous with exponent $\frac{1}{3}$. Thus we can hope that the optimal regularity class is $C^{1,\frac{1}{3}}_{loc}$.

4 $C^{1,1}$ extension

In this Section, let Ω be a subset of the Euclidean space $\mathbb{R}^n$ and F be a 1-field on Ω such that $f : \Omega \mapsto \mathbb{R}$, $Df : \Omega \mapsto \mathbb{R}^n$ are mappings associated with F.

4.1 The minimal Lipschitz extension for 1-field

Let us review the Kirszbraun extension theorem (see Theorem 2.1): Let $\Omega \subset \mathbb{R}^n$ and let $u : \Omega \to \mathbb{R}^m$ be a Lipschitz function. Then there exists a total Lipschitz extension v of u such that $\text{Lip}(v, \mathbb{R}^n) = \text{Lip}(u, \Omega)$. As a consequence, we have $\text{Lip}^*(u) = \text{Lip}(u, \Omega)$ where

$$\text{Lip}^*(u) := \inf\{\text{Lip}(v, \mathbb{R}^n) : v \text{ total extension of } u\}.$$

The natural question is that what is

$$\text{Lip}^*(F) = \inf\{\text{Lip}(Dg, \mathbb{R}^n) : G \text{ is an extension of } F\} \ ?$$

In 2009, Le Gruyer [20] proved that $\text{Lip}^*(F) = \Gamma^1(F, \text{dom}(F))$. Moreover, the constant Γ^1 of 1-field plays a role in the extension problem similarly to the constant Lip of continuous function:

Theorem 4.1 *[20] Let F be a 1-field. The functional* $\Gamma^1 : F \to \Gamma^1(F, \mathrm{dom}(F)) \in \mathbb{R}^+ \cup \{+\infty\}$ *is the unique one satisfying*

(P0) Γ^1 *is increasing, that is, G extends F implies that*

$$\Gamma^1(G, \mathrm{dom}(G)) \geq \Gamma^1(F, \mathrm{dom}(F)).$$

(P1) If G has total domain satisfying $\Gamma^1(G, \mathrm{dom}(G)) < +\infty$, *then the total function g defined by* $g(x) := G(x)(x)$ *is differentiable and its derivative* ∇g *is Lipschitz.*

(P2) If g is a differentiable function of total domain with ∇g *Lipschitz, then*

$$\Gamma^1(G, \mathbb{R}^n) = \mathrm{Lip}(\nabla g, \mathbb{R}^n),$$

where G is the 1-field associate to g, i.e. $G(x)(a) := g(x) + \langle \nabla g(x); a - x \rangle, \forall x, a \in \mathbb{R}^n$.

(P3) For any F such that $\Gamma^1(F, \mathrm{dom}(F)) < +\infty$, *F extends to a total 1-field G satisfying*

$$\Gamma^1(G, \mathrm{dom}(G)) = \Gamma^1(F, \mathrm{dom}(F)).$$

This theorem holds not only in $\mathbb{R}^n$ but in fact in any Hilbert space. Therefore, it generalizes the $C^{1,1}$ version of the classical Whitney extension theorem [29].

To compute the norm Γ^m of the minimal extension on C^m which generalize Le Gruyer's work on minimal C^1 extensions is a very difficult problem and the main thrust is some attempts to guess the natural norm for which one can obtain the minimal extension.

Problem 4.2 *Let* Ω *be a subset of Euclidean space* $\mathbb{R}^n$. *Let* $\mathcal{P}^m(\mathbb{R}^n, \mathbb{R})$ *be the set of m−degree polynomials mapping* $\mathbb{R}^n$ *to* $\mathbb{R}$. *Let us consider a m-field* $T : \Omega \to \mathcal{P}^m(\mathbb{R}^n, \mathbb{R})$. *Find a functional* $\Gamma^m : T \to \Gamma^m(T, \mathrm{dom}(T)) \in \mathbb{R}^+ \cup \{+\infty\}$ *satisfying*

(P0) Γ^m *is increasing, that is, U extends T implies that*

$$\Gamma^m(U, \mathrm{dom}(U)) \geq \Gamma^m(T, \mathrm{dom}(T)).$$

(P1) If U has total domain satisfying $\Gamma^m(U, \mathbb{R}^n) < +\infty$, *then the total function u defined by* $u(x) := U(x)(x)$ *is in* C^m *and* $D^m u$ *is Lipschitz.*

(P2) If $u \in C^m(\mathbb{R}^n)$ *with* $D^m u$ *Lipschitz, then*

$$\Gamma^m(U, \mathbb{R}^n) = \mathrm{Lip}(D^m u),$$

where U is the m−field associate to u.

(P3) For any T such that $\Gamma^m(T, \mathrm{dom}(T)) < +\infty$, *T extends to a total m−field U satisfying*

$$\Gamma^m(U, \mathbb{R}^n) = \Gamma^m(T, \mathrm{dom}(T)).$$

4.2 Relationships between $\Gamma^1(F; \Omega)$ and $\mathrm{Lip}(Df; \Omega)$

It is worth asking what is it the relationship between 1-field Lipschitz constant $\Gamma^1(F; \Omega)$ and $\mathrm{Lip}(Df, \Omega) := \sup\limits_{x \neq y \in \Omega} \frac{\|Df(x) - Df(y)\|}{\|x-y\|}$? The two are equal when $\Omega = \mathbb{R}^n$, but in general one only has $\mathrm{Lip}(Df, \Omega) \leq \Gamma^1(F, \Omega)$ and the strict inequality may happen even when Ω is convex (see [20, 21]). It remains a very interesting question whenever the equality holds. It turns out that the answer depends heavily on the boundary of Ω. In [21], Le Gruyer and Phan show that:

Theorem 4.3 *Let* Ω *be an open set in* $\mathbb{R}^n$. *We have*

$$\Gamma^1(F; \Omega) = \max\left\{\mathrm{Lip}(Df; \Omega), \Gamma^1(F; \partial\Omega)\right\}.$$

Moreover, they prove that $\mathrm{Lip}(Df, \Omega) = \Gamma^1(F, \Omega)$ when either the supremum in the definition of $\Gamma^1(F, \Omega)$ is attained, or $\Gamma^1(F, \Omega) = \Gamma^1(F, \Omega')$ for some compact subset Ω' of Ω.

4.3 Extremal MLEs and AMLEs for 1-field

Suppose that $\Gamma^1(F,\Omega) < +\infty$. From Theorem 4.1, there exists $g \in C^{1,1}(\mathbb{R}^n,\mathbb{R})$ such that $g|_\Omega = f$, $\nabla g|_\Omega = Df$ and the 1-field G associated to $(g,\nabla g)$ satisfies $\Gamma^1(F,\Omega) = \Gamma^1(G;\mathbb{R}^n)$. Such a map G is called a *minimal Lipschitz extension* (MLE) of F.

In 2015, Le Gruyer and Phan [21] constructed explicitly the sup-inf formulas for two *extremal MLEs* U^+ and U^-, namely

$$U^-(x)(x) \le G(x)(x) \le U^+(x)(x),\ \forall x \in \mathbb{R}^n$$

for every MLE G of F. Moreover, they proved that if Ω is finite, then $U^\pm$ are *absolutely minimal Lipschitz extensions* (AMLE) of F, namely

$$\Gamma^1(U^\pm, D) = \Gamma^1(U^\pm, \partial D)$$

for any open, bounded set D satisfying $\overline{D} \subset \mathbb{R}^n\backslash\Omega$.

This result give the existence of AMLEs of F when Ω is finite. In general, we have not uniqueness, since it may happen $u^- < u^+$. In fact, we may even have infinity AMLE of F.

Recently, in 2018, Azagra, Le Gruyer and Mudarra [6] provide necessary and sufficient conditions for the 1-field F to admit an extension G associated to $(g,\nabla g)$ with g convex and of class $C^{1,1}$. As a consequence, they obtain a simple explicit formulas for U^+ and U^-.

When Ω is infinite, U^+ and U^- are extremal MLEs, but in general are not AMLE of F (see [21]). The question on the existence of an AMLE for a 1-field when Ω is *infinite* remains a very difficult *open problem.*

Problem 4.4 *Whether there exists an AMLE of 1-field F ?*

References

[1] Armstrong, S.N., Smart, C.K.: *As easy proof of Jensen's theorem on the uniqueness of infinity harmonic functions.* Calc. Var. Partial Differential Equation. **37**(3-4), 381-384 (2010).

[2] Aronsson, G.: *Minimization problems for the functional* $sup_x\ F(x, f(x), f'(x))$. Ark. Mat. **6**, 33-53 (1965).

[3] Aronsson, G.: *Minimization problems for the functional* $sup_x\ F(x, f(x), f'(x))$ *II* . Ark. Mat. **6**, 409-431 (1966).

[4] Aronsson, G.: *Extension of functions satisfying Lipschitz conditions.* Ark. Mat. **6**, 551-561(1967).

[5] Aronsson, G.: *On certain singular solutions of the partial differential equation* $u_x^2 u_{xx} + 2u_x u_y u_{xy} + u_y^2 u_{yy} = 0$. Manuscripta Math. **47**, no. 1-3, 133-151 (1984).

[6] Azagra, D., Le Gruyer, E., Mudarra, C.: *Explicit formulas for* $C^{1,1}$ *and* $C^{1,\omega}_{conv}$ *extensions of* 1*−jets in Hilbert and superreflexive spaces* Journal of Functional Analysis **274** (10), 3003-3032, (2018).

[7] Aronsson, G., Crandall, M.G, Juutinen, P.: *A tour of the theory of Absolutely Minimizing Functions.* Bull. Am. Math. Soc. **41**(4), 439-505 (2004).

[8] Benyamini, Y., Lindenstrauss, J.: *Geometric Nonlinear Functional Analysis*, Vol 1. American Mathematical Society Colloquium Publications, vol 48. American Mathematical Society, Providence (2000).

[9] Bezdek, K., Connelly, R. *Pushing disks apart- the kneser- poulsen conjecture in the plane*, J. reine angew. Math. **553**, 221-236 (2002).

[10] Champion, T., Pascale, L.D. : *Principles of comparison with distance functions for absolute minimizers*, J. Convex Anal. **14**, no. 3, 515-541 (2007).

[11] Crandall, M.G., Evans, L.C., Gariepy, R.F. : *Optimal Lipschitz extensions and the infinity laplacian*, Calc. Var. Partial Differential Equation. **13**, no. 2, 123-139 (2001).

[12] Crandall, M.G., Lions, P.L.: *Viscosity solutions of Hamilton-Jacobi equations*, Trans. Amer. Math. Soc. **277**, no. 1, 1-42 (1983).

[13] Evans, L.C. and Smart, C. K.: *Everywhere differentiability of infinity harmonic functions* , Calc. Var. **42**, 289-299 (2011)

[14] Hirn, M.J., Le Gruyer, E.L.: *A general theorem of existence of quasi absolutely minimal Lipschitz extensions.* Math. Ann. **359**(3), 595-628 (2014).

[15] Glaeser, G.: *Études de quelques algèbres tayloriennes.* J. d' Analyse math. **6**, 1-124 (1958).

[16] Jensen, R.: *Uniqueness of Lipschitz extension: minimizing the sup-norm of gradient.* Arch. Rational Mech. Anal. **123**(1), 57-74(1993) .

[17] Juutinen, P.:, *Absolutely minimizing Lipschitz extension on a metric space.* Ann. Acad. Sci. Fenn. Math. **27**, 57-67 (2002).

[18] Kirszbraun,M.D.: *Über die zusammenziehende und Lipschitzsche Transformationen.* Fund. Math. **22**, 77-108 (1934).

[19] V.M. Kneser, *Einige bemerkungen uber das minkowskische flachenmass*, Archiv der Mathematik **6**, 382-390 (1955).

[20] Le Gruyer, E.: *Minimal Lipschitz extensions to differentiable functions defined on a Hilbert space.* Geom. Funct. Anal. **19**(4), 1101-1118 (2009).

[21] Le Gruyer, E., Phan, T.V. *Sup–Inf explicit formulas for minimal Lipschitz extensions for 1–fields on* $\mathbb{R}^n$. J. Math. Anal. Appl. **424**(1-2), 1161-1185 (2015).

[22] McShane, E.J.: *Extension of range of functions.* Bull. Amer. Math. Soc. **40**, 837-842 (1934).

[23] Peres, Y., Schramm, O. , Sheffield, S., Wilson, D.B.: *Tug-of-war and the infinity Laplacian.* J. Am. Math. Soc. **22**(1), 167-210 (2009).

[24] Poulsen, E.T. *Problem 10*, Mathematica Scandinavica **2** , 346 (1954).

[25] Savin, O.: C^1 *regularity for infinity harmonic functions in two dimensions.* Arch. Rational Mech. Anal. **176**(3), 351-361 (2006).

[26] Schwartz, J. T. *Nonlinear functional analysis*, New York: Gordon and Breach Science (1969), 100-108.

[27] Sheffield, S., Smart, C.K .: *Vector-valued optimal Lipschitz extension.* Commun. Pure. Appl. Math. **65**(1), 128-154 (2012).

[28] Valentine, F.A. *On the extension of a vector function so as to preserve a Lipschitz condition*, Bull. Am. Math. Soc. **49**, 100-108 (1943).

[29] Whitney, H.: *Analytic extensions of differentiable functions defined in closed sets.* Trans. Amer. Math. Soc. **36**, 63-89 (1934).

Gradient bounds for solutions of nonlinear strictly elliptic equations, applications and extensions

Vinh Duc Nguyen[1,*]

[1]Faculty of Mathematics and Statistics, Ton Duc Thang University, Ho Chi Minh City, Vietnam

Abstract. In this short manuscript, we briefly recall some well-known methods for obtaining gradient bounds of viscosity solutions for elliptic and parabolic equations. The two methods we focus here are the idea coming from Ishii-Lions' method applies for strictly elliptic equations and the other one is so-called weak Bernstein's method developed by Barles applies for degenerate equations. We present some results based on these methods and their applications. We also discuss some promising extensions in the future.

1 introduction

In the theory of nonlinear elliptic or parabolic equations, having gradient bounds is very important for proving the existence of solutions. It is even more useful in many applications, for example to singular perturbations problems, long time convergences.

The classical Bernstein's method is the well-known tool to obtain these bounds but the method only applies for smooth solutions.

The key idea of the classical Bernstein's method (cf. Gilbarg and Trudinger[4] and Lions[5]) is the following: if $u : \Omega \subset \mathbb{R}^N \to \mathbb{R}$ is a smooth solution of

$$-\Delta u = 0 \ \text{ in } \Omega,$$

After performing a change of function $v = |Du|^2$, v solves

$$-\Delta v \leq 0 \ \text{ in } \Omega,$$

The strong maximum principle for elliptic equation says that

$$\sup_{x\in\Omega} v \leq \sup_{x\in\partial\Omega} v \quad \text{hence } \sup_{x\in\Omega} |Du(x)| \leq \sup_{x\in\partial\Omega} |Du(x)|.$$

From the above inequality, we have the bound for $|Du(x)|$, $x \in \Omega$ if we have the bound for $|Du(x)|$ with $x \in \partial\Omega$. With suitable structure conditions on F, we can perform the above arguments to obtain gradient bounds for nonlinear elliptic equations:

$$F(x, u, Du, D^2u) = 0, x \in \Omega. \tag{1.1}$$

Disadvantage of the classical Bernstein's method: in order to the strong maximum principle holds, v needs to be C^2 and hence u needs to be C^3.

*e-mail: nguyenducvinh@tdtu.edu.vn

For making the presentation easier, from now we work on quasi-linear elliptic equations. We do not consider boundary conditions to concentrate on structure conditions

$$-\mathrm{trace}(A(x)D^2u) + H(x, Du) + \epsilon u = 0, x \in \mathbb{T}^N. \tag{1.2}$$

The weak Bernstein's method applies for continuous viscosity solutions developed by Barles [4], is another way to obtain gradient bounds. Instead of the idea of performing a change of function $v = |Du|^2$, Barles [4] develops a so called weak Bernstein's method by considering the test function

$$\Phi(x, y) = u(x) - u(y) - L|x - y|. \tag{1.3}$$

The goal is to prove that for L large enough, $\max_{x,y\in\overline{\Omega}} \Phi(x, y) \leq 0$ and hence $||Du||_\infty \leq L$. Surprisingly, the structure conditions of F obtained in Barles [4] are almost the same with the ones in the classical Bernstein's method.

Disadvantage of weak Bernstein's method: the strength of ellipticity is not totally used.

The Ishii-Lions' method introduced in [15], see also [5, 12] allows us to take profit from the strict ellipticity of the equation to control the first order terms. In Ishii-Lions [15] and Barles [3], weak regularity assumptions are assumed over H, merely a kind of balance between some Hölder continuity in x and the growth size of H with respect to the gradient, more precisely,

$$|H(x, p) - H(y, p)| \leq C|x - y||p|^3 + C(1 + |p|^2) \quad \text{in [3, Assumption (3.4)]}, \tag{1.4}$$

where $x, y \in \mathbb{T}^N$, $p \in \mathbb{R}^N$, $\tau \in [0, 1]$, $C > 0$.

These assumptions are designed for subquadratic Hamiltonians(whose growth is less than $|p|^2$) . This is not surprising since it is known that, in general, the ellipticity is not powerful enough to control the first order terms whose growth is more than quadratic [10]. Under these assumptions, the authors obtained gradient bounds which depend on the L^∞ norm of the solution.

Disadvantage of Ishii-Lions's method: the strength of first order terms is not totally used.

Let us mention a recent work of Barles [7] devoted to the study of weak Bernstein's method to **local** gradient bounds with more general structure conditions for both elliptic and parabolic equations.

2 Some regularity results

In [16], we combine the two above methods in some sense. We use the strength from the ellipticity as well as from the strong coercivity of H to allow arbitrary growth of H with respect to the gradient.

In Theorem 2.1, the coercivity assumption (2.1) is the one needed to obtain the Hölder regularity with exponent $\frac{k-2}{k-1}$ in [10]. The main idea is the following:

- showing that the solution is γ-Hölder continuous *for any* $\gamma \in (\frac{k-2}{k-1}, 1)$.
- improving the regularity up to Lipschitz continuity.

Starting with a Hölder exponent equal to $\frac{k-2}{k-1}$ seems crucial to be able to improve the regularity for the class of H we consider which has a strong growth with respect to the gradient.

Theorem 2.1. *Assume $A \geq \nu Id$, $\nu > 0$ and H satisfies*

$$\text{there exist constants } k > 2, C > 0 \text{ such that } H(x,p) \geq \frac{1}{C}|p|^k - C \tag{2.1}$$

and

$$\begin{cases} \text{there exist a modulus of continuity } \omega \text{ and constants } \alpha \in [0,1], \beta < k-1 \\ \text{such that for all } x, y \in \mathbb{T}^N, p \in \mathbb{R}^N, \\ |H(x,p) - H(y,p)| \leq \omega\Big((1+|p|^\beta)|x-y|\Big)\, |x-y|^\alpha |p|^{(k-1)\alpha+k} + o(|p|^k), \end{cases} \tag{2.2}$$

where $o(|p|^k)/|p|^k \to 0$ as $|p| \to +\infty$, uniformly with respect to $x \in \mathbb{T}^N$. Then, there exists $K > 0$ such that for all $\epsilon > 0$, any continuous solution v^ϵ of (1.2) satisfies

$$\|Dv^\epsilon\|_\infty \leq K. \tag{2.3}$$

.

Assume there exists $k > 2$ such that

$$H(x,p) = a(x)|p|^k + K(x,p),$$

With $a \in C(\mathbb{T}^N)$ and $K(x,p) = o(|p|^k)$, this Hamiltonian satisfies the assumptions of Theorem 2.1.

Gradient bounds are very useful in singular perturbations problems, long time convergences. Let us state a well-known fact when (2.3) holds as a consequence of strong maximum principle for viscosity solutions proved in [13].

Theorem 2.2. *Assume $A \geq \nu Id$, $\nu > 0$ and (2.3) holds. Then any bounded continuous solution of*

$$\begin{cases} \dfrac{\partial u}{\partial t} - \mathrm{trace}(A(x)D^2u) + H(x, Du) = 0, & (x,t) \in \mathbb{T}^N \times (0, +\infty), \\ u(x,0) = u_0(x), & x \in \mathbb{T}^N. \end{cases} \tag{2.4}$$

converges uniformly to a continuous function v as $t \to \infty$. Moreover, v is a solution of the equation

$$-\mathrm{trace}(A(x)D^2v) + H(x, Dv) = 0.$$

3 Future extensions

We wonder if (2.3) and Theorem 2.2 still hold for the following m-laplace equation

$$\epsilon v^\epsilon - \mathrm{div}(|Dv^\epsilon|^{m-2}Dv^\epsilon) + H(x, Dv^\epsilon) = 0, \quad x \in \mathbb{T}^N. \tag{3.1}$$

with similar structure conditions. While (2.3) is quite clear to proceed, Theorem 2.2 is quite unclear since we know that the m-laplacian term $\mathrm{div}(|Dv^\epsilon|^{m-2}Dv^\epsilon)$ is degenerate as $Dv^\epsilon = 0$. Let us state the main idea of the proof for (2.3) taken from our work in preparation [21].

- From the super-quadratic structures on first order terms, in [10], the authors proved that

$$v^\epsilon \text{ is } \frac{k-m}{k-m+1}.$$

• Although the m-laplacian term $\mathrm{div}(|Dv^\epsilon|^{m-2}Dv^\epsilon)$ is degenerate as $Dv^\epsilon = 0$ but when Dv^ϵ big, it is strictly elliptic.
• The proof of gradient bounds consists of improving the Holder exponent.

Obtain local gradient bounds for elliptic equations associated with a nonlinear Neumann boundary condition

$$\begin{cases} -\mathrm{trace}(A(x)D^2u) + H(x, Du) = 0, & x \in \Omega, \\ G(x, u, Du) = 0, & x \in \partial\Omega. \end{cases} \tag{3.2}$$

with similar structure conditions on H. Local gradient bounds for superquadratic Hamiltonians are the best we can expect since it is known that the gradients of solutions can blow up at boundary, see [9] and references therein for more results about this phenomenon.

References

[1] Olivier Alvarez and Martino Bardi. Ergodicity, stabilization, and singular perturbations for Bellman-Isaacs equations. *Mem. Amer. Math. Soc.*, 204(960):vi+77, 2010.

[2] M. Arisawa and P.-L. Lions. On ergodic stochastic control. *Comm. Partial Differential Equations*, 23(11-12):2187–2217, 1998.

[3] G. Barles. Interior gradient bounds for the mean curvature equation by viscosity solutions methods. *Differential Integral Equations*, 4(2):263–275, 1991.

[4] G. Barles. A weak Bernstein method for fully nonlinear elliptic equations. *Differential Integral Equations*, 4(2):241–262, 1991.

[5] G. Barles. $C^{0,\alpha}$-regularity and estimates for solutions of elliptic and parabolic equations by the Ishii & Lions method. In *International Conference for the 25th Anniversary of Viscosity Solutions*, volume 30 of *Gakuto International Series, Mathematical Sciences and Applications*, pages 33–47. Gakkotosho, Tokyo, Japan, 2008.

[6] G. Barles. A short proof of the $C^{0,\alpha}$-regularity of viscosity subsolutions for superquadratic viscous Hamilton-Jacobi equations and applications. *Nonlinear Anal.*, 73(1):31–47, 2010.

[7] G. Barles. Local Gradient Estimates for Second-Order Nonlinear Elliptic and Parabolic Equations by the Weak Bernstein's Method. *Submitted.*

[8] G. Barles and P. E. Souganidis. Space-time periodic solutions and long-time behavior of solutions to quasi-linear parabolic equations. *SIAM J. Math. Anal.*, 32(6):1311–1323 (electronic), 2001.

[9] Guy Barles and Francesca Da Lio. On the generalized Dirichlet problem for viscous Hamilton-Jacobi equations. *J. Math. Pures Appl. (9)*, 83(1):53–75, 2004.

[10] I. Capuzzo Dolcetta, F. Leoni, and A. Porretta. Hölder estimates for degenerate elliptic equations with coercive Hamiltonians. *Trans. Amer. Math. Soc.*, 362(9):4511–4536, 2010.

[11] P. Cardaliaguet and L. Silvestre. Hölder continuity to Hamilton-Jacobi equations with superquadratic growth in the gradient and unbounded right-hand side. *Comm. Partial Differential Equations*, 37(9):1668–1688, 2012.

[12] M. G. Crandall, H. Ishii, and P.-L. Lions. User's guide to viscosity solutions of second order partial differential equations. *Bull. Amer. Math. Soc. (N.S.)*, 27(1):1–67, 1992.

[13] F. Da Lio. Remarks on the strong maximum principle for viscosity solutions to fully nonlinear parabolic equations. *Commun. Pure Appl. Anal.*, 3(3):395–415, 2004.

[14] David Gilbarg and Neil S. Trudinger. *Elliptic partial differential equations of second order*. Classics in Mathematics. Springer-Verlag, Berlin, 2001. Reprint of the 1998 edition.

[15] H. Ishii and P.-L. Lions. Viscosity solutions of fully nonlinear second-order elliptic partial differential equations. *J. Differential Equations*, 83(1):26–78, 1990.

[16] O. Ley and V. D. Nguyen. Lipschitz regularity results for nonlinear strictly elliptic equations and applications. *Journal of Differential Equations 263, 4324-4354, 2017.*

[17] O. Ley and V. D. Nguyen. Gradient bounds for nonlinear degenerate parabolic equations and application to large time behavior of systems. *Nonlinear Anal.*, 130:76–101, 2016.

[18] N. V. Krylov. *Lectures on elliptic and parabolic equations in Hölder spaces*, volume 12 of *Graduate Studies in Mathematics*. American Mathematical Society, Providence, RI, 1996.

[19] P.-L. Lions. *Generalized solutions of Hamilton-Jacobi equations*. Pitman (Advanced Publishing Program), Boston, Mass., 1982.

[20] P.-L. Lions, B. Papanicolaou, and S. R. S. Varadhan. Homogenization of Hamilton-Jacobi equations. *Unpublished*, 1986.

[21] V. D. Nguyen. Regularity results and long time behavior for solutions of p-laplace equations. *In preparation.*

High Performance Energy Prediction using Hadoop with Spark

Hung Duong-Ngoc[1,3,*], *Hoan* Nguyen-Thanh[2], and *Tam* Nguyen-Minh[1]

[1]Ho Chi Minh City University of Technology and Education, 1 Vo Van Ngan, Linh Chieu Ward, Thu Duc District, Ho Chi Minh City,Vietnam.
[2]Ho Chi Minh City University of Technology, 268 Ly Thuong Kiet, District 10, Ho Chi Minh City, VietNam
[3]Tien Giang University, 119 Ap Bac, Ward 5, My Tho City, Tien Giang, Vietnam

Abstract. With the increasingly modern development of the power system. Along with that is the data source collected from them is huge. Combined with other systems such as GIS, MDMS-AMR (Automatic Meter Reading), weather forecast and socio-economic indicators. We consider performing an effective analysis of the data sources in order to understand the evolution, characteristics, and modeling of the power consumption system. Thereby predict future energy trends and build bases for the system model. To implement the issues raised, we appreciate using Hadoop platform for storage and segmentation data, enabling better handle large amounts of data initially. Then, the data was analyzed using the scalable machine learning algorithms - MLib was supported and developed on the Spark/SPARKNET platform. The Hadoop framework has recently evolved to the standard framework implementing the MapReduce model. In this paper, we evaluate Hadoop with Mlib/Sparknet performance in both the traditional model of collocated data and compute services as well as consider the impact of separating out the services. Energy modeling from multiple data sources such large may help to understand the change of the system according to consumer demand for practical, predictable trends of energy in the future and provide the basis for building energy models for similar systems.

1 Introduction

Modeling and forecasting electrical energy consumption has been a research area interested for more than a decade. Today, with climate change and the focus on environment, it is even more important to model and forecast electricity consumption accurately in pursuit of backup opportunities. The importance of measuring and collecting electricity data, together with recent advances in sensor technology, have led to the proliferation of smart meters that measure and communicate electricity consumption. These smart meters measure electricity at intervals of an haft of hour or less in Ho Chi Minh City, whereas some sensor devices can measure consumption in real time. These Big Data have created opportunities to develop new ways of analyzing energy consumption, identifying potential savings, and

* Corresponding author: ngochungduong@gmail.com

measuring energy efficiency. Sensor-based approaches to energy forecasting rely on readings from sensors or smart meters and contextual information such as meteorological information or work schedules to infer future energy behaviour. Typically, historical data such as temperature, day of the week, time of day, and energy consumption are fed into a machine learning model that learns from them and consequently can forecast future energy consumption. The accuracy of these sensor based approaches is comparable or superior to traditional approaches based on modeling in depth the properties of a building [3].

The growing faster for handle high volume and complex data has resulted in the development of a broad set of tools for data processing and analytics. One of the most popular solution is Hadoop [4], an open-source implementation of the MLIB Spark framework [12] which is first introduced by Amazon to deliver timely and cost-effective data processing on a large cluster of commodity machines. It has a simple but powerful functional programming interface to abstract computations and ease the programming efforts from users. The parallelization and distribution of computing tasks are automatically managed by the framework to achieve higher system throughput as well as to support fault tolerance. Today, a Hadoop ecosystem has formed by composing a software stack around Hadoop to address various challenges in Big Data, including unstructured data storage [5], stream processing [8], graph processing [3], query analytics [25], workflow management [7], machine learning [6], etc.

While it is rather easy to write a program running on Hadoop, to manage the resource allocation and job execution of a Hadoop cluster can be a different story. As shown from the previous literature [11], [20], [21], [28], Hadoop has more than hundreds of runtime configurations to control job executions, such as number of reducers, shuffle buffer size, sorting percentage. The setting of these configurations can have significant impact to the execution time. For instance, a factor of 2 to 8 speed up in job execution time has been observed [11] while being given the same amount of resources. But finding a better Hadoop configuration to reduce job execution time requires a substantial amount of knowledge and experiences in the application and Hadoop, because the execution time depends on various factors, including the size and structure of input data, the complexity of computation code, the environment of system, etc. On the other hand, in reality, Hadoop jobs are often written by users from different application domains with little or no knowledge of the framework. The problem is further exacerbated by the fact that modern computing clusters are often consisted of heterogeneous hardware and managed through a virtualization layer. Thus, the resource pool is shared among users and allocated to them in the unit of different types of resource instances (e.g., virtual machine). Therefore, in this paper, we aim to answer a fundamental question as below:

Besides, energy consumption time series are recorded by smart meters at the regular interval of an hour or fewer. Smart meters read the detailed energy consumption in a real-time or near real-time fashion, which provides the opportunity to monitor timely unusual events or consumption behaviors. However, the enabling detection technologies combining smart meters typically using data mining technologies, which require large amounts of training data sets, and significantly complex systems. In a typical application of data mining to anomaly detection, the detection models are produced off-line because the learning algorithms have to process tremendous amounts of data [26]. The produced models are naturally used by off-line anomaly detections, i.e., analyzing consumption data after being loaded into an energy management system (EMS). However, we argue that effective anomaly detection should happen real-time in order to minimize the compromises to the use of energy. The efficiency of updating the detection model and the accuracy of the detection results are the important consideration for constructing a real-time anomaly detection system.

We are interested in the time prediction because it is one of the most critical information for managing systems and resources. With the prediction answer, we could search for a better

Hadoop configuration to minimize execution time; we could determine the minimum resource requirement of a job; we could develop an execution time aware scheduling algorithm to improve resource usage fairness. We propose a statistical anomaly detection algorithm to detect the anomalous daily electricity consumption. The anomaly detection is based on consumption patterns, which are usually quite similar for a customer, such as in weekdays, or in weekends/holidays. Then machine learning methods, such as linear regression, are used to determine the proper coefficient values in the equation for different class of applications or jobs. Promising results (i.e., more than 90% of prediction accuracy) have been reported from the literature [14].

On the other hand, as shown by the recent work in various application domains [2], [15], Deep Learning (DL) technique based on neural network model can further improve the prediction accuracy over traditional machine learning methods without requiring complicated system modeling. But little attention has been paid to apply such technique to job time prediction. Therefore, in this paper, we made an attempts to predict Hadoop job execution time using DL technique, and built a system to optimize Hadoop system performance in a shared resource cloud environment. To support big data capability, we choose Spark Streaming as the speed layer technology for detecting anomalies on a large amount of data streams, Spark as the batch layer technology for computing anomaly detection models, and PostgreSQL as the serving layer for saving the models and detected anomalies; and sending feedbacks to customers. The proposed system can be integrated with smart meters to detect anomalies directly.

2 Related work

Figure I provides a high-level overview of the two deployment models we consider in this paper. The two models are: 1) traditional model in which data and compute services are collocated and 2) alternate model in which data and compute services are separated. This section reviews related work in machine learning with Big Data, and in electricity consumption prediction.

The specific scenario considered is electricity consumption prediction for event venues such as sports arenas, theatres, and conference centres. In this scenario, consumption patterns are not as strongly related to hours of the day and days of the week, as is the case with office buildings, but are driven by event schedules and event attributes such as event type (basketball, hockey, ...) and seating capacity. In addition to electricity readings, the following attributes are considered:

Day of the year: 1 to 365

Day of the week: 1 to 7

Hour of the day: 1 to 24 Event day: indicates whether there was an event on the day of the reading

Event type: category of events, such as basketball and hockey. Three input features are used, one for each of basketball, hockey, and other.

Seating configuration: captures seating capacity for an event.

The data consist of one sample for each electricity reading, and the event schedule is captured through date/time attributes (day of the year, day of the week, hour of the day) and the event type. Samples corresponding to non-event periods have 0 for the event type, whereas those corresponding to time periods during events have an event type describing the category of event, such as basketball or hockey.

2.1 Implementation

In this paper, we consider the time prediction and optimization problem of elastic data processing by running on-demand Hadoop jobs in a cloud environment as shown in Figure 1.

The objective of this paper is to evaluate various suggested approaches for Big Data processing with respect to accuracy and time, not necessarily to create a completely new approach. Hence, prediction with local SVR as described in this section relies mostly on already available components, but it combines them in a way that enables efficient energy prediction.

Fig. 1 describes the training and testing process for energy prediction using local SVR. First, in step 1, the data set is divided into a training and a testing set. Because energy prediction is a time series problem in which older data are used to predict newer data, a portion of the data at the end of the time series is reserved solely for testing. The remainder of the set is used for training and parameter optimization.

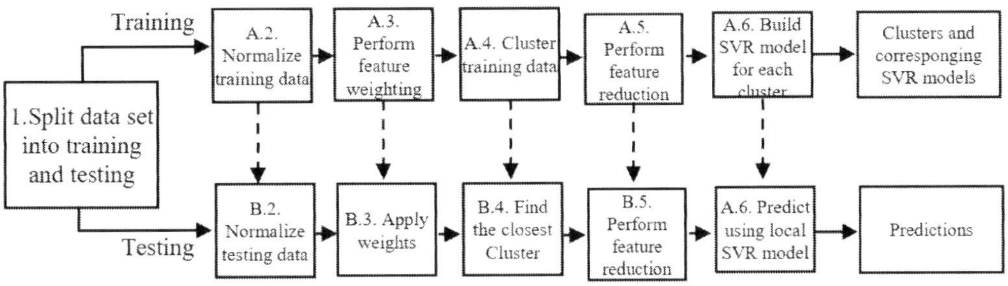

Fig. 1. Local SVR process.

In this system model, users submit their data analytic jobs as a Hadoop workflow. Each Hadoop job in a workflow runs on a on-demand virtual cluster which only has the same lifetime as the job. To improve resource utilization, computing resources are virtualized and offered to users in terms of different size of VM instance type. Hadoop also provides several configurations that could be tuned for individual job execution, such as number of reducers. Hence, our goal is to predict the execution time of a Hadoop job under the setting of a given Hadoop configuration and resource configuration, where Hadoop configuration refers to the parameters listed in Table I, and resource configuration refers to the type of VM and the number of VMs. Once the time prediction knowledge is obtained, a set of optimization techniques can be developed to achieve higher system performance, lower execution cost, and other potential benefits.

Table 1. Hadoop configuration and the range of its parameter values considered in the paper.

Configuration	default	min	max
job.reduces	1	1	128
tasktracker.map.tasks.maximum	2	2	12
tasktracker.reduce.tasks.maximum	2	2	12
task.io.sort.mb	100	100	150
task.io.sort.factor	10	10	100
reduce.shuffle.parallelcopies	5	5	20
jobtracker.handler.count	10	10	100
tasktracker.http.threads	40	40	60
map.cpu.vcores	1	1	4
reduce.cpu.vcores	1	1	4
map.memory.mb	1024	1024	2048
reduce.memory.mb	1024	1024	2048

app.Spark.am.resource.mb	512	1024	2048
app.Spark.am.resource.cpu-vcores	1	1	4
child.java.opts	-Xmx200m	-Xmx200m	-Xmx1024m

In addition to energy consumption, the data set included event-related data as described in Section IV.A. 80% of the data were used for parameter selection and training, and 20% were used for testing. The training set contained readings for all of 2013, thus accounting for all seasons. The testing set included data for the first 3 months of 2014.

Three prediction approaches were implemented: SVR, local SVR as presented in Section IV.B, and H2O deep learning. Each implementation uses the grid search approach with five-fold cross validation for the parameter selection:

- **SVR**: Implemented in R language [31] using the "e1071" package. Two parameters were tuned: 10 values for the cost C from 1e-6 to 1e+3 with exponential increments and 10 values for the radial basis parameter from 1e-8 to 1e+1. This makes for a total of 100 configurations.
- **Local SVR**: Implemented in R language [31] using the "stats" package for k-means clustering and the "e1071" package for SVR. The SVR model for each cluster was tuned using the same approach as in standalone SVR; 10 values for the cost C and 10 values for the parameter . In addition, ten values of the number of clusters k (from 20 to 110 by increments of 10) were considered.

2.2 Data Driven Approach

Our time prediction method is driven by analyzing historical job execution data. We believe this approach is particularly

Table 2. Data collections of job profile

Category	**configuration**
Job information	job name
	VM type
	VM number
	job status (succeeded/failed)
Hadoop counters	mapreduce.job.reduces
	Spark.task.io.sort.mb
	Spark.task.io.sort.factor
	...
System metrics	CPU usage {sum,avg,max,min,stddev} in each time period
	Memory usage {sum,avg,max,min,stddev} in each time period
	Disk {read, write} {sum,avg,max,min} in each time period
	Network {input, output} {sum,avg,max,min} in each time period
Data information	input data size
	number of input data file
	#word/KB
	#line/KB
	#word/line in {avg,stddev}
	#char/line in {avg,stddev}
	#char/word in avg,stddev

suitable for Hadoop jobs and data analytic applications because most users use similar tools or programs to analyze their data. For instance, Mahout is one of the famous library built on top of Hadoop to support scalable machine learning applications. There are only a handful of machine learning algorithms that are implemented in Mahout, such as KMeans, Stochastic

SVD, etc. Hence, users essentially keep running the same set of programs but with their own data input. Not to mention that most data analytic applications, such as OLAP (On-Line Analytical Processing), tend to periodically perform the same operations over their updated datasets. Therefore, it is rather easy to collect sufficient amount of job execution history of a program for program behavior analysis and execution time prediction.

As shown in Figure 1, our system collects four types of data with respect to a job execution, naming job information, Hadoop counter, system metrics and data information. They are combined together into a job profile, and used by the methods introduced in Section III for time prediction and optimization. Table II summarizes the attributes in a job profile, and their brief descriptions are as follows. 1) Job information is the data collected from the job submission system. The data contains the overall job execution information, such as job status, job submission/start/finish time, and also some program or user information, such as executable filename and userID, etc. 2) Hadoop counters are the statistical numbers collected by the Hadoop framework during job execution. More than tens of built-in counters are supported in Hadoop, and custom counters could be added too. This set of data is critical for characterizing the runtime behavior of a job, especially in terms of the data size generated from each map phase or reduce phase. 3) System metrics are the resource usage information collected from monitoring system, such as CPU utilization, memory usage, disk read/write size and network I/O bandwidth, etc. Each metric contains a set of time-series measurements from compute nodes. To limit the size of our job profile, our prediction analysis only uses the aggregate statistic numbers (i.e., max/min/mean) over 4 time periods for each system metric. 4) Data information is getting from analyzing a small subset of random sampled data files, such as the number of lines, the number of words per line, etc. This information allows us to characterize each data file besides just knowing its file size.

2.3 Implementation

To validate our approach, we build our proposed system based on an open source cloud platform software, OpenStack [24], which has been widely used to build private and public cloud platforms. OpenStack has a modular architecture to allow open source community develop its components in separate projects. The core components provide IaaS (Infrastructure-as-a-Service) service to support widely available virtualization technologies. For our work, we also installed several other components and integrate them together to realize the system shown in Figure 1. Sahara is a Hadoop cluster deployment service. It plays the role of a submission system to launch Hadoop clusters using the orchestration template managed by Heat. The user data is stored in Swift, an object store service. Originally, Hadoop configuration and VM instance type must be specified by users when jobs are submitted to Sahara. But in our implementation, we added a prediction and optimization plug-in component to determine proper configurations for the users automatically. To collect the profile of any running jobs in the system, we implemented a daemon running on the Hadoop master node to collect job information from Sahara, system metrics from Ceilometer monitoring service, Hadoop counters from Hadoop log files, and finally the data information from analyzing several 1KB sample data chunks of job input files. Then all the job profile data information is stored into a MongDB database, and periodically pulled out to build the prediction models as described in the next section.

3 Prediction & optimization methods

Our prediction and optimization methods are summarized in Figure 2. The input of our problem is a given Hadoop job, and the output of our solution is a pair of recommended settings for its Hadoop configuration, denoted as HCopt, and the resource configuration, denoted as RC_{opt}. Our approach is based on deep learning technique, so the solution can be divided into online and offline two parts. The offline part indicated by the dotted box in the figure is to build models using the training data collected from historical job execution data (i.e., job profile). On the other hand, the online part is to determine {HCopt,RCopt} based on the training models, and it is consisted of four components: Profiler, Classifier, Predictor, Optimizer. We introduce the methods used by each of these components in the rest of section.

3.1 Profiler

In this paper, we consider the time prediction and optimization problem of elastic data processing by running on-demand Hadoop jobs in Our approach is driven by the job execution information collected from the system. Upon receiving a job from users, Profiler immediately takes the responsibility to generate its job profile. The profile is used by Classifier and Predictor for online job classification and execution time prediction. The profile is also stored as the training datasets for building the models offline. As detailed in Section II-B, profile is consisted of four types of data collections, and most of them can only be collected during job execution time. Therefore, the main challenge of profiler is to get these detailed job execution information without introducing too much profiling overhead. Our approach is to create a small sample job which runs with the same code requested by users under a default setting of Hadoop configuration and resource configuration.

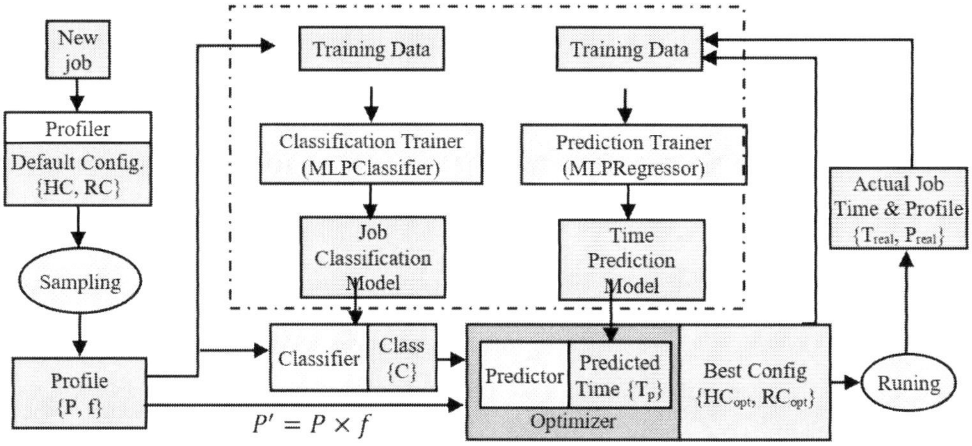

Fig. 2. Our learning-based prediction and optimization methods. HC is a Hadoop configuration.

RC is a resource configuration, P is a job profile, f is a scaling vector, C is an application class, and T is job execution time. The input data of sample jobs is downsized by randomly selecting a subset records from the original input data, so the execution time of sampling jobs can be short. To capture the influence of downsizing data, we profile a job with two different sampling data sizes to determine an scaling vector, f, so that the profiling values of the original data size can be estimated as $P \times f$.

3.2 Classifier

A variety of applications can run on Hadoop. Although they all have to follow the Spark programming model, their execution behaviors could still be widely varied. Therefore, we decide to classify jobs into several classes, and then build a time prediction model for each class individually. As shown by our evaluation in Section V-A, classification is necessary to achieve higher prediction accuracy. To build the classification model, we use the collected profiles of sampled jobs as the training dataset, and label the datasets according the type of applications in our system. Thus, give the profile of a job, the model identifies it to a most similar class. We apply the neural network model by using Scikit-Learn MLPClassifier (multilayer perceptron classifier) in our implementation. It's a fully connected neural network (CNN). After our tuning, we set the number of hidden layer to 3, and the number of neural in each layer to 100. We adopt adam as the optimizer, and the activation function is ReLU. We also set batch size to 125, alpha to 0.003 and the initial value of learning rate to 0.002.

3.3 Anomaly Detection Model

Generally speaking, residential electricity consumption is highly correlated to temperature. For example, in winter, electricity consumption increases since the temperature decreases because of the heating needs. In summer, electricity consumption increases when the temperature is higher because of cooling loads. The daily consumption pattern of a customer typically demonstrates the periodic characteristics, due to the living habit of the customer, e.g., the morning peak appears between 7 and 8 o'clock if a customer usually gets up at 7 o'clock; and evening peak appears between 17 and 20 o'clock (due to cooking and washing) if the customer gets home at 5 o'clock after work.

The PARX model, thus, uses a daily period, taking 24hours of the day as the seasons, i.e., t = 0...23, and uses the previous p days' consumptions at the hour at t for auto-regression. The PARX model at the s-th season and at the n-th period is formulated as

$$Y_{s,n} = \sum_{i=1}^{p} \alpha_{s,i} Y_{s,n-i} + \beta_{s,1} XT1 + \beta_{s,2} XT2 + \beta_{s,3} XT3 + \varepsilon_s, \quad s \in t \tag{1}$$

where Y is the data point in the consumption time-series; p is the number of order in the auto-regression; XT1,XT2 and XT3 are the exogenous variables accounting for the weather temperature, defined in the equations of (2); α and β are the coefficients; and ε is the value of the white noise.

$$XT1 = \begin{cases} T-20 & if\ T>21 \\ 0 & otherwise \end{cases} \quad XT2 = \begin{cases} 16-T & if\ T<14 \\ 0 & otherwise \end{cases} \quad XT3 = \begin{cases} 5-T & if\ T<6 \\ 0 & otherwise \end{cases} \tag{2}$$

The variables represent the cooling (temperature above 20◦), heating (temperature below 16◦), and overheating (temperature below 5◦), respectively. The anomaly detection algorithm is of using unique variate Gaussian distribution, described in the following. Given the training data set, X = {xand the vari-1,x2,...,xn} whose data points obey the normal distribution with the mean μ ance δ2, the detection function is defined as

$$p(x;\mu,\delta) = \frac{1}{\delta\sqrt{2\pi}} e^{-\frac{(x-\mu)^2}{2\delta^2}} \tag{3}$$

where $\mu = \frac{1}{n}\sum_{i=1}^{n} x_i$, $\delta^2 = \frac{1}{n}\sum_{i=1}^{n}(x_i - \mu)^2$ and . For a new data point, x, this function computes its probability density. If the probability is less than a user-defined threshold, i.e., p(x) < ε, it is classified as an anomaly, otherwise, it is a normal data point. In our model training process, we compute the L1 distance between the actual and predict consumptions, i.e., ||Yt − Y^t||, where Yt is the actual hourly consumption at the time t, and Yt is the predict hourly consumption at the time t. The predict hourly consumption, Y^t, is computed using the PARX model in Eq.1.

We now describe the implementation of the anomaly detection system. We choose Spark Streaming, Spark, and PostgreSQL as the speed layer, batch layer and serving layer technology, respectively (see Fig.3). The system employs Spark to compute the models for anomaly detection, which reads the data from the Hadoop distributed file system (HDFS) in the batch layer. The batch job runs at a regular time interval, computes, and updates the detection models to the table in PostgreSQL database. Spark Streaming is used to process real-time data streams, e.g., directly reads the readings from smart meters, and detects abnormal consumption with the detection algorithm. The detection algorithm always uses the latest models getting from the PostgreSQL database. Spark Stream writes the detected anomalies back to the PostgreSQL database, which will be used for the notification of customers.

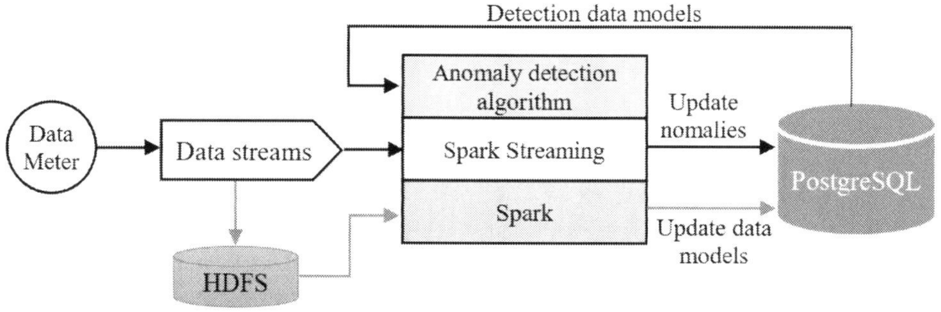

Fig. 3. The nomaly detection system

3.4 Predictor

The goal of Predictor is to predict the execution time of a job under a given Hadoop configuration and resource configuration. Therefore, the input data of our prediction model includes the job profile as well as the Hadoop configuration and resource configuration, and the output result from the prediction model is the job execution time. Since the sampled jobs only run with the default configurations, the training dataset for prediction model is collected from the actual job execution, including {HCopt,RCopt,Preal,Treal} as shown in Figure 2. Again, We apply neural network as our time prediction model, and implement it using Scikit-Learn MLPRegressor(multi-layer perceptron regressor). Similar to the classification model we use, it's a fully connected neural network. The difference between MLPClassifier and MLPRegressor is that MLPRegressor doesn't have activation function in the output layer, so the output value is a continuous value which can be used for regression. After our tuning, we set the number of hidden layer to 8, and the number of neural in each layer to 100. We adopt lbfgs as the optimizer, which is an optimizer in the family of quasi-Newton methods, and the activation function is ReLU.

3.5 Optimizer

Optimizer aims to determine the best job running configuration {HCopt,RCopt} with respect to an optimization objective function using the predicted execution time information. In this work, we focus on minimizing the execution time or the execution cost of individual jobs. Same as the pricing model used in most cloud services, the execution cost is the execution time multiplied by the unit price of resource instances per time interval. Since the resource configuration decides the type of resource instance, and the number of provisioned resources, the predicted execution cost can also be derived from our predictor. In the future, we also plan to explore the opportunity of optimizing other system level performance requirements as well, such as guarantee service quality of jobs with deadline, minimize average response time of job scheduler, increase the utilization and fairness of resource allocation, etc.

To apply our time prediction model for optimization, a general approach is to search through all the possible configurations. But even in the setting of our experiments, the number of dimension of configurations is 22, and the total number of possible configurations are more than 1029. Hence, brute force search is not feasible. Therefore, we adapt a 2-step approximate search approach to address this problem which consists of a global searching step and a local searching step. In the global searching step, a search of the entire global solution space is performed by randomly generating a set of configurations. The parameter values in these configurations are uniformly sampled from their corresponding valid value ranges. Then all these sampled configurations are evaluated by

Table 3. Types of vm instances.

Type	#CPU	Memory	Disk	Normalized Cost
Small	1	2	100GB	1
Medium	3	6	300GB	3
Large	6	12	600GB	6

the optimization objective function, and the one with the best result among them is selected as the global best. In the local searching step, the global best found from the first step is input as the initial search position. Then the gradient boosted trees algorithm is used to search the local minima with respect to global best position. In the implementation, we use the python optimization package, Scikit-Optimize, and call the functions of dummy minimize and gbrt minimize for global searching and local searching, respectively.

4 Experimental setup

4.1 Environment & Setup

We conducted experiments on an OpenStack cloud platform built by in-house cluster with 8 physical nodes. Each node has 2 Intel Xeon X5670 2.93GHz CPUs and 96GB memory. Three types of VM instances, Small, Medium, and Large, are offered in the cloud platform as shown in Table III. The large size VM is the default configuration in our experiments.

The application workloads in our evaluation were generated from HiBench [17], which is a well-known workload generators for Hadoop. As summarized in Table IV, we consider 7 applications from 4 different categories, includes text and graph processing, sort, and machine learning. All of them are commonly-used applications in Hadoop framework. Noted, for PageRank, Kmeans Clustering, and TriangleCount, their executions are consisted of two phases running different type of jobs. Hence, there are a total of 10 types of jobs in our experiments.

To evaluate the accuracy of our job classification model and the error rate of our time prediction model, we use the following metrics:

Job Classification

$$AvgAccu = \frac{1}{N}\sum_1^N \begin{cases} 0, & if\ Predict(n) \neq Actual(n) \\ 1, & if\ Predict(n) = Actual(n) \end{cases} \tag{4}$$

where N is the number of test cases. Time Prediction

$$AvgErr = \frac{1}{N}\sum_1^N \frac{Predict(n) - Actual(n)}{Actual(n)} \tag{5}$$

where N is the number of test cases.

4.2 State-of-the-art Comparison

To show the advantage of deep learning technique, we compared our prediction results to three other state-of-art prediction methods proposed in previous literature. We refer them as "EQ", "ML1", and "ML2" in the rest of paper, and their methods are detailed in below.

EQ, proposed in [14], is a white-box model based prediction method. It constructs an equation to compute the execution

Table 4. Types of applications in the evaluation.

Category	Application	Goal
Text	WordCount	Word frequency counter
	InvertedIndex	Computing inverted index of a set of documents
Graph	PageRank	Rank web pages
	TriangleCount	Triangle counting in a graph
Sort	Sort	Sort data
	TeraSort	
Machine Learning	KMeans Clustering	Group objects with similar

time according to the running behavior of a Spark job. It is known that Spark execution is consisted of three phases: map phase, data shuffle phase, and reduce phase. So this method models each of these phases separately. For instance, the time of map phase is proportional to the number of Map waves and input data size, where the number of Map waves is the number of map tasks divided by the number mapper slots (i.e., computing cores). In addition, because the execution time of first Reduce wave can be overlapped with the time of the last Map wave, the time of the first Reduce wave, and the time of subsequent Reduce waves are also modelled separately in this method. Finally, a linear regression method is used to tune the coefficients in the model according to different types of workloads. Comparing to other white-box approaches, this method already builds a more sophisticate equation to model the execution time. However, it is impossible to construct equations to explicitly model the effect of every parameters in Hadoop configuration. Therefore, this approach can only predict the time under different resource configurations, but not under different Hadoop configurations.

Tree-based, proposed in [11], is a gray-box modeling approach using tradition machine learning techniques. It first classifies configuration parameters into the parameter sets for map and reduce tasks using the random forest feature importance method. Then it builds a 2-level prediction model by tree-based regression such as random forest regressor, extra trees regressor and gradient boosting regressor as the ensemble method. The 2-level prediction model means that there is an additional prediction step on the intermediate features (such as the median task execution time) between the input features and the final time results. The idea of this 2-level modeling is similar to deep learning. But instead of learning these intermediate features automatically from training dataset, here it requires human knowledge to define these features properly.

SVM, proposed in [21], is also a gray-box modeling approach using tradition machine learning techniques. It applies a stepwise regression for feature selection and uses Supported Vector Machine(SVM) regression for building the time prediction model. Both machine learning techniques, ML1 and ML2, can be used to predict the time under any given Hadoop configuration and resource configuration because all the parameters in the configurations can be treated as a feature in model.

5 Experimental evaluation

5.1 Job Classification

Here we evaluate the benefit and accuracy our job classification method. As mentioned in Section IV-A, there are 10

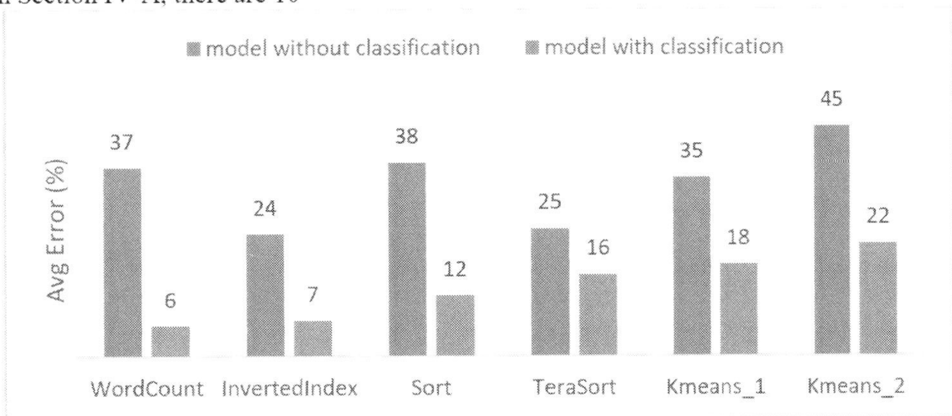

Fig. 4. The comparison of error rates between the time prediction model built with job classification and without job classification.

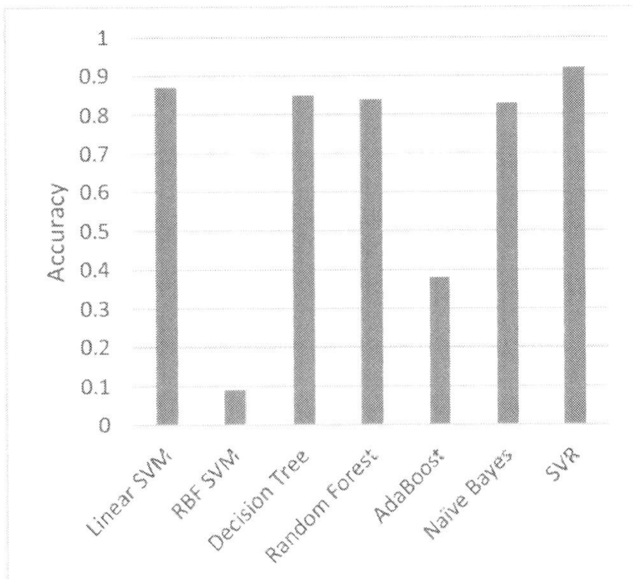

Fig. 5. The comparison of accuracy among different classification algorithms. Our method model is SVR.

types of jobs in our experiments. Without classification, all jobs share the same time prediction model regardless their job type. On the contrary, with classification, a prediction model is built for each type of jobs individually. Figure 3 is the comparison of prediction error rates between the two approaches. As shown by the results, the error rates without classification are much higher than the error rates with classification across all types of jobs. The differences range from a factor of 1.5(i.e., Terasort) ~ 11.7 (i.e., Pagerank _2, 2nd execution phase of Pagerank) times higher. Therefore, classification is necessary to ensure our prediction accuracy.

In this work, we use deep learning (neural network) to build the classification model and identify the type of a new arrival job in our system. To validate our choice, Figure 4 compares the accuracy of our classification method to several other popular classification algorithms, including: 1) Nearest Neighbors classified by a majority vote of its neighbors, 2) Supported Vector Machine(SVM) with both linear and radial basis function (RBF) as the kernel function, 3) Decision Tree and Random Forest are tree-based model learning from separating and pruning, 4) Adaptive Boosting with the iterative algorithms, and 5) Naive Bayes applying the probabilistic theorems. As shown, our method, named Neural Network in the figure, did achieve the highest classification accuracy among all.

5.2 Time Prediction

For the evaluation of prediction accuracy, we compare the accuracy of our approach, NN, to three other methods, EQ, Tree-based, and SVM, as introduced in Section IV-B. EQ represents the class of traditional approaches that tries to construct explicit equations for capturing the effects of some key factors, such as the number of cores, tasks, slots, and input data size. The equation coefficients are determined by a linear regression method under a given Hadoop configuration. However, there could be tens or hundreds of parameters in Hadoop configuration. Hence, its prediction accuracy starts to deviate when the jobs are running with varied Hadoop configurations. To show this limitation, Figure 5 compares the prediction error rate when jobs are only generated with the default Hadoop configuration versus the results when jobs are generated with a randomly selected Hadoop configuration. As shown, for jobs with default Hadoop configuration, the error rate is less than 25% for the 3 types of job we tested. But for jobs with random selected Hadoop configuration, the error rate quickly grows 5%~10%. This result indicates the fact that this class of prediction method is unsuitable to address our prediction problem with large numbers of features with complex inter-dependency relationship.

Unlike EQ, Tree-based and SVM are similar to our neural network approach. They both can be easily generalized to consider the large number of features in Hadoop configuration.

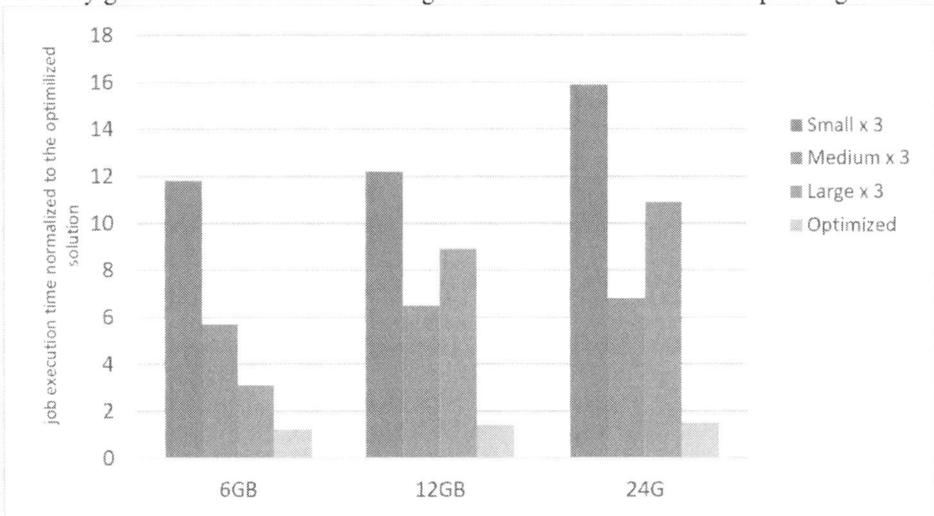

Fig. 7. Optimization results of the normalized execution time for Terasort.

But they rely on different techniques to train their models. Figure 6 compares the error rates of these methods for each of 10 types of jobs in our experiments. We found that Treebased is more likely suffering from the over-fitting problem because it has a simpler model comparing to the others. Take Kmeans1 as an example, we observed that Tree-based has a training error rate less than 8%, but its actual error rate is up to 23%. As a result, Tree-based has the highest error rates for all the job types, except PageRank1 (i.e., the iterated value propagation in PageRank algorithm). This is likely because the data input size of each iteration could change, but our profiler only records the initial input size in the job profile. The results between SVM and NN are close. But, clearly, NN still performs better than SVM in most cases, especially for PageRank2 and WordCount. The overall prediction error rate of our method is only 12.3%.

5.3 Time & Cost Optimization

To demonstrate the benefits of our auto-tuned system based on job time prediction information, we evaluate the time and cost saving of running a Terasort application with three different input data sizes (6GB, 12GB, and 24GB). In the experiment, the available size of VM is Small, Medium, and Large as shown in Table III, and the number VMs can be either 3 or 6. Without deep knowledge of Hadoop, we consider jobs can only be run with the default Hadoop configuration but with varied VM sizes and VM numbers. In the contrary, our approach tends to choose the best resource and Hadoop configuration for optimizing the given objectives.

The comparison result of execution time and cost are shown in Figure 7 and Figure 8, respectively. We normalize the experiment results with the optimized one, which is set to 1. For time optimization, having more computing resources, with larger and more number of VMs, often results in shorter execution time, but not always. For instance, in the case of 12GB and 24GB input data size, we observed that Medium size VMs had better performance than Large size VM. This is because the default number of reducer is 1, so the I/O contention on the reducer grows larger when more intermediate data is generated with a faster rate. Therefore, it is not trivial to find the optimal setting. But as shown by the plot, our method did achieve the best execution time by reducing the execution time by at least a factor of 10.5 comparing to other settings. The resource configurations found by our optimizer are 6 Large VMs in all three test cases, but the Hadoop configurations are different for each cases. Similarly, for cost optimization, our auto-tuned method also achieved the best results by reducing the cost by at least a factor of 1.12 comparing other settings. Again, the resource configurations found by our optimizer are 3 Small VMs in all three test cases, but the Hadoop configurations are still different for each cases.

6 Conclusion

In this paper, we developed a method to predict the execution time of Hadoop using deep learning technique. This paper explores the ability of recently developed Big Data approaches with respect to energy consumption prediction. The focus is on evaluating if and how findings from machine learning with Big Data can benefit consumption prediction. An approach based on local learning with support vector regression is presented. The approach takes advantage of parameter reduction to increase training speed. Local SVR outperformed H2O in both accuracy and training time. The presented local SVR was evaluated on the energy consumption prediction for event venues, but it could be applied for other energy consumption prediction scenarios.

Future work will evaluate the same approaches on much larger data sets to determine their performance on truly Big Data. Comparison with other distributed Big Data algorithms such as those supported by Spark will be performed. To partition the data, locality-sensitive hashing (LSH) will be evaluated as a potential replacement for k-means. Comparing to three other previous proposed methods, our evaluation showed that our prediction method achieved the highest time prediction accuracy in most test cases and significantly reduced job running cost and time. Our study results provided strong evidences that deep learning technique is well suited to predict the execution time of a job, especially when the job is managed by a complex computing framework and dynamic provisioned resource environment. In the future, we will keep exploring the opportunity of using such technique to address other resource management problems in cloud systems, such as fail prediction and scaling control, etc.

Acknowledgement

This study is conducted under the "Big Data Technologies and Applications Project (3/4)" of the Institute for Information Industry which is subsidized by the Ministry of Economic Affairs of the Republic of China. They also would like to thank Budweiser Gardens for providing valuable data for this project.

References

1. Large synoptic survey telescope. https://www.lsst.org/.
2. H. Altwaijry, E. Trulls, J. Hays, P. Fua, and S. Belongie. Learning to match aerial images with deep attentive architectures. In 2016 IEEE Conference on Computer Vision and Pattern Recognition (CVPR), pages 3539–3547, June 2016.
3. H. Zhao and F. Magoulès, "A review on the prediction of building energy consumption," *Renewable and Sustainable Energy Reviews*, vol. 16, no. 6, pp. 3586–3592, 2012.
4. Apache. Apache hadoop. [Online]. Available: http://hadoop.apache.org/.
5. Apache. Apache mahout. [Online]. Available: http://mahout.apache.org/.
6. Apache. Apache storm. [Online]. Available: http://sotrm.apache.org/.
7. AWS. Aws. [Online]. Available: https://aws.amazon.com/tw/.
8. K. J. Bowers, B. J. Albright, L. Yin, B. Bergen, and T. J. T. Kwan. Ultrahigh performance three-dimensional electromagnetic relativistic kinetic plasma simulation. Physics of Plasmas, 15(5):7, 2008.
9. C. O. Chen, Y. Q. Zhuo, C. C. Yeh, C. M. Lin, and S. W. Liao. Machine learning-based configuration parameter tuning on hadoop system. In 2015 IEEE International Congress on Big Data, pages 386–392, June 2015.
10. J. Dean and S. Ghemawat. Mapreduce: Simplified data processing on large clusters. Commun. ACM, 51(1):107–113, Jan. 2008.
11. "Apache Mahout." http://mahout.apache.org/.
12. "Apache Spark." http://spark.apache.org/.
13. K. Grolinger, A. L'Heureux, M. A. M. Capretz, and L. Seewald, "Energy forecasting for event venues: Big Data and prediction accuracy," Energy and Buildings, vol. 112, pp. 222–233, 2016.
14. Oxdata, "H2O." http://www.h2o.ai/2016.
15. V. N. Vapnik, Estimation of Dependences Based on Empirical Data, New York: Springer-Verlag. 1982.
16. L. Bottou and V. Vapnik, "Local learning algorithms," Neural Computation, vol. 4, no. 6, pp. 888–900, 1992.
17. I. W. Tsang, J. T. Kwok, and P.-M. Cheung, "Core vector machines: Fast SVM training on very large data sets," Journal of Machine Learning Research, pp. 363–392, 2005.
18. F. Niu, B. Recht, R. Christopher, and S. J. Wright, "Hogwild!: A lockfree approach to parallelizing stochastic gradient descent," Advances in Neural Information Processing Systems, pp. 693–701, 2011.

19. H. Zhang, G. Chen, and B. Ooi, "In-memory Big Data management and processing: A survey," IEEE Transactions on Knowledge and Data Engineering, vol. 27, no. 7, pp. 1920–1948, 2015.
20. X.-W. Chen and X. Lin, "Big Data deep learning: challenges and perspectives," IEEE Access, vol. 2, pp. 514–525, 2014.
21. M. M. Najafabadi, F. Villanustre, T. M. Khoshgoftaar, N. Seliya, R. Wald, and E. Muharemagic, "Deep learning applications and challenges in big data analytics," Journal of Big Data, vol. 2, no. 1, pp. 1–21, 2015.
22. R. K. Jain, K. M. Smith, P. J. Culligan, and J. E. Taylor, "Forecasting energy consumption of multi-family residential buildings using support vector regression: investigating the impact of temporal and spatial monitoring granularity on performance accuracy," Applied Energy, vol. 123, pp. 168–178, 2014.
23. L. Suganthi and A. Samuel, "Energy Models for Demand ForecastingA Review," Renewable and Sustainable Energy Reviews, vol. 16, no. 2, pp. 1223–1240, 2012.
24. A. S. Ahmad, M. Y. Hassan, M. P. Abdullah, H. A. Rahman, F. Hussin, H. Abdullah, and R. Saidur, "A review on applications of ANN and SVM for building electrical energy consumption forecasting," Renewable and Sustainable Energy Reviews, vol. 33, pp. 102–109, 2014.
25. H. C. Jung, J. S. Kim, and H. Heo, "Prediction of building energy consumption using an improved real coded genetic algorithm based least squares support vector machine approach," Energy and Buildings, vol. 90, pp. 76–84, 2015.
26. Lee, W., Stolfo, S.J., Chan, P.K., Eskin, E., Fan, W., Miller, M., Zhang, J.: Realtime data mining-based intrusion detection. In: DARPA Information Survivability Conference and Exposition II, DISCEX 2001, vol. 1, pp. 89–100. IEEE Press, New York (2001)

Unknown parameter identification of mobile heating source by using the sensitivity of sensor network

Thanh-Phong Tran[1,*]

[1]Tien Giang University, 119 Ap Bac, Ward 5, My Tho City, Tien Giang, Vietnam

Abstract. This communication focuses on an inverse problem of the physical phenomena which can be described by the partial differential equation. Especially, an unknown parameter identification process of the heat conduction problem in the thermal domain related to mobile heating source tracking is investigated. To do this, an iterative minimization of a quadratic cost-function based on the conjugate gradient method is widespread since this algorithm provides regularization properties for an estimation of the trajectory of a mobile heating source. Furthermore, another identification method is proposed by using the conjugate gradient method combined with the sensitivity problem of the sensor network. A quasi-online adaptation of this numerical method aims to identify the unknown characteristics with a small delay after the required temperature measurements by selecting the better sensors locations, this could also reduce the computational time. This approach offers the opportunity to move mobile sensors in order to increase the overall sensitivity and thus to improve the quality of the identification which is a shorter delay and better accuracy.

Keywords: heat equation, inverse problem, modeling, parametric identification, partial differential equation.

1 Introduction

In engineering science, it is necessary to model many physical phenomena or technical system which can be modeled by partial differential equation (PDE) in order to predict and to act on their effect. However, to determine the value of the parameters is not easy when building their mathematical models. To do this, an inverse problem can be formulated as follows: find the unknown parameters such that the simulated output provided by the resolution of direct problem are closed to the measurements provided by sensors. It is usual to investigate such inversion as a minimization problem where a quadratic cost-function has to be minimized. Especially, we use the heat conduction problem which is also described by a parabolic partial differential equation such as an example. The goal aims to identify the trajectory of the mobile heating source. It is well known that inverse heat conduction problems are ill-posed in Hadamard sense and that it is still possible to find accurate solutions using regularization techniques [1]. Iterative minimization based on conjugate gradient method (CGM) is known as a stable algorithm for this case. Stabilizing effect during the iterative minimization is highlighted in [2] and it is shown that this method acts like a sequential filtering mechanism capable of rejecting random perturbations in measurements during the identification process.

*e-mail: tranthanhphong@tgu.edu.vn

In this research context, the quasi online identification was proposed and developed in [3], it is based on successive time intervals for identification which slide on the total time horizon in order to take into account the continuous updating of observations. The disadvantage was necessary to use all of sensors. A sensors positioning method is investigated in order to improve it. To do this, several mobile pointwise sensors are considered. Thus sensitivity functions are determined throughout the process and are analysed in order to determine the better position for each sensor. Such methodology has been developed in [4] for a set of fixed sensor in order to select a subset of relevant sensors among a large number of candidate sensors. The main problem of mobile sensors positioning is to avoid observations redundancy and to deal with the constraints and conflicts related to the trajectories. In order to illustrate the proposed approach a demonstrator is built in the laboratory aided by the simulated experimentation. A numerical software dedicated to quasi online identification based on conjugate gradient method is implemented. Adaptation to sensors positioning is discussed considering this validated numerical tool.

The content of this article consists of four following parts. In the next part, physical system context and mathematical model are briefly described. Thus direct problem is stated. In the third part, the methodology of inverse problem (identification of the trajectory) is proposed and quasi online identification method using the sensitivity problem of sensor network is presented. Some numerical results of this method is considered in order to discuss about sensors positioning in the fourth section. The concluding remarks are presented in the last section.

2 Problematic

Let us consider that a small mobile heating source is moving on a plane surface. The main objective is to determine the source trajectory considering temperatures measurements provided by five mobile sensors. In the following several elements of the experimental device are described. Then partial differential equations system is proposed in order to provide an efficient numerical predictive tool.

A mobile heating source is moving on a square plate. In this study, the hypothesis of a 2D geometry was considered in order to reduce the computing times. This assumption is valid only if heat transfers in the plate thickness are neglected. Then, a thin metallic plate (with a high thermal conductivity) is considered. Aluminum plate was chosen (square plate with a size equal to 3m) and horizontally set on a support providing insulation.

2.1 Physical system context

To construct an experimental model for verifying the method proposed in this paper, we assume that there is one heat source S moving on the surface of a square aluminum sheet $\Omega \in \mathbb{R}^2$ of lateral dimension L and thickness e described as shown in Figure 1. The boundary of the studying domain is denoted by $\partial\Omega \in \mathbb{R}$. The variable space of the system $(x, y) \in \Omega = \left]-\frac{L}{2}, +\frac{L}{2}\right[\times \left]-\frac{L}{2}, +\frac{L}{2}\right[$ are measured in meter and the variable time $t \in \mathrm{T} = \left[0, t_f\right]$ is measured in second. This metal sheet is heated by a heat source having the thermal flux density functions $\phi(t)$ in Wm^{-2} which are assumed to be a homogeneous disk D with center $I\left(x(t), y(t)\right)$ and radius r. The temperature distribution function of a metal plate is $\theta(x, y, t)$ in Kelvin is a continuous function in space and time. Assuming that the values of parameters of a system used to construct the experimental model $\Psi\left(\Omega, \rho, c, \lambda, h, \phi(t), I(t), \theta_0\right)$ are known and listed in Table 1 with the unit of measure of the quantities in the unit of measure in the International System of Units (SI).

In particular, the metal plate is heated by a heat source traveling on its surface (Cartesian coordinate system xOy) to allow us to investigate the heat transfer on the surface and inside the plate. At the same time, the heat density densities of the sources are given by the function $\phi(t) = \phi_{\max} \exp\left(-(t-\alpha)^2/\beta^2\right)$ in (W/m^2) with $\alpha = 600$ and $\beta = 175$. The expression of the total thermal power density function of source $\Phi(x, y, z; t)$ is used to heat the experimental metal plate as follows:

$$\Phi(x, y, z; t) = \begin{cases} \phi(t) & if\ (x, y, z) \in D(I(t), r) \\ 0 & otherwise \end{cases} \tag{1}$$

This expression can be represented continuously and differentially as a function of the component density functions in time variables and in spatial coordinates as follows:

$$\Phi(x, y, z, t) = \frac{\phi(t)}{\pi} arccotan\left(\eta\sqrt{(x - x(t))^2 + (y - y(t))^2} - r\right) \tag{2}$$

The regularization parameter $\eta \in \mathbb{R}^+$ has been chosen to describe with precision the heat flux discontinuity. The time interval $\left[0, t_f\right]$ can be divided into N_t segments and defined using piecewise continuous linear functions: $\left[0, t_f\right] = \bigcup_{i=0}^{N_t-1} [t_i, t_{i+1}]$ with $t_i = \tau i$ and a discretization step defined by $\tau = t_f/N_t$. In order to avoid losing of generality, the orbital equation of all positioning of heat source $I(x(t), y(t), z(t))$ were re-established as discrete functions linearly and rewritten using basic triangle function $s^i(t)$ with $i = 0, 1, \ldots, N_t$:

$$s^i(t) = \begin{cases} 1 + t/\tau - i & if & t \in [t_{i-1}, t_i] \\ 1 - t/\tau + i & if & t \in [t_i, t_{i+1}] \\ 0 & otherwise & \end{cases} \tag{3}$$

Then, the density function of heat flow is expressed as follows $\phi(t) = \sum_{i=0}^{N_t} \phi^i s^i(t) = \left(\overline{\phi}\right)^{tr} \overline{s(t)}$ and the equation of motion of the heat source is expressed as follows $x(t) = \sum_{i=0}^{N_t} x^i s^i(t) = (\overline{x})^{tr} \overline{s(t)}$ and $y(t) = \sum_{i=0}^{N_t} y^i s^i(t) = (\overline{y})^{tr} \overline{s(t)}$. It denotes that "$tr$" is the symbol of the transposition matrix.

To evaluate the reliability of the proposed mathematical model to simplify the heat transfer equation into two-dimensional space, a set of heat sensors are fixed on the metal plate in order to collect temperature data from the sensor locations during the experiment. Or, the temperature value on the sheet metal is heated by each heat source recorded by sensors. Furthermore, to assess the effect of errors during the measurement process, it is assumed that the temperature collected from the sensors has been affected by the noise. These disturbances are followed by Gaussian probability distributions $\mathcal{N}(\mu, \sigma^2)$ with mean $\mu = 1$ and standard deviation $\sigma = 0$.

2.2 Mathematical modelling of studied system

Temperature $\theta(x, y, t)$ is described by the following partial differential equation system as follow where density ρ in kgm^{-3}, specific heat capacity c in $Jkg^{-1}K^{-1}$, thermal conductivity λ in $Wm^{-1}K^{-1}$, natural heat convection coefficient h in $Wm^{-2}K^{-1}$ and ambient temperature θ_0 in $K(Kevin)$ are assumed to be constant.

The direct problem of this system is expressed by equation (4), when the source term $\frac{\Phi(x,y;t) - 2h(\theta(x,y;t) - \theta_0)}{e}$, expressed in Wm^{-3}, is relevant if the thickness e is small enough. In the studied configuration, according to the heat fluxes range and to the parameters taken into account in the studied case, a previous numerical study has shown that the 2D model is valid in comparison with the 3D domain.

$$\begin{cases} \forall(x,y,t) \in \Omega \times T & \rho c \dfrac{\partial\theta(x,y,t)}{\partial t} - \lambda\Delta\theta(x,y,t) = \dfrac{\Phi(x,y;t) - 2h(\theta(x,y,t) - \theta_0)}{e} \\ \forall(x,y) \in \Omega & \theta(x,y;0) = \theta_0 \\ \forall(x,y,t) \in \partial\Omega \times T & -\lambda \dfrac{\partial\theta(x,y,t)}{\partial\vec{n}} = 0 \end{cases} \quad (4)$$

Usually, such mathematical problem is solved using numerical method such as the finite element method [5][6][7] and a numerical software such as Comsol Multiphysics TM interfaced with Matlab® [8][9]. The previous direct problem is well-posed since few noises on parameters Ψ introduce small disturbances on state estimation $\theta(x,y,t)$. Several illustrations for direct problem solution are shown in the following figures for a mobile source on a small area of the plate (about 4 square meters, see figure 1).

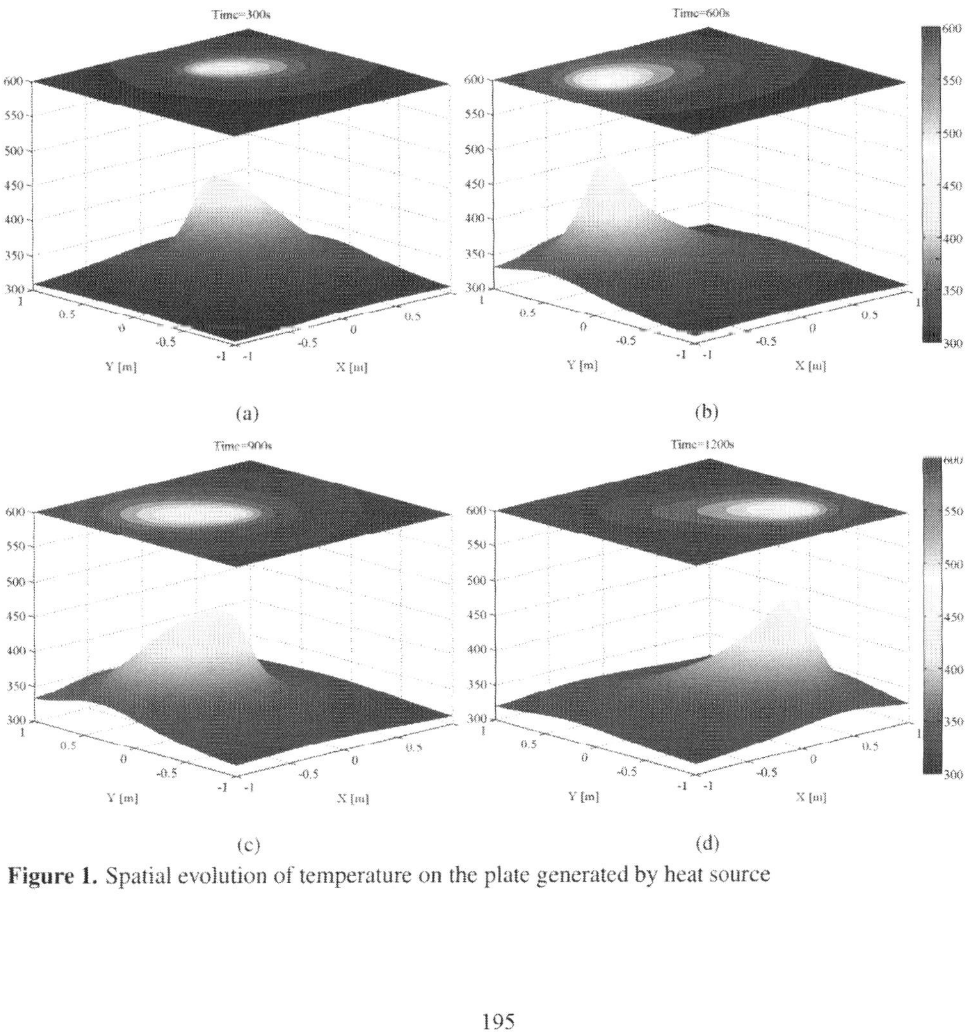

Figure 1. Spatial evolution of temperature on the plate generated by heat source

Temperatures are observed by a virtual network of 4 mobile sensors located on the plate. These "observed" temperatures are arbitrarily disturbed by a Gaussian noise. Once a parameter is unknown, mathematical formulations of the inverse problem considered in order to identify the unknown parameters are proposed in the following section.

3 Methodology of inverse problem

3.1 Inverse problem

Let us consider in the following that the source trajectory $\{x_I(t), y_I(t)\}$ is not known and has to be identified. In such an aim sensors measurements are available: $\hat{\theta}_m(t)$ provided by sensors C_m (for m=1 to 5). Even if observations are provided by pyrometers on a small surface, sensors C_m are assumed to be pointwise.

Inverse problem can be expressed as follows: Find $\{x_I(t), y_I(t)\}$ such that $J(\theta, I) = \frac{1}{2}\sum_{m=1}^{5}\int_0^{t_f}\left(\theta(C_m, t; I) - \hat{\theta}_m(t)\right)^2 dt$ is minimum with $theta(x, y; t)$ solution of (3).

Without loss of generalities, a discrete formulation is considered;

Find $\overline{I} = (\overline{x}_I, \overline{y}_I)$ such that $J\left(\theta, \overline{I}\right) = \frac{1}{2}\sum_{m=1}^{5}\sum_{j=1}^{n}\left(\theta\left(C_m, t^j; \overline{I}\right) - \hat{\theta}_m^j\right)^2$ is minimum with $\theta(x, y; t)$ solution of (3).

where n is related to the time sample for sensors measurements. CGM is implemented to identify the unknown parameters [10][11]. This algorithm requires iterative resolution of three well-posed problems: the direct problem (3) to calculate the cost-function $J\left(\theta, \overline{I^k}\right)$ and estimate the quality of the estimate $\overline{I^k}$ at iteration k; the sensitivity problem to calculate the descent depth (in the descent direction); the adjoint problem to determine the gradient of the cost-function $J\left(\theta, \overline{I}\right)$ and thus to define the next descent direction [12][13].

In the following, the mathematical formulation of the problems will be presented in detail in order to calculate the intermediate parameters of the identification method.

3.1.1 *Sensitivity problem*

Let us consider the variation of temperature $\delta\theta(\omega)$ with $(\omega) = (x, y, t)$ induced by a variation of the total heating flux noted: $\Phi^+(x, y; t) = \Phi(\omega) + \varepsilon\delta\Phi(\omega)$ with $\delta\theta(\omega) = \lim_{\varepsilon\to 0}\left(\frac{\theta^+(\omega) - \theta(\omega)}{\varepsilon}\right)$.

It's combined with the equation (4), the sensitivity problem is described by the following system:

$$\begin{cases} \rho c \dfrac{\partial\delta\theta(x,y,t)}{\partial t} - \lambda\Delta\delta\theta(x,y,t) = \dfrac{\delta\Phi(x,y,t) - 2h\delta\theta(x,y,t)}{e} & \forall(x,y,t) \in \Omega\times T \\ \delta\theta(x,y;0) = 0 & \forall(x,y) \in \Omega \\ -\lambda\dfrac{\partial\delta\theta(x,y,t)}{\partial\vec{n}} = 0 & \forall(x,y,t) \in \partial\Omega\times T \end{cases} \tag{5}$$

where: $\xi(\omega) = \sqrt{(x - x_I(t))^2 + (y - y_I(t))^2}$

$$\begin{aligned} \delta\Phi(\omega) &= \frac{\partial\Phi(\omega)}{\partial x_I(t)}\delta x_I(t) + \frac{\partial\Phi(\omega)}{\partial y_I(t)}\delta y_I(t) \\ &= \frac{\eta\phi(t)}{\pi\xi(\omega)\left(1 + \eta^2\left(\xi(\omega) - r\right)^2\right)}\left((x - x_I(t))\delta x_I(t) + (y - y_I(t))\delta y_I(t)\right) \end{aligned} \tag{6}$$

The sensitivity problem solution $\delta\theta(\omega)$ is useful to calculate the descent depth γ^{k+1} for each iteration, such as:

$$\overrightarrow{\Phi}^{k+1} = \overrightarrow{\Phi}^{k} - \gamma^{k+1}\overrightarrow{d}^{k+1} \tag{7}$$

Descent depth γ^{k+1} minimizes the criterion $J\left(\theta;\overrightarrow{\Phi}^{k+1}\right)$: $[\gamma^{k+1} = \underset{\gamma\in}{\text{Argmin}}\, J\left(\theta;\overrightarrow{\Phi}^{k+1}\right)$ or $\dfrac{\partial J\left(\theta;\overrightarrow{\Phi}^{k} - \gamma^{k+1}\overrightarrow{d}^{k+1}\right)}{\partial\gamma^{k+1}} = 0$. This implies that descent depth calculated at each iteration is:

$$\gamma^{k+1} = \frac{\int\limits_0^{t_f}\sum\limits_{n=1}^{N_c}\left(\theta_{C_n}\left(t;\overrightarrow{\Phi}^{k+1}\right) - \hat{\theta}_{C_n}(t)\right)\delta\theta\left(C_n,t;\overrightarrow{\Phi}^{k}\right)dt}{\int\limits_0^{t_f}\sum\limits_{n=1}^{N_c}\left(\delta\theta\left(C_n,t;\overrightarrow{\Phi}^{k}\right)\right)^2 dt} \tag{8}$$

In order to solve the sensitivity problem, the descent direction $\overrightarrow{d}^{k+1}$ $k \in {}^{6}\mathbb{N}_t$ has to be known. In such an aim the cost function gradient has to be computed *(according to the following problem)*.

3.1.2 Adjoin problem

In order to determine the gradient $\nabla\overrightarrow{J} = \left(\dfrac{\partial J}{\partial x_I};\dfrac{\partial J}{\partial y_I}\right)$ with $\forall i = 1,2,....,N_t$ at each iteration, a Lagrangian formulation $\ell(\theta(\omega),\Phi,\psi)$ is introduced such that:

$$\begin{aligned}\ell(\theta(\omega),\Phi,\psi) &= J(\theta(\omega),\Phi)\\ &+\int\limits_0^{t_f}\int\limits_\Omega\left[\rho c\frac{\partial\theta(\omega)}{\partial t} - \lambda\Delta\theta(\omega) - \frac{\Phi(\omega)}{e} + \frac{2h(\theta(\omega)-\theta_0)}{e}\right]\psi d\Omega dt\end{aligned} \tag{9}$$

If $\theta(\omega)$ is solution of the heat equation (4) then $\ell(\theta(\omega),\Phi,\psi) = J(\theta(\omega),\Phi) \rightarrow \delta\ell(\theta(\omega),\Phi,\psi) = \delta J(\theta(\omega),\Phi)$. When $\psi(x,y;t)$ is fixed, we denote $d\theta(\omega) = \theta_{C_n}\left(\omega;\overrightarrow{\Phi}^{k+1}\right) - \hat{\theta}_{C_n}(\omega)$, the Lagrangian variation can be written as:

$$\begin{aligned}&\delta\ell(\theta(\omega),\Phi,\psi)\\ &= \delta J(\theta(\omega),\Phi) + \int\limits_0^{t_f}\int\limits_\Omega\left[\rho c\frac{\partial\delta\theta(\omega)}{\partial t} - \lambda\Delta\delta\theta(\omega) - \frac{\delta\Phi - 2h(\delta\theta(\omega))}{e}\right]\psi(\omega)d\Omega dt\\ &= \int\limits_0^{t_f}\int\limits_\Omega\sum_{n=1}^{N_c} d\theta(t)\delta\theta(\omega)\delta_D(C_n)d\Omega dt + \int\limits_0^{t_f}\int\limits_\Omega\rho c\frac{\partial\delta\theta(\omega)}{\partial t}\psi(\omega)d\Omega dt\\ &- \int\limits_0^{t_f}\int\limits_\Omega\lambda\Delta\delta\theta(\omega)\psi(\omega)d\Omega dt + \int\limits_0^{t_f}\int\limits_\Omega\frac{2h\delta\theta(\omega)}{e}\psi(\omega)d\Omega dt - \int\limits_0^{t_f}\int\limits_\Omega\frac{\delta\Phi}{e}\psi(\omega)d\Omega dt\end{aligned} \tag{10}$$

Let us denote by $E(\omega) = \sum\limits_{n=1}^{N_c} d\theta(\omega)\delta_D(x(t) - x_{C_n}(t))\delta_D(y(t) - y_{C_n}(t))$ with $\delta_D(x - x_{C_n})\delta_D(y - y_{C_n})$ the Dirac distribution related to sensor $C_n(x_{C_n}, y_{C_n})$. Then it comes:

$$\begin{aligned}\delta\ell(\theta(\omega),\Phi,\psi) &= \int\limits_0^{t_f}\int\limits_\Omega\left[E(\omega) - \rho c\frac{\partial\psi(\omega)}{\partial t} - \lambda\Delta\psi(\omega) + \frac{2h\psi(\omega)}{e}\right]\delta\theta(\omega)d\Omega dt\\ +\rho c\int\limits_\Omega\delta\theta\left(x,y;t=t_f\right)\psi(\omega)d\Omega &+ \lambda\int\limits_0^{t_f}[\delta\theta(\omega)\nabla\psi(\omega)]_{\partial\Omega}dt - \int\limits_0^{t_f}\int\limits_\Omega\delta\Phi\frac{\psi(\cdot)}{e}d\Omega dt.\end{aligned} \tag{11}$$

Since $\frac{\partial \ell(\theta(x,y;t),\Phi,\psi)}{\partial \theta(x,y;t)}\delta\theta(\cdot)=0 \quad \forall \delta\theta(x,y;t)$, then the terms multiplied by $\delta\theta(x,y;t)$ have to be equal to zero, then it is necessary that the adjoin function $\psi(x,y;t)$ is solution of the following problem:

$$\begin{cases} \rho c \frac{\partial \psi(x,y,t)}{\partial t} + \lambda \Delta \psi(x,y,t) = E(x,y,t) + \frac{2h\psi(x,y,t)}{e} & \forall (x,y,t) \in \Omega \times T \\ \psi(x,t;t_f) = 0 & \forall (x,y) \in \Omega \\ -\lambda \frac{\partial \psi(x,y,t)}{\partial \vec{n}} = 0 & \forall (x,y,t) \in \partial\Omega \times T \end{cases} \tag{12}$$

Finally, when $\psi(x,y,t)$ is solution of the previous problem then: $\delta\ell(\theta(x,y,t),\Phi,\psi) = \frac{\partial \ell(\theta(x,y,t),\Phi,\psi)}{\partial \Phi}\delta\Phi = -\int_0^{t_f}\int_\Omega \delta\Phi \frac{\psi(x,y,t)}{e} d\Omega dt = \delta J(\theta(x,y,t),\Phi)$.

After several mathematical developments which cannot be presented due to the limited number of pages of this communication, cost function gradients can be formulated as follows:

$$\begin{aligned} \nabla J_{x_l^i} &= -\int_0^{t_f}\int_\Omega \frac{\eta\phi(t)}{\pi} \frac{(x-x_l(t))s^i_{x_l}(t)}{\pi\xi(x,y,t)\left(1+\eta^2(\xi(x,y,t)-r)^2\right)} \frac{\psi(x,y,t)}{e} d\Omega dt \\ \nabla J_{y_l^i} &= -\int_0^{t_f}\int_\Omega \frac{\eta\phi(t)}{\pi} \frac{(x-y_l(t))s^i_{y_l}(t)}{\pi\xi(x,y,t)\left(1+\eta^2(\xi(x,y,t)^{\text{`}}-r)^2\right)} \frac{\psi(x,y,t)}{e} d\Omega dt \end{aligned} \tag{13}$$

The descent direction can be estimated at each new iteration $k+1$ from the previous gradient formula, as follow:

$$\vec{d}^{k+1} = -\vec{\nabla J}(\theta;\Phi^k) + \frac{\left\|\vec{\nabla J}(\theta;\Phi^k)\right\|^2}{\left\|\vec{\nabla J}(\theta;\Phi^{k-1})\right\|^2}\vec{d}^k \tag{14}$$

where $\|\cdot\|$ is the Euclidean norm.

In the following section, the application of the online identification method based on the iterative regularization method (CGM) considering a network of fixed sensors is numerically implemented in order to identify the flux density as well as the trajectory of two sources.

3.2 Identification method based on CGM using sensitivity of sensor network

In recent works [3] authors have investigated the interest of CGM adaptation to quasi on line identification. Indeed, the main inconvenient for CGM is the convergence time which can be very important according to the problem complexity. Several strategies have been tested and compared to the reference strategy which is the offline approach: identification algorithm is launched when all the measurements are available. Quasi online identification is based on time windows for observation which are sliding according to the identification process.

- constant offset with constant time window size;
- adaptive overlap with constant time window size;
- time window size related to a priori information;
- adaptive time window size and constant offset;
- adaptive time window size and adaptive overlap.

In the configuration studied in [3] the last strategy has provided identification result in a computational time divided by 200 in comparison with the reference strategy (offline identification). From the results obtained from the method of choice of sensors in offline mode proposed above, we develop another method of choice of sensors in quasi-online mode. Near-line identification is performed using the adaptation of the conjugate gradient method (predictive with adaptive window size) using the on-line sensor selection algorithm.

Algorithm of selectioning sensor method based on the sentivity problem	
Step 1	- initialisation of parameter: k, L^2
	- time interval $\tau_m = [\tau_m^-, \tau_m^+]$
Step 2	- solve the sensitivity problem
	- extract values of $\theta(C_k, t)$
Step 3	- calculate $L_k^2 = \|\delta\theta(C_k, t)\|$
	- choose a sensor having the largest value of L^2
	- remove this sensor in the list
Step 4	- update index $k = k + 1$
	- ***if*** $(k > k_{max})$
	$\rightarrow$ end of algorithm
	- return to step 1.

This algorithm allows us to select the most relevant sensors over the time interval $\tau_m = [\tau_m^-, \tau_m^+]$ in which to deploy the robot-sensors. In addition, the online identification method makes it possible to "continue" during the experiment the mobile sources. This method relies on finding the most sensitive sensors by maximizing the following standard over slippery time intervals τ_m:

$$L_k^2 = \|\delta\theta(C_k, t)\| = \sqrt{\sum_{\tau_m} (\delta\theta(C_k, t))^2} \; (\forall k) \qquad (15)$$

First of all, it is necessary to determine the time interval using the method of Determination of the interval proposed and to initialize the parameters of the algorithm. By comparing these sensor standard values, the most sensitive sensors are selected. These are the optimal positions (in the sense of sensitivity) for robot-sensors in the next interval of the identification window. The application of this online sensor selection algorithm makes it possible to develop optimal sensor deployment strategies.

4 Numerical results and discussion

By applying proposed identification method based on CGM for an inverse problem combined with the sensitivity of sensor network, we can build an example to identify trajectory of a mobile heat source. After realizing the algorithm of this method, we obtained some results of identification process (see figure 2).

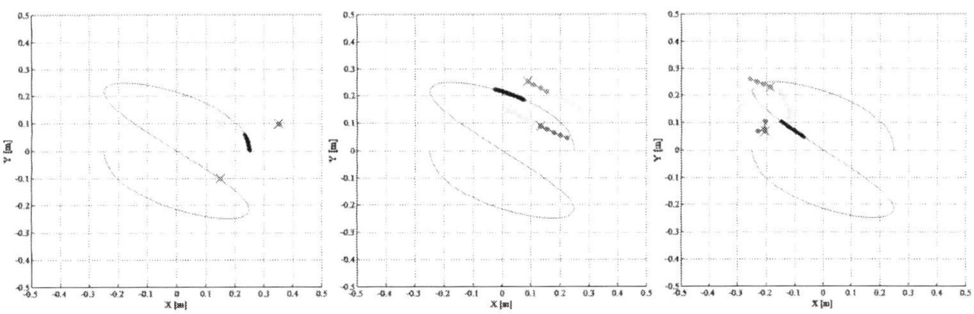

(a) Time interval 0 to 71 (b) Time interval 240 to 321 (c) Time interval 776 to 847

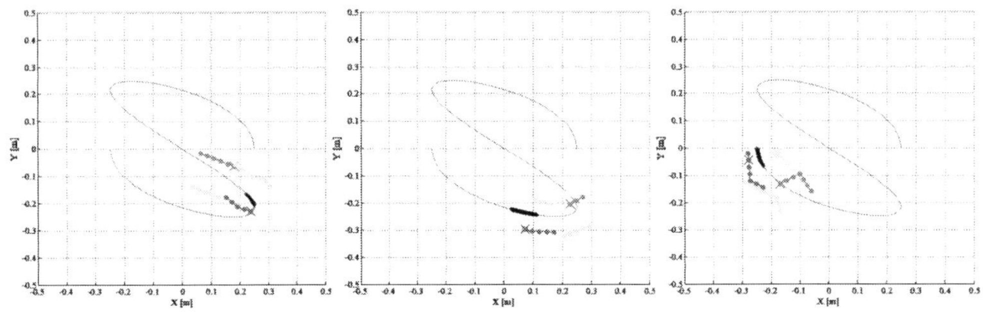

(d) Time interval 0 to 71 (e) Time interval 240 to 321 (f) Time interval 776 to 847

Figure 2. Results of identification process

In these above figures, a set of four better sensors were chosen in order to track the movement of the heat source according to its trajectory. These sensors provide the measured temperatures for an acquisition which is used in the proposed identification method based on CGM. It allows identifying the trajectory of this mobile heat source. It is now necessary to identify for each source the position every 15 seconds.

This represents an experiment of 1800 seconds is 121 parameters to identify. The acquisition of temperatures at different points on the plate will be carried out by a pack of 4 mobile sensors.

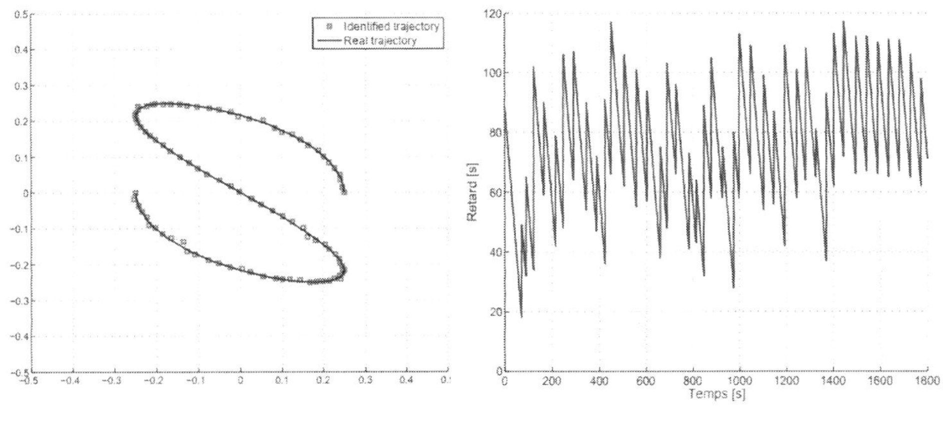

(a) Identified trajectory (b) Time of delay

Figure 3. Results of identified trajectory and time of relay

In figure 3, the figure (a) shows the identified trajecroty of the mobile heat source by using the proposed method. While the duration of the campaign is 1800 seconds, the estimation procedure converged in 1886 seconds, 86 seconds after the end of the experiment, the average identification delay is 85 seconds. We also obtain the following results (identification time, average temperature residues, the standard deviation of residues):

Table 1. Results of quasi-online identification process

$t_{identification}$ (s)	$\mu_{residues}$ (K)	$\sigma_{residues}$ (K)
1886	-0.19	0.64

5 Conclusion and outlooks

In this paper, a quasi-online unknown parameter identification method is proposed by using an iterative minimization based on the conjugate gradient method. Considering a mobile radiative heating source on a plane metallic surface, several mobile sensors are used in order to determine the unknown source trajectory. This ill-posed inverse problem is solved quasi-online by means of adapted iterative regularization method by minimizing a quadratic cost-function. This iterative regularization technique was used for an inverse problem in the thermal domain. Furthermore, the sensitivity problem of the sensor network was investigated to propose a method to choose the better position of each sensor. With the numerical results of this research, some future prospects are related to developing the whole methodology and to proposing the identification of new strategies for sensor moving and source tracking (in order to take into account redundancy and several heating sources). A long-term goal is the management of a fleet of embedded sensors, e.g. in the automobile, in order to track polluting cloud in a city.

References

[1] OM Alifanov,*"Inverse Heat Transfer Problems"*, Springer Science and Business Media (1994).

[2] M Prud'homme, TH Nguyen, *"On the iterative regularization of inverse heat conduction problems by conjugate gradient method"*, International Communications in Heat and Mass Transfer, 25 (1998).

[3] A Vergnaud, L Perez, L Autrique, *"Quasi-online parametric identification of moving heating devices in a 2D geometry"*, International Journal of Thermal Sciences, 102 (2016).

[4] L Perez, A Vergnaud, *"Observation strategies for mobile heating source tracking"*, High Temperatures-High Pressures, 45-1 (2016).

[5] DW Pepper, JC Heinrich, *"The finite element method – Basic concepts and Aplications"*, Taylor and Francis, Group, pp. 240 (1992).

[6] L Edsberg, *"Introduction to computation and modeling for differential equations"*, Wiley-Interscience, pp. 256 (2008).

[7] AJ Baker, *"Finite elements: computational engineering sciences"*, Wiley, pp. 288 (2012).

[8] WBJ Zimmerman, *"Multiphysics modeling with finite element methods"*, World Scientific Publishing, pp. 432. (2006).

[9] RW Pryor, *"Multiphysics modeling using Comsol v.4 – A first principles approach"*, Mercury Learn Inform, 2012.

[10] Y Jarny, MN Ozisik, JP Bardon, *"A general optimization method using adjoint equation for solving multidimensional inverse heat conduction"*, International Journal of Heat and Mass Transfer, 34-11, pp. 2911-2919 (1991).

[11] KA Woodbury, *"Inverse Engineering Handbook"*, Ed. CRC Press, pp. 480 (2002).

[12] C Huang, W Chen, *"A 3D inverse forced convection problem in estimating surface heat flux by conjugate gradient method"*, International Journal of Heat and Mass Transfer, 43, pp. 317-3181 (2000).

[13] S Beddiaf, L Perez, L Autrique, JC Jolly, *"Parametric identification of a heating mobile source in a three dimensional geometry"*, Inverse Problems in Science and Engineering, 23-1, pp. 93-111 (2015).

A proposal method for reducing the order of general heat conduction equation

Thanh-Phong Tran[1,*]

[1]Tien Giang University, 119 Ap Bac, Ward 5, My Tho, Tien Giang, Vietnam

Abstract. In the context of investigating methods dedicated to identifying unknown parameters of the system described by partial differential equations, particularly in the field of heat transfer, it has been realized that the heat transfer process in particular three-dimensional features is really complex and takes longer to calculate. Therefore, an equivalent mathematical model which is simpler proposed to reduce the calculation time and the costs of experimental activities. We observe that the mathematical models of the diffusion equation can be minimized in three-dimensional space into a similar two-dimensional pattern within certain limits did not change the physical properties of heat transfer process. A mathematical model and the numerical results of simulation experiments in order to prove effectiveness the proposed method will be presented in detail in this article.

Keywords: heat transfer equation, mathematical model, minimization, multidimensional space, partial differential equation.

1 Introduction

Nowadays, it's necessary to model the physical phenomena mathematically dedicated to research activities and purpose decision or prediction about their acts on our life, environment and production activities [1][2]. Evidently, there are many physical problems could be described by partial differential equation (PDE), such as the spread of oil spills on the seawater surface, the movement of contaminated water from the plant or forest fires, the movement of polluted clouds, etc [3][4].

Physical processes often have mathematical models similar to the ones mentioned above. Instead of developing algorithms or solutions for specific phenomena, we can build similar problems. Especially, the heat transfer in the conductor, it can also be modeled by the second order partial differential equation. In addition, simulation and experimentation can be made easier and cheaper to reproduce real-world phenomena.

However, during the study of heat transfer in the conductor, it was found that solving the problem of heat transfer equation in three-dimensional (3D) is not simple and takes too long when the computer was calculating these equations [5]. Instead, this general equation can be minimized into a two-dimensional (2D) space in some specific applications but without losing the physical properties of the heat transfer. This proposed model makes the calculation simpler, so it helps to find solutions faster and more efficient. That is the main objective of this article. Accordingly, mathematical models will in turn be introduced, compared and

[*]e-mail: tranthanhphong@tgu.edu.vn

concluded through the simulation results on the software. In addition, the results of this study will be useful for research and development of algorithms for problems described by the partial differential equation.

The content of this article consists of four following sections. The next section will briefly the problematic and mathematical models describing the heat transfer process in 3D space is shown in third part. The third section will focus on the proposed method for minimizing heat conduction equation in 3D into 2D. The numerical results obtained from the simulation and some discussions will be addressed in the fourth section. Finally, the concluding remarks will be represented in the last section.

2 Problematic and mathematical models

2.1 Problematic on physical phenomena

To construct an experimental model for verifying the method proposed in this paper, we assume that there are two heat sources, in case more general, $S_{j=1,2}$ moving on the surface of a square aluminum sheet $\Omega \in \mathbb{R}^3$ of lateral dimension L and thickness e described as shown in Figure 1. The boundary of studying domain is denoted by $\partial\Omega \in \mathbb{R}^2$ [6][7][8]. The variable space of the system $(x, y, z) \in \Omega = \left]-\frac{L}{2}, +\frac{L}{2}\right[\times \left]-\frac{L}{2}, +\frac{L}{2}\right[\times \left]-\frac{e}{2}, +\frac{e}{2}\right[$ are measured in meter and the variable time $t \in \mathrm{T} = \left[0, t_f\right]$ is measured in second. This metal sheet is heated by two heat sources having the thermal flux density functions $\phi_{j=1,2}(t)$ in Wm^{-2} which are assumed to be a homogeneous disk D_j with center $I_j\,(x(t), y(t), z(t))$ and radius r_j. The temperature distribution function of a metal plate $\theta(x, y, z, t)$ in Kelvin is a continuous function in space and time.

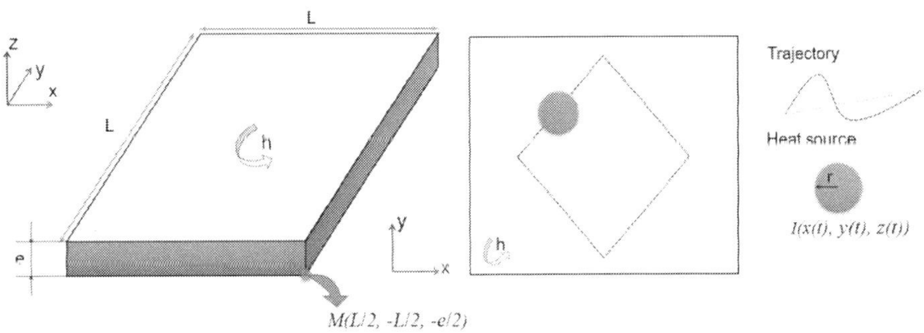

Figure 1. Representation of studied modeled system

Assuming that the values of parameters of a system used to construct the experimental model $\Psi\,(\Omega, \rho, c, \lambda, h, \phi(t), I(t), \theta_0)$ are known with the unit of measure of the quantities in the unit of measure in International System of Units (SI). In particular, the metal plate is heated by a heat source traveling on its surface (Cartesian coordinate system xOy) to allow us to investigate the heat transfer on the surface and inside the plate. The expression of the total thermal power density function of the sources $\Phi\,(x, y, z; t)$ is used to heat the experimental metal plate as follows:

$$\Phi\,(x, y, z; t) = \begin{cases} \phi_1(t) & if\,(x, y, z) \in D_1\,(I_1(t), r_1) \\ \phi_2(t) & if\,(x, y, z) \in D_2\,(I_2(t), r_2) \\ 0 & otherwise \end{cases} \qquad (1)$$

In other words, this expression can be represented continuously and differentially as a function of the component density functions in time variables and in spatial coordinates as follows:

$$\Phi(x, y, z, t) = \sum_{j=1}^{2} \frac{\phi_j(t)}{\pi} arccotan\left(\eta \sqrt{\left(x - x_j(t)\right)^2 + \left(y - y_j(t)\right)^2 + \left(z - z_j(t)\right)^2} - r_j\right) \quad (2)$$

The regularization parameter $\eta \in \mathbb{R}^+$ has been chosen to describe with precision the heat flux discontinuity. The time interval $\left[0, t_f\right]$ can be divided into N_t segments and defined using piecewise continuous linear functions: $\left[0, t_f\right] = \bigcup_{i=0}^{N_t-1} [t_i, t_{i+1}]$ with $t_i = \tau i$ and a discretization step defined by $\tau = t_f/N_t$. In order to avoid losing of generality, the orbital equation of all positioning of heat source $I(x(t), y(t), z(t))$ were re-established as discrete functions linearly and rewritten using basic triangle function $s^i(t)$ with $i = 0, 1, \ldots, N_t$:

$$s^i(t) = \begin{cases} 1 + t/\tau - i & if \quad t \in [t_{i-1}, t_i] \\ 1 - t/\tau + i & if \quad t \in [t_i, t_{i+1}] \\ 0 & otherwise \end{cases} \quad (3)$$

Then, the density function of heat flow is expressed as follows $\phi_j(t) = \sum_{i=0}^{N_t} \phi^i_j s^i(t) = \left(\overline{\phi_j}\right)^{tr} \overline{s(t)}$ and the equation of motion of the heat source is expressed as follows $x_j(t) = \sum_{i=0}^{N_t} x^i_j s^i(t) = \left(\overline{x_j}\right)^{tr} \overline{s(t)}$, $y_j(t) = \sum_{i=0}^{N_t} y^i_j s^i(t) = \left(\overline{y_j}\right)^{tr} \overline{s(t)}$ and $z_j(t) = \sum_{i=0}^{N_t} z^i_j s^i(t) = \left(\overline{z_j}\right)^{tr} \overline{s(t)}$. It denotes that "*tr*" is the symbol of the transposition matrix.

To evaluate the reliability of the proposed mathematical model to simplify the heat transfer equation into two-dimensional space, 10 heat sensors are fixed on the metal plate as shown in Figure 3 in order to collect temperature data of the sensor locations during the experiment. Or, the temperature value on the sheet metal is heated by each heat source recorded by five sensors. Furthermore, to assess the effect of errors during the measurement process, it is assumed that the temperature collected from the sensors has been affected by the noise. These disturbances are followed by Gaussian probability distributions $\mathcal{N}(\mu, \sigma^2)$ with mean $\mu = 1$ and standard deviation $\sigma = 0$.

2.2 Mathematical modelling of studied system

In the context of this research, it hypothesizes that characteristic parameters of the system $\Psi(\Omega, \rho, c, \lambda, h, \phi(t), I(t), \theta_0)$ are well-known and that the assumptions of parameters ρ, c, λ, h are constants. If the used material is heterogeneous that means thermal conductivity is a function of other parameters $\lambda(t) = \lambda(\zeta(t))$, then the solution of this problem becomes more complex even though they do not affect its results. In addition, the natural heat transfer coefficient is difficult to determine because it is highly dependent on the particular environment surrounding the heat conductor and has a great influence on the heat exchanger. However, these issues will be addressed in other articles.

The heat transfer equation in the aluminum sheet mentioned above is modeled by the mathematical equation system in the form of partial differential equations in a general way, as follows:

$$\begin{cases} \rho c \dfrac{\partial \theta (x, y, z, t)}{\partial t} - \lambda \Delta \theta (x, y, z, t) = 0 & \forall (x, y, z, t) \in \Omega \times \Gamma \\ \theta (x, y, z, 0) = \theta_0 & \forall (x, y) \in \Omega \\ -\lambda \dfrac{\partial \theta (x, y, z, t)}{\partial \vec{n}} = h (\theta(x, y, z, t) - \theta_0) - \Phi (x, y, z, t) & \forall (x, y, z, t) \in \partial\Omega \times \Gamma \end{cases} \tag{4}$$

where $\Delta\theta (x, y, z, t) = \left(\dfrac{\partial^2\theta(x, y, z, t)}{\partial x^2} + \dfrac{\partial^2\theta(x, y, z, t)}{\partial y^2} + \dfrac{\partial^2\theta(x, y, z, t)}{\partial z^2} \right)$ is the Laplace operator of the temperature variable $\theta(x, y, z, t)$ in space and time.

In the above equation, the initial condition of the differential equation is considered as the temperature of the surrounding environment. The boundary conditions of this equation depend on the natural heat convection coefficient h and the total heat flux density function $\Phi (x, y, z, t)$. In that, the natural heat convection coefficient is difficult to measure directly, it is usually determined by experience or model and chosen so that the sensitivity of the temperature distribution function changes at least compared to the variation of other parameters of system.

3 Minimization of heat conduction equation in 3D into 2D

3.1 Hypothesis

The idea of this article is to propose a thermal model in 2D geometry that has been considered and used as a good approximation of thermal transfers in a thin plate. The hypothesis of a 2D geometry (infinitely fine plate) was considered in order to reduce the computation time. This hypothesis is valid if we can neglect heat transfers in the thickness of the plate.

To do this, it is necessary to consider a thin plate of a metal of high thermal conductivity. For financial reasons, aluminum was retained. This square plate of 3 m is placed horizontally on a support providing insulation. The latter is composed of rock wool, thermal insulation supporting high temperatures. In order to the validate the model for the experimental prototype, a study aimed at studying the errors generated by an approximation in 2D geometry has been investigated and is presented in this section.

To reduce the general heat conduction equation in three-dimensional space into bi-dimensional, we fix $z(t) = e$ and this value is enough small. From the equation (4), we calculate boundary conditions of this system.

- Boundary condition of lateral surface:

$\partial\Omega_{lat}$ is the boundary of lateral surface of the aluminum plate and the thickness is too small.

$$-\lambda \frac{\partial \theta (x, y, z, t)}{\partial \vec{n}} = 0 \quad with \quad \forall (x, y, z; t) \in \partial\Omega_{lat} \times T \tag{5}$$

- Boundary condition of superior face:

$\partial\Omega_{sup}$ is the boundary of superior face of the aluminum plate.

$$-\lambda \frac{\partial \theta (x, y, z, t)}{\partial \vec{n}} = -e\Phi(x, y, z, t) \quad with \quad \forall (x, y, z; t) \in \partial\Omega_{sup} \times T \tag{6}$$

- Boundary condition of inferior face:

$\partial\Omega_{sup}$ is the boundary of inferior face of the aluminum plate.

$$-\lambda \frac{\partial \theta (x, y, z, t)}{\partial \vec{n}} = \frac{1}{R} (\theta(x, y, z, t) - \theta_0) \quad with \quad \forall (x, y, z; t) \in \partial\Omega_{inf} \times T \tag{7}$$

From the equations (5) to (7), we can rewrite the formulation of the heat transfer equation in 3D with a combination of the boundary conditions of (4) in this case, as follow:

$$\begin{cases} \forall (x, y, z; t) \in \Omega \times T & \rho c \dfrac{\partial \theta(x, y, t)}{\partial t} - \lambda \Delta \theta(x, y, t) = 0 \\ \forall (x, y, z) \in \Omega & \theta(x, y, t)(x, y; 0) = \theta_0 \\ \forall (x, y, z; t) \in \partial\Omega_{lat} \times T & -\lambda \dfrac{\partial \theta(x, y, t)}{\partial \vec{n}} = 0 \\ \forall (x, y, z; t) \in \partial\Omega_{sup} \times T & -\lambda \dfrac{\partial \theta(x, y, t)}{\partial \vec{n}} = -e\Phi(x, y, t) \\ \forall (x, y, z; t) \in \partial\Omega_{inf} \times T & -\lambda \dfrac{\partial \theta(x, y, t)_p}{\partial \vec{n}} = \dfrac{1}{R}(\theta(x, y, t) - \theta_0) \end{cases} \tag{8}$$

The coefficient R represents thermal resistance, also called thermal isolation coefficient of surface, is expresses itself in m^2KW^{-1}. We can also reuse the equation (8) to build another form of the partial differential equation system in order to obtain a new formulation. In the next section, a proposed PDE in second order which can replace this one in 3D will be represented.

3.2 Formulation of models

In this study, the hypothesis of a geometry 2D *(infinitely fine plate)* was considered in order to reduce the computing times. This assumption is valid only if we can neglect the heat transfer in the thickness of the plate. To do this, it is necessary to consider a fine plate of a metal having a high thermal conductivity. For financial reasons, aluminum was chosen (gold, silver and copper not being possible). This square plate 3m of the each side posed horizontally on a support providing insulation. This last is composed of Rockwool thermal insulation to high temperatures.

$$\begin{cases} \forall (x, y, t) \in \Omega \times T & \rho c \dfrac{\partial \theta(x, y, t)}{\partial t} - \lambda \Delta \theta(x, y, t) = \dfrac{\Phi(x, y, t) - 2h(\theta(x, y, t) - \theta_0)}{e} \\ \forall (x, y) \in \Omega & \theta(x, y; 0) = \theta_0 \\ \forall (x, y, t) \in \partial\Omega \times T & -\lambda \dfrac{\partial \theta(x, y, t)}{\partial \vec{n}} = 0 \end{cases} \tag{9}$$

The heat flux density function can be rewritten as follows:

$$\Phi(x, y; t) = \begin{cases} \phi(t) & if \quad (x, y) \in D(I(t), r) \\ 0 & otherwise \end{cases}$$

and the total flux can also be expressed continuously and differentially as: $\Phi(x, y, t) = \dfrac{\phi(t)}{\pi} \text{arcoctan}\left(\eta\left(\sqrt{(x - x_I(t))^2 + (y - y_I(t))^2} - r\right)\right)$

4 Numerical results and discussion

4.1 Numerical results

In order to verify the effectiveness of proposed method or verify that the heat transfers are surfacing on the aluminum plate, two mathematical models were compared, a modeled simulation of a system was built which the parameters are in Table 1. The systems of partial differential equations (4) and (9) are solved using the finite element method [3] implemented by the Comsol Multiphysics code interfaced with Matlab software [9][10][11][12].

Table 1. Input parameters of proposed models

$\rho c = 2.4 \times 10^6$	$\lambda = 237$	$h = 15$	$R = 1.2$
$\phi_{heat}(x, y, t) = 10^5 e^{-10^3 d^2} sin\left(\frac{2\pi t}{1200}\right)$			$\theta_0 = 291$

In Table 2, $d(x, y)$ is the distance between the point (x, y) and the center of the top face of the aluminum plate. It can be noted that the 3D configuration studied is axisymmetric. The following table indicates at various points the mean absolute error between two models, as well as the absolute average deviation of temperature $emod[K]$ between 2 points located on both sides of the plate (information $ecart[K]$ obtained with the model (2)).

Table 2. Numerical results of comparison of two models 2D and 3D

d [m]	0	0.05	0.10	0.15	0.20	0.25
emod [K]	1.4	1.2	0.7	0.6 K	0.4	0.3
ecart [K]	0.3	0.02	0.003	0.002	0.001	0.0008

In Figure 2, the temperature changes at different points of the upper face of the plate are presented for a plate thickness $e = 2mm$. It shows that the 2D model is very satisfactory for points near the heating source. For points far away from the source, weak errors of model will be present.

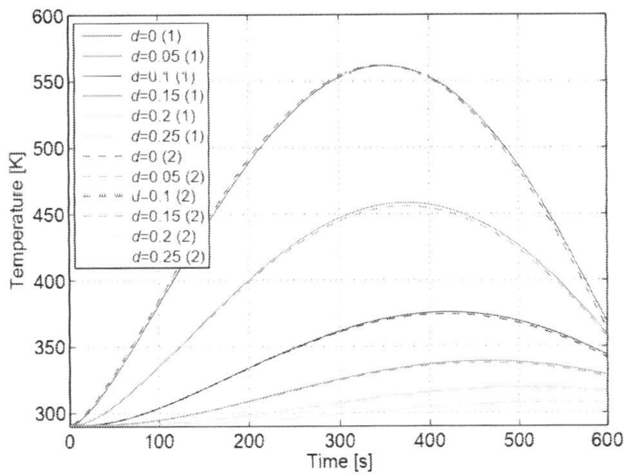

Figure 2. Comparison of the 2D and 3D models

Moreover, when we focus on the computational time, the response time of model 2D is $t_{2D} = 5.32(s)$. While it can find the results of model 3D after $t_{3D} = 11.04(s)$, it means t_{3D}/t_{2D} is 2.08 times. It's shown that the goal of this research to reduce the mathematical complexity and the computational time is satisfied.

4.2 Application in modelling of mobile heat sources by using proposed method

In this section, the proposed method is applied to model a heat conduction problem generated by two mobile heat sources. The trajectory of these moving heat sources is described in Figure 3. At the same time, the heat density densities of the sources are given by the function $\phi_1(t) = \phi_{\max}\exp\left(-(t-\alpha)^2/\beta^2\right)$ and $\phi_2(t) = \phi_{\max} - \phi_1(t)$ in (W/m^2) with $\alpha = 600$ and $\beta = 175$, the graph of these functions is shown in Figure 2 (b).

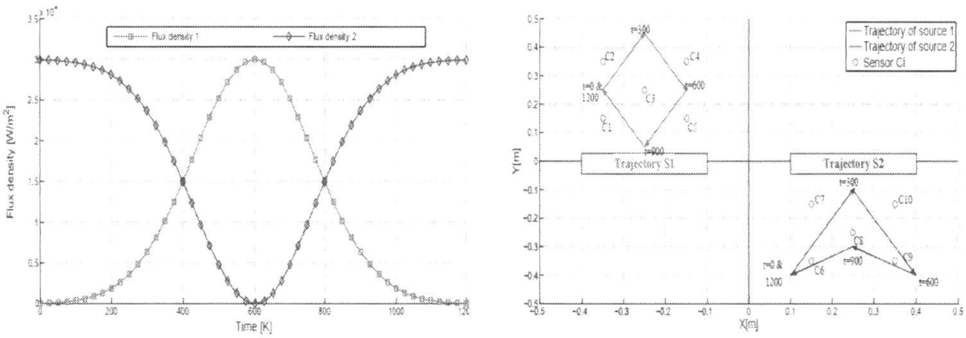

(a) (b)

Figure 3. Heat flux density (a) & Trajectory (b) of two heat sources

Solving the direct problem (9) with PDE in the second order, can be expressed the problem in 3D, yields the spatiotemporal distribution of temperature within the domain, see Figure 4 a to d, respectively correspond with the moments $t = \{300, 600, 900, 1200\}$ (s).

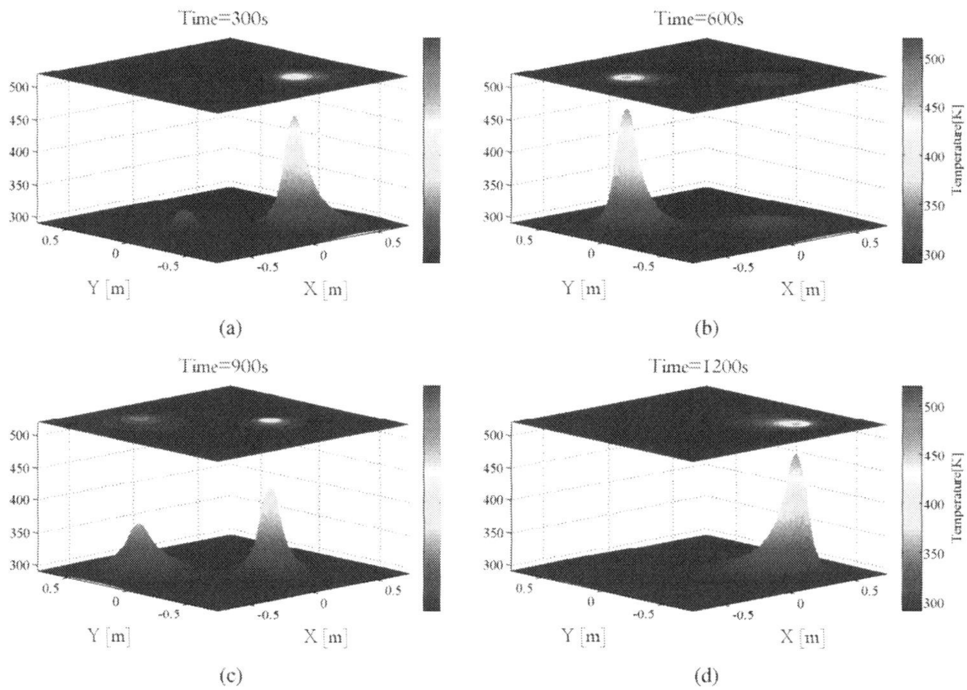

Figure 4. Spatial evolution of temperature on the plate generated by two heat sources

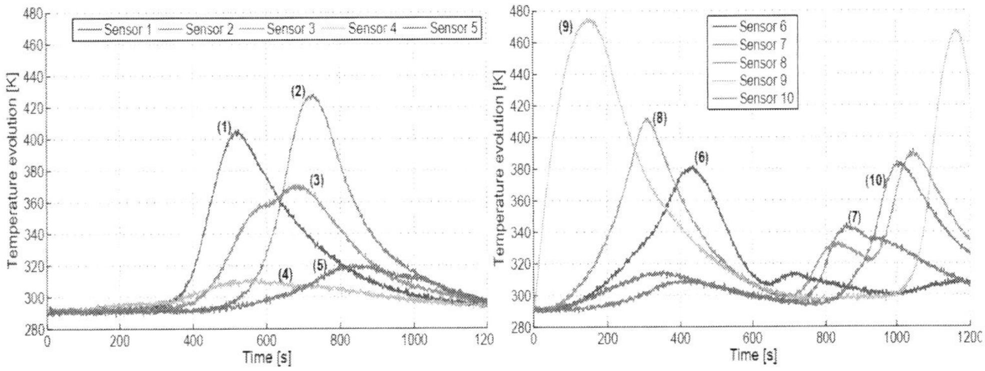

(a) Temperature evoltion of sensors C_{1-5} (b) Temperature evoltion of sensors C_{6-10}

Figure 5. Heat flux density (a) and Trajectory (b) of two heat sources

These numerical data are considered as "measures" of temperatures to test if the identification procedure makes it possible to determine what is the density of flow which is the cause, for an example. A network of fixed "temperature sensors" is considered (see Figure 3 (b)). The numerical results are noisy with a realistic measurement noise: disturbances followed by Gaussian probability distributions $\mathcal{N}(\mu, \sigma^2)$ with mean $\mu = 1$ and standard deviation $\sigma = 0$. The temperature evolution at these sensors is shown in Figure 5.

5 Conclusion and perspective

This paper proposes a mathematical model of heat transfer in two-dimensional space that can replace heat transfer in three-dimensional space but does not change the physical properties of the system. The purpose of the proposed alternative model is to simplify the mathematical model to make it easier to solve mathematical problems associated with this mathematical model. From there, it makes the response time of the system built from this model faster and makes the associated algorithms more robust. Through the statistics of the results from the simulation, it was shown that the mathematical model describing the heat transfer process in the proposed two-dimensional space could replace the heat equation of the reaction in complex three-dimensional space. more than The results of this study will be of great help to other studies related to the use of mathematical models describing heat transfer in later thermal conductivity.

References

[1] Vergnaud, A., Perez, L., Lucidarme, P., Autrique, L., *"DARC-EDP : Conception du prototype"*. In 4ième Journées des Démonstrateurs en Automatique. Université d'Angers (2013).

[2] Vergnaud, A., Tran, T. P., Perez, L., Lucidarme, P., Autrique, L., *"Deployment strategies of mobile sensors for monitoring of mobile sources: method and prototype"*. In Control Architectures of Robots 2015, 10th National Conference - CAR2015, (2015).

[3] Beddiaf, S., *"Identification paramétrique de systèmes d'EDP paraboliques non linéaires en géométrie 3D par une méthode de régularisation itérative"*. Thèse de doctotat, Université d'Angers (2013).

[4] Vergnaud, A., *"Adaptive deployment of a mobile sensors network to optimize the monitoring of an evolutionary phenomenon described by partial differential equations"*. Doctoral thesis, University of Angers (2015).

[5] Beddiaf, S., Perez, L., Autrique, L., Jolly, J.-C., *"Parametric identification of a heating mobile source in a three-dimensional geometry"*. Inverse Problems in Science and Engineering, 23(1), 93–111, (2015).

[6] Lefèvre, F., Le Niliot, C., *"A boundary element inverse formulation for multiple point heat sources estimation in a diffusive system: Application to a 2D experiment"*. Inverse Problems in Engineering, 10(6), 539–557, (2002).

[7] Martin, T. J., Dulikravich, G. S.,*"Inverse Determination of Boundary Conditions and Sources in Steady Heat Conduction With Heat Generation"*. Journal of Heat Transfer, 118(3), 546–554, (1996).

[8] Gillet, L. P. and L. A. and M., *"Implementation of a conjugate gradient algorithm for thermal diffusivity identification in a moving boundaries system"*. Journal of Physics: Conference Series, 135(1), 12082, (2008).

[9] DW Pepper, JC Heinrich, *"The finite element method – Basic concepts and Aplications"*, Taylor and Francis, Group, pp. 240 (1992).

[10] L Edsberg, *"Introduction to computation and modeling for differential equations"*, Wiley-Interscience, pp. 256 (2008).

[11] AJ Baker, *"Finite elements: computational engineering sciences"*, Wiley, pp. 288 (2012).

[12] WBJ Zimmerman, *"Multiphysics modeling with finite element methods"*, World Scientific Publishing, pp. 432. (2006).

Non-linear failure rate: A comparison of the Bayesian and frequentist approaches to estimation

Tien Thanh Thach[1,*] and *Radim* Briš[1]

[1]Department of Applied Mathematics, VŠB – Technical University of Ostrava, The Czech Republic

Abstract. In this article, a new generalization of linear failure rate called non-linear failure rate is developed, analyzed, and applied to a real dataset. A comparison of Bayesian and frequentist approaches to the estimation of parameters and reliability characteristics of non-linear failure rate is investigated. The maximum likelihood estimators are obtained using the cross-entropy method to optimize the log-likelihood function. The Bayes estimators of parameters and reliability characteristics are obtained via Markov chain Monte Carlo method. A simulation study is performed in order to compare the proposed Bayes estimators with maximum likelihood estimators on the basis of their biases and mean squared errors. We demonstrate that the proposed model fits a well-known dataset better than other mixture models.

1 Introduction

The Weibull failure rate,

$$h(t) = \frac{a}{b}\left(\frac{t}{b}\right)^{a-1} \tag{1}$$

equals 0 at time $t = 0$. This failure rate model might be only appropriate for modeling some physical systems that do not suffer from random shocks. The model, where initial failure rate is negligible might be inappropriate for modeling some physical systems that require initial positive failure rate. That is the physical systems that suffer from random shocks and also from wear out failures. The linear failure rate (LFR),

$$h(t) = a + bt \tag{2}$$

provides partly a remedy for this problem.

The LFR was originally proposed in [1], and has been studied in [2] as a special case of polynomial failure rate model for type II censoring. It is a generalization of an exponential distribution model ($b = 0$) and the Rayleigh distribution model ($a = 0$). The LFR model can also be considered as a mixture of failure rates of an exponential distribution and a Rayleigh distribution. Because the Rayleigh distribution also has a lot of limitations, as well as the LFR, that is not flexible enough to represent the most common types of failure rates, new generalizations of LFR must be developed.

In our study, we bring a new non-linear failure rate (NLFR) model, which is a generalized version of the LFR. We introduce it as a mixture of failure rates of the exponential and Weibull

*e-mail: tien.thach.thanh@vsb.cz

distributions. One of the first attempt to solve the non-linear failure rate model provides [3] but solutions were hard to find due to computational difficulties. In addition, the non-linear failure rate generalized in [3] differs from our new model in other details.

The remainder of this article is organized as follows. Section 2 introduces the intended non-linear failure rates model including basic characteristics of the corresponding lifetime distribution as well. Section 3 brings maximum likelihood estimators for the three unknown parameters of the model on one hand and the adaptive Markov chain Monte Carlo (MCMC) algorithm based on MCMC method on the other hand. In addition, Bayesian methodology resulting from the special likelihood function of the mixture model is demonstrated bringing formulas for alternative estimators in comparison with MLE. Section 4 reports simulation study on specially selected situations where both MCMC and MLE are investigated. Section 5 shows an illustrative example to demonstrate our developed estimation procedures on a real dataset. Finally, Section 6 brings conclusions.

2 The model

2.1 Non-linear failure rate model

Let the failure rate function of the system be in either of the following two states: (1) Initially, it experiences a constant failure rate model. (2) When the system wears out, it also experiences a wear out failure model. That is

$$h(t) = a + bt^{k-1}, \quad a, b, k, t > 0 \tag{3}$$

This model allows an initial positive failure rate, $h(0) = a$, whereas $h(0) = 0$ for most other increasing failure rate function, such as with the Weibull distribution. As mentioned in [2] for the LFR model, this type of situation would exist if failures result at random and also from wear out or deterioration.

Figure 1 shows that non-linear failure rate in equation (3) characterizes three of the most common types of failure rate functions which have been described in [4]:

1. *Increasing failure rate (IFR)*: the instantaneous failure rate increases as a function of time. We expect to see an increasing number of failures for a given period of time.
2. *Decreasing failure rate (DFR)*: the instantaneous failure rate decreases as a function of time. We expect to see the decreasing number of failures for a given period of time.
3. *Constant failure rate (CFR)*: the instantaneous failure rate is constant for the observed lifetime. We expect to see a relatively constant number of failures for a given period of time.

2.2 Characteristics of the lifetime distribution

Using the relationship between $h(t)$ and $R(t)$ we have

$$\begin{aligned} R(t) &= \exp\left\{-\int_0^t h(s)ds\right\} \\ &= \exp\left\{-\int_0^t \left(a + bs^{k-1}\right)ds\right\} \\ &= \exp\left\{-\left(at + \frac{b}{k}t^k\right)\right\} \end{aligned} \tag{4}$$

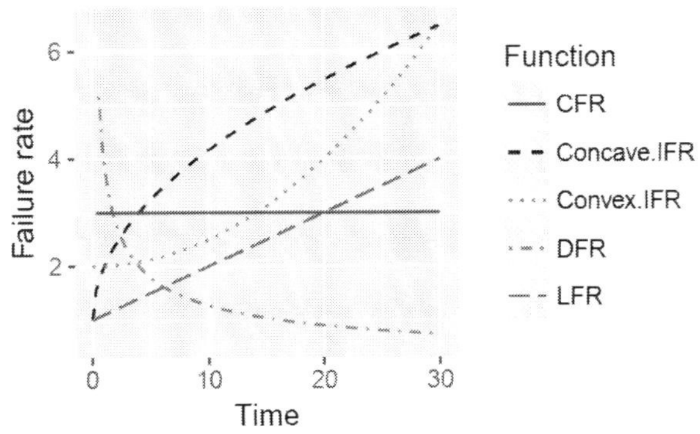

Figure 1. LFR and NLFR models

Then, the probability density function is given as

$$\begin{aligned} f(t) &= h(t)R(t) \\ &= \left(a + bt^{k-1}\right)\exp\left\{-\left(at + \frac{b}{k}t^k\right)\right\} \end{aligned} \tag{5}$$

The cumulative failure rate is given by

$$\begin{aligned} H(t) &= \int_0^t h(s)ds \\ &= at + \frac{b}{k}t^k \end{aligned} \tag{6}$$

And the mean time to failure (MTTF) is given by

$$\begin{aligned} MTTF &= \mathbb{E}(T) \\ &= \int_0^\infty R(t)dt \\ &= \int_0^\infty \exp\left\{-\left(at + \frac{b}{k}t^k\right)\right\} dt \end{aligned} \tag{7}$$

This integral can be obtained by using some suitable numerical methods.

3 Estimation of parameters and reliability characteristics

Let $D : t_1, ..., t_n$ be the random failure times of n devices under test whose failure time distribution is given as in equation (5) and $\boldsymbol{\theta} = (a, b, k)$. If there is no censoring, the likelihood function takes the general form

$$L(D|\boldsymbol{\theta}) = \prod_{i=1}^{n} f(t_i|\boldsymbol{\theta}) \tag{8}$$

The log-likelihood function can be written as

$$\log L(D|\boldsymbol{\theta}) = \sum_{i=1}^{n} \log f(t_i|\boldsymbol{\theta}) \tag{9}$$

If some observations are right censored, we have to make an adjustment to this expression. For an observation of an observed failure, we put in the pdf as above. But for a right-censored observation, we put in the reliability function, indicating that observation is known to exceed a particular value. The likelihood function in general then takes the form

$$L(D|\boldsymbol{\theta}) = \prod_{i=1}^{n} f(t_i|\boldsymbol{\theta})^{\delta_i} R(t_i|\boldsymbol{\theta})^{1-\delta_i} \tag{10}$$

This expression means that when t_i is an observed failure, the censoring indicator is $\delta_i = 1$, and we enter a pdf factor. When t_i is a censored observation, we have $\delta_i = 0$, and we enter a reliability factor [5].

The log-likelihood function for censoring case is

$$\log L(D|\boldsymbol{\theta}) = \sum_{i=1}^{n} \delta_i \log f(t_i|\boldsymbol{\theta}) + \sum_{i=1}^{n} (1-\delta_i) \log R(t_i|\boldsymbol{\theta}) \tag{11}$$

3.1 The cross-entropy method for continuous multi-extremal optimization

The main idea for the cross-entropy (CE) method for optimization can be stated as follows: Suppose we wish to maximize some "performance" function $S(\mathbf{x})$ over all elements/states $\mathbf{x}$ in some set $\mathcal{X} \subset \mathbb{R}^n$. Let us denote the maximum by γ^*, thus

$$S(\mathbf{x}^*) = \gamma^* = \max_{\mathbf{x}\in\mathcal{X}} S(\mathbf{x}) \tag{12}$$

To proceed with CE, we first randomize our deterministic problem by defining a family of pdfs $\{f(\cdot;\mathbf{v}), \mathbf{v} \in \mathcal{V}\}$ on the set $\mathcal{X}$. Next, we associate with equation (12) the estimation of

$$\ell(\gamma) = \mathbb{P}_u(S(\mathbf{X}) \geq \gamma) = \mathbb{E}_u I_{\{S(\mathbf{X})\geq\gamma\}} \tag{13}$$

the so-called associated stochastic problem (ASP). Here, $\mathbf{X}$ is a random vector with pdf $f(\cdot;\mathbf{u})$, for some $\mathbf{u} \in \mathcal{V}$ (for example, $\mathbf{X}$ could be a normal random vector) and γ is a known or unknown parameter. Note that there are in fact two possible estimation problems associated with equation (13). For a given γ we can estimate ℓ, or alternatively, for a given ℓ we can estimate γ, the root of equation (13). Let us consider the problem of estimating ℓ for a certain γ close to γ^*. Then, typically $\{S(\mathbf{X}) \geq \gamma\}$ is a rare event, and estimation of ℓ is a non-trivial problem. The CE method solves this efficiently by making adaptive changes to the probability density function according to the Kullback-Leibler CE, thus creating a sequence $f(\cdot;\mathbf{u}), f(\cdot;\mathbf{v}_1), f(\cdot;\mathbf{v}_2), \ldots$ of pdfs that are "steered" in the direction of the theoretically optimal density $f(\cdot;\mathbf{v}^*)$ corresponding to the degenerate density at an optimal point. In fact, the CE method generates a sequence of tuples $\{(\gamma_t, \mathbf{v}_t)\}$, which converges quickly to a small neighborhood of the optimal tuple $(\gamma^*, \mathbf{v}^*)$. More specifically, we initialize by setting $\mathbf{v}_0 = \mathbf{u}$, choosing a not very small quantity ϱ, say $\varrho = 10^{-2}$, and then we proceed as follows:

Adaptive updating of γ_t. For a fixed $\mathbf{v}_{t-1}$, let γ_t be the $(1-\varrho)$-quantile of $S(\mathbf{X})$ under $\mathbf{v}_{t-1}$. That is, γ_t satisfies

$$\mathbb{P}_{\mathbf{v}_{t-1}}(S(\mathbf{X}) \geq \gamma_t) \geq \varrho \tag{14}$$

$$\mathbb{P}_{\mathbf{v}_{t-1}}(S(\mathbf{X}) \leq \gamma_t) \geq 1-\varrho \tag{15}$$

where $\mathbf{X} \sim f(\cdot;\mathbf{v}_{t-1})$

Algorithm Generic CE Algorithm for Optimization

1: Choose some $\hat{\mathbf{v}}_0$. Set $t = 1$.

2: Generate a sample $\mathbf{X}_1, \ldots, \mathbf{X}_n$ from the density $f(\cdot; \hat{\mathbf{v}}_{t-1})$ and compute the sample $(1 - \varrho)$-quantile $\hat{\gamma}_t$ of the performances according to equation (16).

3: Use the same sample $\mathbf{X}_1, \ldots, \mathbf{X}_n$ and solve the stochastic program equation (18). Denote the solution by $\tilde{\mathbf{v}}_t$.

4: Apply equation (19) to smooth out the vector $\tilde{\mathbf{v}}_t$.

5: Repeat steps 2-4 until a pre-specified stopping criterion is met.

A simple estimator of γ_t, denoted $\hat{\gamma}_t$, can be obtained by drawing a random sample $\mathbf{X}_1, \ldots, \mathbf{X}_n$ from $f(\cdot; \mathbf{v}_{t-1})$ and evaluating the sample $(1 - \varrho)$-quantile of the performances as

$$\hat{\gamma}_t = S_{\lceil (1-\varrho)N \rceil} \tag{16}$$

Adaptive updating of $\mathbf{v}_t$. For fixed γ_t and $\mathbf{v}_{t-1}$, derive $\mathbf{v}_t$ from the solution of the program

$$\max_{\mathbf{v}} D(\mathbf{v}) = \max_{\mathbf{v}} \mathbb{E}_{\mathbf{v}_{t-1}} I_{\{S(\mathbf{X}) \geq \gamma_t\}} \ln f(\mathbf{X}; \mathbf{v}) \tag{17}$$

The stochastic counterpart of equation (17) is as follows: for fixed $\hat{\gamma}_t$ and $\hat{\mathbf{v}}_{t-1}$ (the estimate of $\mathbf{v}_{t-1}$), derive $\hat{\mathbf{v}}_t$ from the following program

$$\max_{\mathbf{v}} \hat{D}(\mathbf{v}) = \max_{\mathbf{v}} \frac{1}{N} \sum_{i=1}^{N} I_{\{S(\mathbf{X}_i) \geq \hat{\gamma}_t\}} \ln f(\mathbf{X}_i; \mathbf{v}) \tag{18}$$

Instead of updating the parameter vector $\mathbf{v}$ directly via the solution of equation (18) we use the following smoothed version

$$\hat{\mathbf{v}}_t = \alpha \tilde{\mathbf{v}}_t + (1 - \alpha) \hat{\mathbf{v}}_{t-1}, \quad i = 1, \ldots, n \tag{19}$$

where $\tilde{\mathbf{v}}_t$ is the parameter vector obtained from the solution of equation (18), and α is called the smoothing parameter, with $0.7 < \alpha \leq 1$.

Using normal updating, the sample $\mathbf{X}_1, \ldots, \mathbf{X}_n$ are sample from an n-dimensional multivariate normal distribution with independent components, $\mathcal{N}(\hat{\boldsymbol{\mu}}_{t-1}, \hat{\boldsymbol{\sigma}}^2_{t-1})$. While applying CE algorithm, the mean vector $\hat{\boldsymbol{\mu}}_t$ should converge to $\mathbf{x}^*$ and the vector of standard deviations $\hat{\boldsymbol{\sigma}}_t$ converge to the zero vector. In short, we should obtain a degenerated pdf with all mass concentrated in the vicinity of the point $\mathbf{x}^*$. More detail for this explanation can be found in [6, 7], and a short introduction can also be found in [8].

3.2 Maximum likelihood estimation

In our study, we maximize the log-likelihood function given in equation (9) in case of non-censored data or equation (11) in case of censored data using CE algorithm and obtain the maximizer $\hat{\boldsymbol{\theta}} = (\hat{a}, \hat{b}, \hat{k})$. By using the invariance property of MLE's,

1. The MLE for $R(t)$, say $\hat{R}(t)$, will be

$$\hat{R}(t) = \exp\left\{-\left(\hat{a}t + \frac{\hat{b}}{\hat{k}} t^{\hat{k}}\right)\right\} \tag{20}$$

2. The MLE for $h(t)$, say $\hat{h}(t)$, will be

$$\hat{h}(t) = \hat{a} + \hat{b}t^{\hat{k}-1} \tag{21}$$

3. The MLE for $H(t)$, say $\hat{H}(t)$, will be

$$\hat{H}(t) = \hat{a}t + \frac{\hat{b}}{\hat{k}}t^{\hat{k}} \tag{22}$$

4. The MLE for MTTF will be

$$M\hat{T}TF = MTTF(\hat{a}, \hat{b}, \hat{k}) \tag{23}$$

which can be obtained by installing into formula (7) and integrating.

3.3 The Metropolis-Hastings algorithm

The Metropolis-Hastings algorithm is the most popular MCMC method. The basic idea of the algorithm is to generate data from some generic distribution, say target distribution. Suppose that our target distribution has density π with respect to some reference measure (usually d-dimensional Lebesgue measure). Then, given $\mathbf{X}_n$, a "proposed value" $\mathbf{Y}_{n+1}$ is generated from some pre-specified density $q(\mathbf{X}_n, \mathbf{Y})$, and then accepted with probability

$$\alpha(\mathbf{x}, \mathbf{y}) = \begin{cases} \min\left\{\frac{\pi(\mathbf{y})}{\pi(\mathbf{x})}\frac{q(\mathbf{y},\mathbf{x})}{q(\mathbf{x},\mathbf{y})}, 1\right\}, & \pi(\mathbf{x})q(\mathbf{x}, \mathbf{y}) > 0 \\ 1, & \pi(\mathbf{x})q(\mathbf{x}, \mathbf{y}) = 0 \end{cases} \tag{24}$$

If the proposed value is accepted, we set $\mathbf{X}_{n+1} = \mathbf{Y}_{n+1}$; otherwise, we set $\mathbf{X}_{n+1} = \mathbf{X}_n$. The function $\alpha(\mathbf{x}, \mathbf{y})$ is chosen, of course, precisely to ensure that the Markov chain $\mathbf{X}_0, \mathbf{X}_1, \ldots$ is reversible with respect to target density $\pi(\mathbf{y})$, so that the target density is stationary for the chain. If the proposed is symmetric, then this reduces to

$$\alpha(\mathbf{x}, \mathbf{y}) = \begin{cases} \min\left\{\frac{\pi(\mathbf{y})}{\pi(\mathbf{x})}, 1\right\}, & \pi(\mathbf{x})q(\mathbf{x}, \mathbf{y}) > 0 \\ 1, & \pi(\mathbf{x})q(\mathbf{x}, \mathbf{y}) = 0 \end{cases} \tag{25}$$

This particular case of the Metropolis-Hastings algorithm is called the Metropolis algorithm.

It has long been recognized that the choice of the proposed density $q(\mathbf{x}, \mathbf{y})$ is crucial to the success (e.g. rapid convergence) of the Metropolis-Hastings algorithm. The most common case involves a symmetric random-walk Metropolis algorithm (RWM) in which the proposed value is given by $\mathbf{Y}_{n+1} = \mathbf{X}_n + \mathbf{Z}_{n+1}$, where the increments $\{\mathbf{Z}_n\}$ are i.i.d. from some fixed symmetric distribution (e.g. $N(0, \sigma^2 I_d)$). In this case, the crucial issue become how to scale the proposal (e.g. how to choose σ): too small and the chain will move to slowly; to large and the proposals will usually be rejected. Instead, we must avoid both extremes [9].

3.4 Adaptive MCMC

The search for improved proposal distribution is often done manually, through trial and error, though this can be difficult, especially in high dimensions. An alternative approach is adaptive MCMC, which ask the computer to automatically "learn" better parameter value "on the fly"– that is, while algorithm runs.

Suppose $\{P_\gamma\}_{\gamma \in \mathcal{Y}}$ is a family of Markov chain kernels, each having stationary distribution π. (For example, perhaps P_γ corresponds to RWM algorithm with increment distribution

$N(0, \gamma^2 I_d)$.) An adaptive MCMC would randomly update the value of γ at each iteration, in an attempt to find the best value.

Let Γ_n be the chosen kernel choice at the nth iteration, so

$$\mathbb{P}(\mathbf{X}_{n+1} \in A | \mathbf{X}_n = \mathbf{x}, \Gamma_n = \gamma, \mathbf{X}_{n-1}, \ldots, \mathbf{X}_0, \Gamma_{n-1}, \ldots, \Gamma_0) = P_\gamma(\mathbf{x}, A) \tag{26}$$

for $n = 0, 1, 2, \ldots$. Here the $\{\Gamma_n\}$ are updated according to some adaptive algorithm. In principal, the choice of Γ_n could depend on the entire history $\mathbf{X}_{n-1}, \ldots, \mathbf{X}_0, \Gamma_{n-1}, \ldots, \Gamma_0$, though in practice it is often the case that the pairs process $\{(\mathbf{X}_n, \Gamma_n)\}_{n=0}^{\infty}$ is Markovian. In general the algorithms are quite easy to implement, requiring only moderate amounts of extra computer programming [9].

3.5 Bayesian estimation

For our mixture model, the Bayesian model is constructed by specifying a prior distribution for $\boldsymbol{\theta} = (a, b, k)$, and then multiplying with the likelihood function to obtain the posterior distribution.

Denote the prior distribution of $\boldsymbol{\theta}$ as $\pi(\boldsymbol{\theta})$, the posterior distribution of $\boldsymbol{\theta}$ given $D : t_1, \ldots, t_n$ is given by

$$\pi(\boldsymbol{\theta}|D) = \frac{L(D|\boldsymbol{\theta})\pi(\boldsymbol{\theta})}{\int L(D|\boldsymbol{\theta})\pi(\boldsymbol{\theta})d\boldsymbol{\theta}} \tag{27}$$

Because the denominator in equation (27) is a normalizing constant, Bayes' theorem is often expressed as:

$$\pi(\boldsymbol{\theta}|D) \propto L(D|\boldsymbol{\theta})\pi(\boldsymbol{\theta}) \tag{28}$$

As mentioned in [11], the prior distribution is given beforehand, usually based on prior information of the parameters, such as that from historical data, previous experiences, expert suggestions, even wholly subjective suppositions, or simply from the point of mathematical conveniences.

Here, adopted [12], we assume that a, b, and k each have independent gamma(α_1, β_1), gamma(α_2, β_2), and gamma(α_3, β_3) priors respectively, i.e.

$$\pi_1(a) \propto a^{\alpha_1 - 1} \exp\{-\beta_1 a\}, \; \alpha_1, \beta_1 > 0 \tag{29}$$

$$\pi_2(b) \propto b^{\alpha_2 - 1} \exp\{-\beta_2 b\}, \; \alpha_2, \beta_2 > 0 \tag{30}$$

$$\pi_3(k) \propto k^{\alpha_3 - 1} \exp\{-\beta_3 k\}, \; \alpha_3, \beta_3 > 0 \tag{31}$$

If $\alpha_1 = \alpha_2 = \alpha_3 = 1, \beta_1 = \beta_2 = \beta_3 = 0$ we have a generalized uniform distribution on $\mathbb{R}^+$ [11] or a diffuse prior, and if $\alpha_1 = \alpha_2 = \alpha_3 = \beta_1 = \beta_2 = \beta_3 = 0$, we have a non-informative prior.

Based on the assumption of independence of individual priors, the joint prior of $\boldsymbol{\theta}$ is given by

$$\pi(\boldsymbol{\theta}) = \pi_1(a)\pi_2(b)\pi_3(k) \tag{32}$$

Then, under the square error loss function, the Bayes estimator of a, b, k, failure rate function $h(t)$, reliability function $R(t)$ and cumulative failure rate function are given by

$$a^* = \mathbb{E}(a|D) = \int_{\theta} a\pi(\boldsymbol{\theta}|D)d\boldsymbol{\theta} \tag{33}$$

$$b^* = \mathbb{E}(b|D) = \int_{\theta} b\pi(\boldsymbol{\theta}|D)d\boldsymbol{\theta} \tag{34}$$

$$k^* = \mathbb{E}(k|D) = \int_\theta k\pi(\boldsymbol{\theta}|D)d\boldsymbol{\theta} \quad (35)$$

$$h^*(t) = \mathbb{E}(h(t;\boldsymbol{\theta})|D) = \int_\theta h(t;\boldsymbol{\theta})\pi(\boldsymbol{\theta}|D)d\boldsymbol{\theta} \quad (36)$$

$$R^*(t) = \mathbb{E}(R(t;\boldsymbol{\theta})|D) = \int_\theta R(t;\boldsymbol{\theta})\pi(\boldsymbol{\theta}|D)d\boldsymbol{\theta} \quad (37)$$

$$H^*(t) = \mathbb{E}(R(t;\boldsymbol{\theta})|D) = \int_\theta H(t;\boldsymbol{\theta})\pi(\boldsymbol{\theta}|D)d\boldsymbol{\theta} \quad (38)$$

In our study, we use adaptive MCMC sampling to generate sample $\boldsymbol{\theta}_i = (a_i, b_i, k_i), i = 1, \ldots, n$ from the posterior distribution $\pi(\theta|D)$. Then, Monte Carlo integration estimates a^*, b^*, k^*, $h^*(t)$, $R^*(t)$ and $H^*(t)$ by calculating the means:

$$a^* \approx \frac{1}{n}\sum_{i=1}^{n} a_i \quad (39)$$

$$b^* \approx \frac{1}{n}\sum_{i=1}^{n} b_i \quad (40)$$

$$k^* \approx \frac{1}{n}\sum_{i=1}^{n} k_i \quad (41)$$

$$h^*(t) \approx \frac{1}{n}\sum_{i=1}^{n} h(t;\boldsymbol{\theta}_i) \quad (42)$$

$$R^*(t) \approx \frac{1}{n}\sum_{i=1}^{n} R(t;\boldsymbol{\theta}_i) \quad (43)$$

$$H^*(t) \approx \frac{1}{n}\sum_{i=1}^{n} H(t;\boldsymbol{\theta}_i) \quad (44)$$

4 Monte Carlo simulations

A Monte Carlo simulation study is conducted to compare the performance of CE and MCMC estimators for the parameters of the non-linear failure rate. For each of the following choice of parameters, we simulate 1000 datasets with sample size $n = 25, 50, 100$ and 200, respectively, and based on each dataset we computed the CE and MCMC estimator for the model parameters. The datasets are generated from the failure distribution Eq.(5) using the accept-reject method. A clear explanation and proof of this method can be found in [13]. In order to obtain MCMC estimators, we have implemented the adaptive MCMC algorithm to construct a Markov chain of length 20,000 with burn-in of 1000 and reduced autocorrelation by retaining only every 5 iterations of the chain and obtain 3801 samples. For prior information we used the generalized uniform distribution on $\mathbb{R}^+$, i.e. $\alpha_1 = \alpha_2 = \alpha_3 = 1, \beta_1 = \beta_2 = \beta_3 = 0$. Note here that when $k = 1$, our model coincides with the exponential model with constant failure rate $\lambda = a + b$.

1. $a = 0.03, b = 0.07$, and $k = 0.5$

2. $a = 0.03, b = 0.07$, and $k = 2$

3. $a = 0.03, b = 0.07$, and $k = 3$

Table 1. Comparison of CE and MCMC when $a = 0.03, b = 0.07$, and $k = 0.5$

n	Method	Bias a	MSE a	Bias b	MSE b	Bias k	MSE k
25	CE	−0.0018	0.0004	0.0011	0.0015	0.0546	0.0466
	MCMC	0.0042	0.0002	−0.0021	0.0008	0.0761	0.0260
50	CE	−0.0023	0.0002	0.0014	0.0009	0.0241	0.0225
	MCMC	0.0003	0.0001	−0.0004	0.0005	0.0383	0.0123
100	CE	−0.0025	0.0002	0.0017	0.0006	0.0215	0.0140
	MCMC	−0.0019	0.0001	0.0016	0.0004	0.0302	0.0079
200	CE	−0.0020	0.0001	0.0022	0.0003	0.0170	0.0068
	MCMC	−0.0031	0.0001	0.0037	0.0002	0.0279	0.0054

Table 2. Comparison of CE and MCMC when $a = 0.03, b = 0.07$, and $k = 2$

n	Method	Bias a	MSE a	Bias b	MSE b	Bias k	MSE k
25	CE	−0.0063	0.0013	0.0108	0.0014	−0.0019	0.0781
	MCMC	0.0464	0.0030	−0.0064	0.0005	0.1947	0.1919
50	CE	−0.0084	0.0009	0.0111	0.0009	−0.0253	0.0498
	MCMC	0.0284	0.0013	−0.0079	0.0005	0.1896	0.1260
100	CE	−0.0071	0.0006	0.0084	0.0005	−0.0309	0.0247
	MCMC	0.0178	0.0007	−0.0081	0.0004	0.1502	0.0768
200	CE	−0.0083	0.0004	0.0103	0.0004	−0.0543	0.0187
	MCMC	0.0092	0.0004	−0.0034	0.0003	0.0716	0.0377

Tables 1-3 list the results of the simulation study. Bias is calculated as the mean of 1000 estimates minus the true value, and the mean square error (MSE) is calculated as the mean of the squared differences between 1000 estimators and the true value.

Figure 2-7 visualize the results from the Tables 1-3. From these figures, we see that the biases and MSEs of a and b are very small. So what we are interesting here is the biases and MSEs of the shape parameter k. For all three special cases of the shape parameter k:

- In most of the cases, the CE estimator seem to have less biases than the MCMC estimator (see right panel of Figures 2, 4 and 6).
- In case of decreasing failure rate, i.e. $k < 1$, the MSEs of MCMC estimator seem to be always smaller than MSEs of CE estimator (see right panel of Figure 3).
- In case of increasing failure rate, i.e. $k > 1$, the MSEs of CE estimator seem to be always smaller than MSEs of MCMC estimator (see right panel of Figures 5 and 7).

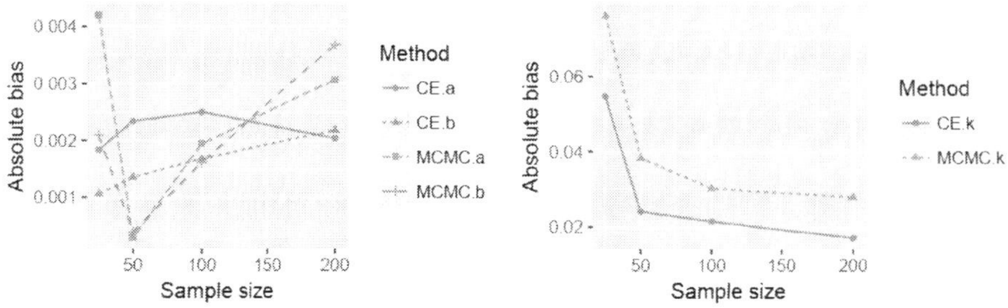

Figure 2. Biases of CE and MCMC estimators for a, b and k when $k = 0.5$

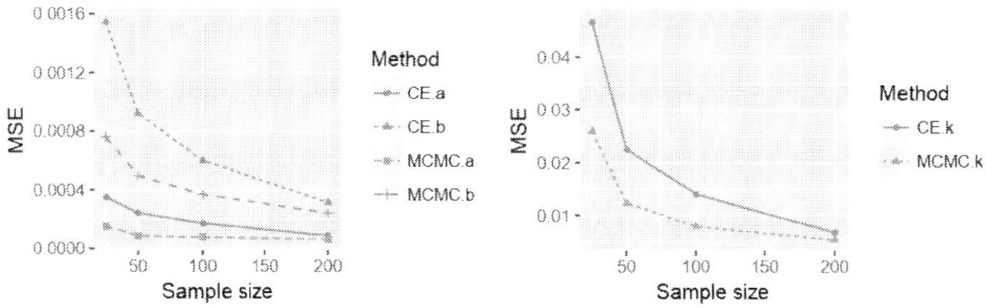

Figure 3. MSEs of CE and MCMC estimators for a, b and k when $k = 0.5$

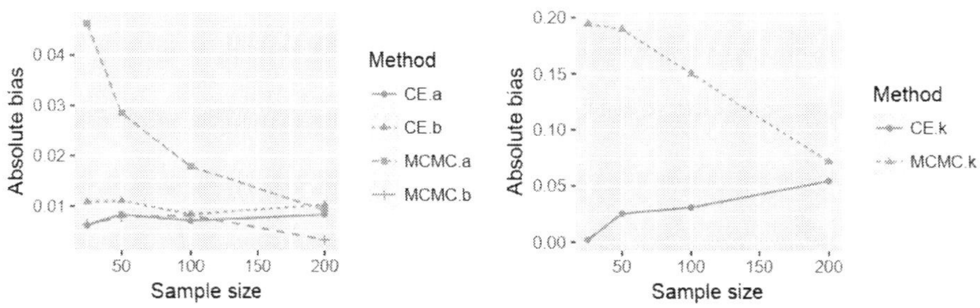

Figure 4. Biases of CE and MCMC estimators for a, b and k when $k = 2$

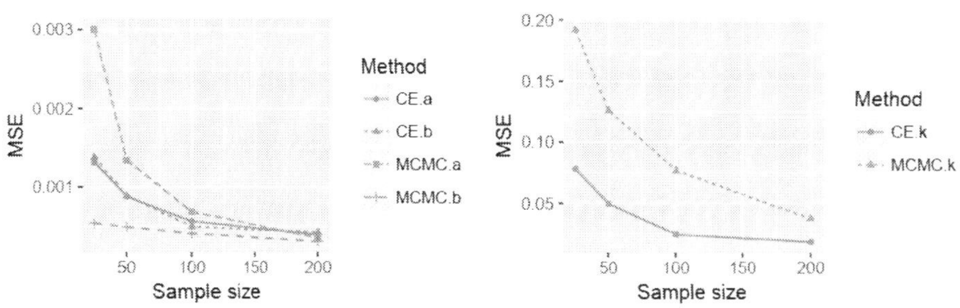

Figure 5. MSEs of CE and MCMC estimators for a, b and k when $k = 2$

Table 3. Comparison of CE and MCMC when $a = 0.03, b = 0.07$, and $k = 3$

n	Method	Bias a	MSE a	Bias b	MSE b	Bias k	MSE k
25	CE	−0.0088	0.0012	0.0211	0.0015	−0.1453	0.1193
	MCMC	0.0389	0.0023	0.0061	0.0011	0.2035	0.3075
50	CE	−0.0098	0.0007	0.0202	0.0010	−0.1602	0.0833
	MCMC	0.0216	0.0010	0.0019	0.0007	0.1591	0.1999
100	CE	−0.0096	0.0005	0.0201	0.0007	−0.1866	0.0694
	MCMC	0.0115	0.0005	0.0010	0.0005	0.0847	0.1072
200	CE	−0.0084	0.0003	0.0183	0.0005	−0.1770	0.0517
	MCMC	0.0064	0.0003	0.0002	0.0003	0.0564	0.0566

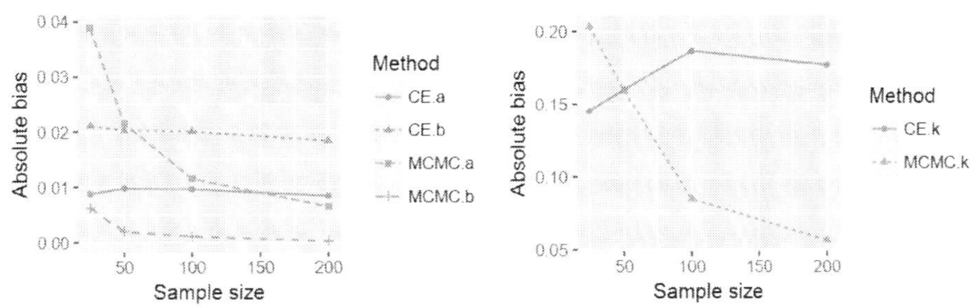

Figure 6. Biases of CE and MCMC estimators for a, b and k when $k = 3$

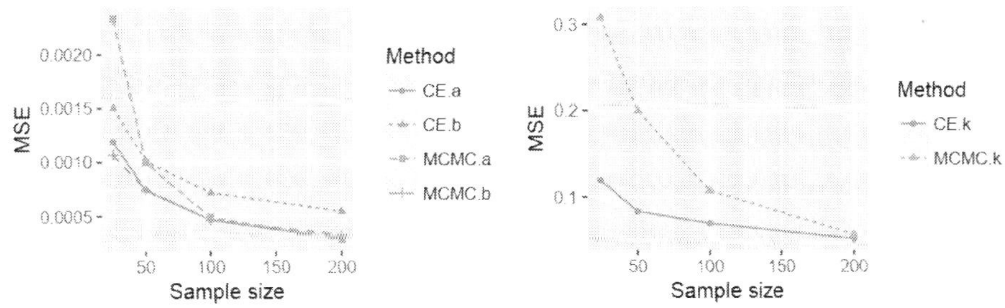

Figure 7. MSEs of CE and MCMC estimators for a, b and k when $k = 3$

5 Illustrative examples

We consider the aircraft windshield failure data given in Table 4. The data consist of 153 observations. Among them 88 are classified as failed windshields, and the remaining 65 are censored time, means that had not failed at the time of observation. The unit for measurement is 1000 hours [14].

In this study, we use both CE and MCMC methods to estimate the parameters and reliability characteristics. In order to obtain MCMC estimators, we have implemented the adaptive

Table 4. Aircraft windshield failure data

Failure Times				Service Times		
0.040	1.866	2.385	3.443	0.046	1.436	2.592
0.301	1.876	2.481	3.467	0.140	1.492	2.600
0.309	1.899	2.610	3.478	0.150	1.580	2.670
0.557	1.911	2.625	3.578	0.248	1.719	2.717
0.943	1.912	2.632	3.595	0.280	1.794	2.819
1.070	1.914	2.646	3.699	0.313	1.915	2.820
1.124	1.981	2.661	3.779	0.389	1.920	2.878
1.248	2.010	2.688	3.924	0.487	1.963	2.950
1.281	2.038	2.823	4.035	0.622	1.978	3.003
1.281	2.085	2.890	4.121	0.900	2.053	3.102
1.303	2.089	2.902	4.167	0.952	2.065	3.304
1.432	2.097	2.934	4.240	0.996	2.117	3.483
1.480	2.135	2.962	4.255	1.003	2.137	3.500
1.505	2.154	2.964	4.278	1.010	2.141	3.622
1.506	2.190	3.000	4.305	1.085	2.163	3.665
1.568	2.194	3.103	4.376	1.092	2.183	3.695
1.615	2.223	3.114	4.449	1.152	2.240	4.015
1.619	2.224	3.117	4.485	1.183	2.341	4.628
1.652	2.229	3.166	4.570	1.244	2.435	4.806
1.652	2.300	3.344	4.602	1.249	2.464	4.881
1.757	2.324	3.376	4.663	1.262	2.543	5.140
1.795	2.349	3.385	4.694	1.360	2.560	

Table 5. Point estimates and two-sided 90% and 95% HPD intervals for a, b, k and MTTF

	MCMC	90% HPD	95% HPD
a	0.0348	[0.0070, 0.0661]	[0.0025, 0.0722]
b	0.0716	[0.0307, 0.1077]	[0.0274, 0.1201]
k	2.9227	[2.4056, 3.4397]	[2.3223, 3.5605]
$MTTF$	3.0351	[2.8177, 3.2302]	[2.7918, 3.2895]

MCMC algorithm to construct a Markov chain of length 50,000 with burn-in of 20,000 and reduced the autocorrelation by retaining only every 5 iterations of the chain and obtain 6001 samples. In fact, it is not necessary to set the long length of burn-in of 20,000 because the adaptive MCMC rapidly converges to the stationary distribution. For prior information we used the generalized uniform distribution on $\mathbb{R}^+$, i.e. $\alpha_1 = \alpha_2 = \alpha_3 = 1, \beta_1 = \beta_2 = \beta_3 = 0$. Table 5 shows our MCMC point estimates and two-sided 90% and 95% highest posterior density (HPD) intervals for a, b, k and MTTF, and Table 6 shows our CE point estimates and two-sided 90% and 95% bootstrap percentile confident intervals (PERC) for a, b, k and MTTF.

Figure 8 shows posterior distributions and trace plots of each parameter of the Bayesian model obtained by MCMC algorithm, and Figures 9-11 show time courses of all relevant functions obtained by both CE and MCMC methods, i.e. reliability, failure rate and cumulative failure rate functions.

Table 6. Point estimates and two-sided 90% and 95% bootstrap percentile confident intervals for a, b, k and MTTF

	CE	90% PERC	95% PERC
a	0.0269	[0.0000, 0.0569]	[0.0000, 0.0626]
b	0.0692	[0.0375, 0.1127]	[0.0321, 0.1244]
k	2.9284	[2.5112, 3.4443]	[2.4331, 3.5554]
$MTTF$	3.0535	[2.8478, 3.2726]	[2.8105, 3.2996]

Table 7. Estimated AIC for nine mixture models for aircraft windshield failure data

Mixture models	Models forms and parameters	AIC
1. Weibull-Weibull	$p \times \text{Weib}(\beta_1, \alpha_1) + (1-p) \times \text{Weib}(\beta_2, \alpha_2)$	350.159
2. Weibull-Exponential	$p \times \text{Weib}(\beta_1, \alpha_1) + (1-p) \times \text{Exp}(\lambda)$	348.260
3. Weibull-Normal	$p \times \text{Weib}(\beta_1, \alpha_1) + (1-p) \times \text{Nor}(\mu, \sigma)$	349.710
4. Weibull-Lognormal	$p \times \text{Weib}(\beta_1, \alpha_1) + (1-p) \times \text{Lnor}(\mu, \sigma)$	351.235
5. Normal-Exponential	$p \times \text{Nor}(\mu, \sigma) + (1-p) \times \text{Exp}(\lambda)$	351.513
6. Normal-Lognormal	$p \times \text{Nor}(\mu, \sigma) + (1-p) \times \text{Lnor}(\mu_2, \sigma_2)$	351.579
7. Lognormal-Exponential	$p \times \text{Lnor}(\mu, \sigma) + (1-p) \times \text{Exp}(\lambda)$	349.301
8. Linear failure rate	$h(t) = a + bt$	359.106
9. Non-linear failure rate	$h(t) = a + bt^k$	347.371

Table 8. MLEs of parameters for nine mixture models

Mixture models	MLEs of parameters
1. Weibull-Weibull	$\hat{\beta}_1 = 1.249, \hat{\alpha}_1 = 0.245, \hat{\beta}_2 = 2.777, \hat{\alpha}_2 = 3.485, \hat{p} = 0.017$
2. Weibull-Exponential	$\hat{\beta} = 2.768, \hat{\alpha} = 3.484, \hat{\lambda} = 4.052, \hat{p} = 0.983$
3. Weibull-Normal	$\hat{\beta} = 7.359, \hat{\alpha} = 4.481, \hat{\mu} = 2.303, \hat{\sigma} = 0.868, \hat{p} = 0.387$
4. Weibull-Lognormal	$\hat{\beta} = 1.246, \hat{\alpha} = 0.395, \hat{\mu} = 1.075, \hat{\sigma} = 0.440, \hat{p} = 0.029$
5. Normal-Exponential	$\hat{\mu} = 3.053, \hat{\sigma} = 1.220, \hat{\lambda} = 24.003, \hat{p} = 0.995$
6. Normal-Lognormal	$\hat{\mu}_1 = 0.302, \hat{\sigma}_1 = 0.182, \hat{\mu}_2 = 1.073, \hat{\sigma}_2 = 0.443, \hat{p} = 0.027$
7. Lognormal-Exponential	$\hat{\mu} = 1.080, \hat{\sigma} = 0.435, \hat{\lambda} = 1.729, \hat{p} = 0.965$
8. Linear failure rate	$\hat{a} = 0.004, \hat{b} = 0.161$
9. Non-linear failure rate	$\hat{a} = 0.027, \hat{b} = 0.069, \hat{k} = 2.928$

 https://doi.org/10.1051/itmconf/20182003001

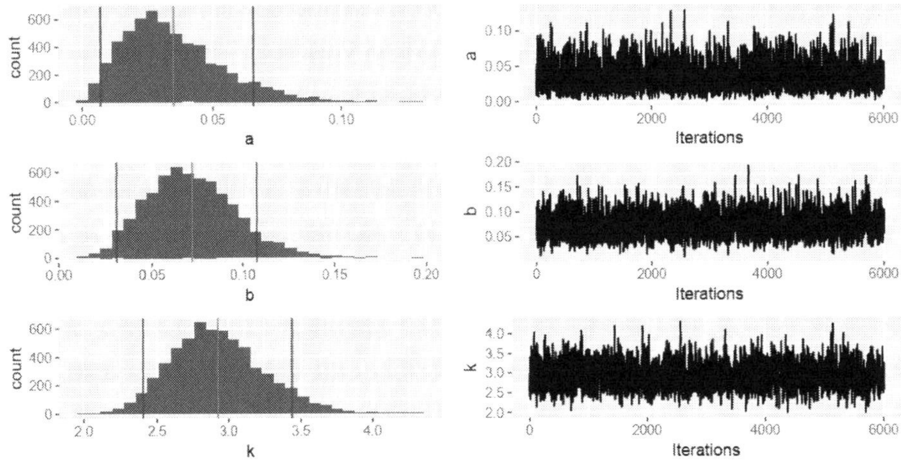

Figure 8. Posterior densities and trace plots of each parameter of the Bayesian model

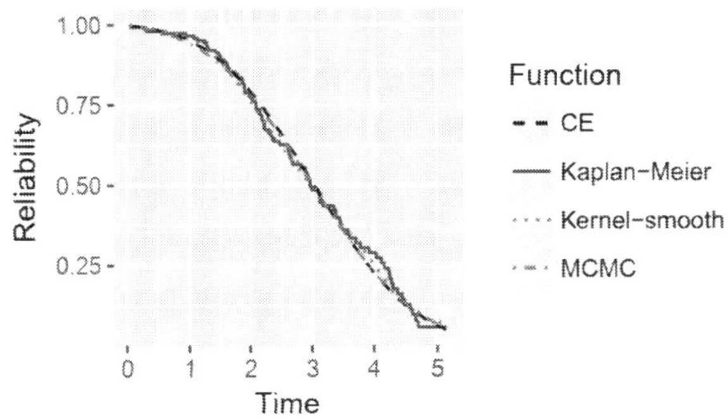

Figure 9. The time courses of the reliability functions.

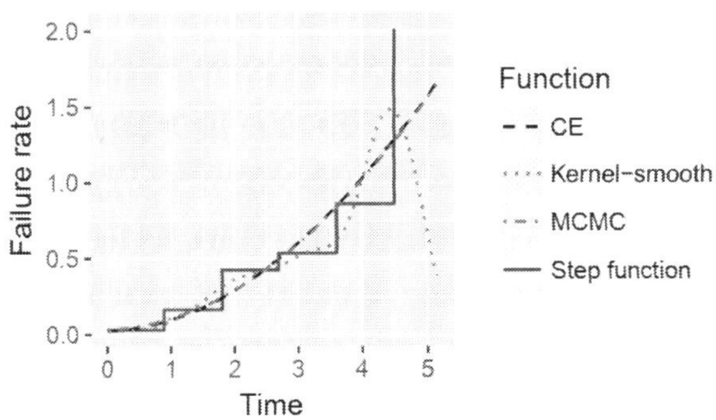

Figure 10. The time courses of the failure rate functions.

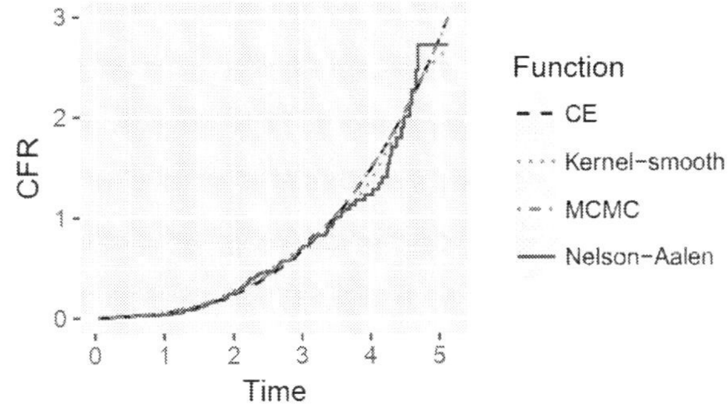

Figure 11. The time courses of the cumulative failure rate functions.

Tables 7-8 represent the Akaike Information Criterion (AIC) values and MLEs of parameters for mixture models, respectively. The AIC values and MLEs from model 1 to 7 have been obtained in [14]. In comparison with these results, Table 7 shows that non-linear failure rate has smallest AIC value. A model for which the AIC is the minimum is considered to be the best approximating model among a set of alternative models.

For this right censored dataset, both CE and MCMC methods give almost the same results. The MLEs are easily obtained by using CE method. This fact is due to the stochastic search of CE method. To improve the result of CE algorithm, one can choose the initial value(s) of (vector) standard deviation $\hat{\sigma}_0$ (sometimes changing the initial value(s) of (vector) mean $\hat{\mu}_0$ is needed) through trial and error until we get the smallest optimum value which yields the optimal estimator(s). Nevertheless, Bayes credible intervals are easily obtained by using MCMC method.

6 Conclusions

The benchmark dataset, the aircraft windshield failure data, shows that the innovative non-linear failure rate might be more appropriate than some mixtures of distributions for modeling the data which might be originated from more than one failure mode.

We recommend using the CE and MCMC methods as the mutual add-in tools for parameter estimation. MLEs are easily obtained by using CE method. The CE method is not so sensitive to the choice of the initial values of the algorithm in comparison with other alternative optimization methods. However, Bayes credible intervals are easily achieved by using MCMC method.

In our study, we consider only the sum of two components in the non-linear failure rate which results from the assumption of two independent competing risks, i.e. the random shock and the wear out. The case of dependent competing risks will be considered next. And we only use the diffuse priors on the parameters for Bayesian estimation, so this study can also be continued with other alternative priors.

Acknowledgement

This work was supported by the European Regional Development Fund in the Research Centre of Advanced Mechatronic Systems project, project number CZ.02.1.01/0.0/0.0/16_019/0000867 within the Operational Programme Research, Development and Education, and partially by the VŠB – Technical University of Ostrava in the SGS project number SP2018/68.

References

[1] Kodlin D., A new response time distribution (Biometrics) **2**, 227-239 (1967).

[2] Bain L. J., Analysis for the linear failure-rate life-testing distribution (Technometrics) **16**, 551-559 (1974).

[3] Salem S. A., Bayesian estimation of a non-linear failure rate from censored samples type II (Microelectronics and Reliability) **32**, 1385-1388 (1992).

[4] Hamada M. S., Wilson A. G., Martz H. F., Reese C. S., *Bayesian reliability* (Springer, 2008).

[5] Moore D. F., *Applied survival analysis using R* (Springer, 2016).

[6] Kroese D. P., Porotsky S., Rubinstein R. Y., The cross-entropy method for continuous multi-extremal optimization (Methodology and Computing in Applied Probability) **8**, 383-407 (2006).

[7] Rubinstein R. Y., Kroese D. P., *The cross-entropy method: a unified approach to combinatorial optimization, Monte Carlo simulation and machine learning* (Springer-Verlag, New York, 2004).

[8] Zio E., *The Monte Carlo simulation method for system reliability and risk analysis* (Springer, 2013).

[9] Brooks S., Gelman A., Jones G., Meng X. L., *Handbook of Markov Chain Monte Carlo* (CRC Press, 2011).

[10] Canfield R., A Bayesian approach to reliability estimation using a loss function (IEEE Transactions on Reliability) **19**, 13-16 (1970).

[11] Jiang H., Xie M., Tang L. C., Markov chain Monte Carlo methods for parameter estimation of the modified weibull distribution (Journal of Applied Statistics) **35**(6), 647-658 (2008).

[12] Kundu D., Howlader H., Bayesian inference and prediction of the inverse Weibull distribution for type-II censored data. (Computational Statistics and Data Analysis) **54**(6), 1547-1558 (2010).

[13] Robert C. P., Casella G., *Introducing Monte Carlo methods with R* (Springer, 2010).

[14] Blischke W. R., Karim M. R., Murthy D. P., *Warranty data collection and analysis* (Springer, 2011).

Identify the correlation between the teaching perspectives and student evaluation: A case study of a sample university in Ho Chi Minh city

Huy Nguyen[1,*] and *Radim* Briš[2]

[1]Faculty of Mathematics and Statistics, Ton Duc Thang University, Ho Chi Minh City, Vietnam
[2]Department of Applied Mathematics, Faculty of Electrical Engineering and Computer Science, VŠB -Technical University of Ostrava, Ostrava, Czech Republic

Abstract. Education is a critical issue in any cultural background. In fact, the quality of teaching shall be considered and linked with student evaluation because there seems to have a strong correlation between them. This study aims to understand the different perspectives on teaching by academic staffs of a university X in Ho Chi Minh City and also is to determine whether there are one or two dominant teaching perspective preferences. In addition, the students' comments were also collected and investigated to identify whether or not there are relationships between these dominant teaching perspectives and student end-of-course evaluations. Finally, the researcher proposed a new form of evaluation to help measure better students' expectations.

1 Introduction

Teaching adults and higher education is a major issue which had been discussed by Kember since 1997 and no further published researches were made beyond that one. Understanding of the teaching perspective is not only to specify what the teachers should and are supposed to do in order to successfully fulfill students' expectations but also to improve the quality of teaching with respect to student evaluations. Even though every educator has different preferred styles of teaching, then they have to be followed by set of standards.

Moreover, in different countries the teaching strategies may vary a little or much depending on their students' needs and expectations. For example, education in Australia tend to be more relaxed than in Vietnam as the Australian teachers allow their students to learn individually, meanwhile Vietnamese students are required to study with disciplines and they are more likely told what to do and learn rather than self-study. Hence, students may not fully understand the reasons for studying some courses and how to apply in the future career and this causes less motivation in their courses.

* Corresponding author: nguyenbachuy@tdtu.edu.vn

2 Literature review

Kember identified five significant teaching perspectives, such as: transmission, apprenticeship, developmental, nurturing and social reform.

Transmission: A good teacher should know how to link his/her teaching to the content as a solid tool in order to effectively transmit message or knowledge to learners.

Apprenticeship: Teachers play an important role in helping their students to conceive the materials with ease and comfort. Different levels of students from not good to excellent need different levels of guidance so teachers have to have lots of experience to satisfy their needs.

Developmental: Content must be appropriately prepared and conducted. Shall teachers choose an easy way to approach their learners to engage in the materials from simplex to complex, also combining asking questions and examples to illustrate their points more comprehensible.

Nurturing: Teaching should come from heart and head. People are more motivated to work without fear or menace. Learners are really educated if they know the outcomes of what they are doing and trying hard for, also their achievement is a combination of their ability and effort.

Social reform: Shall good teachers make their students aware of the values of knowledge and how and when to apply in real life. Each graduate will contribute to society different knowledge which has been taught from classes and from valuable experiences earning from field trip or internship. Each person has different preferred style of learning, however those diverse styles are identified in the classroom as every student is taught in the same way as others (Goodlad, 1984). Students with clear identification of learning style usually achieve better results than those do not. For example: practionners concluded that grade-point averages for those who have clearly identified their preferred learning style much higher than others (Andrews, 1990; Lemmon, 1985; Stone, 1992; Turner, 1993). The best way to teach learners no matter how old they are is firstly to help them deeply understand conceptual backgrounds then more details should be given along with lots of examples so they could systemize the materials themselves at home. In that case they do not fear or worry about what they have just been taught.

'Quality Teaching' and 'Successful Teaching' are distinguished by Fenstermacher and Richardson in 2005 as follows:

A quality teaching is more than just transmitting the knowledge to the learners, but also the way the instructors teach. Fenstermacher and Richardson refer it as by good teaching we mean that the content taught accords with disciplinary standards of adequacy and completeness, and that the methods employed are age appropriate, morally defensible, and undertaken with the intention of enhancing the learner's competence with respect to the content studied (p.191). Meanwhile, 'successful teaching' is evaluated based on student outcomes through assessment and according to these authors it is referred as follows: by successful teaching we mean that the learner actually acquires, to some reasonable and acceptable level of proficiency, what the teacher is engaged in teaching (p.191).

3 Research Methodology

This is a quantitative study; we have conducted survey approach in order to collect students' perceptions and comments. The sets of questions consist of two types: yes/no and open-ended. The students from different schools and departments were selected to answer the survey, 150 surveys were given out and only 140 were found as complete data that can be used for analysis. Surveys comprised 15 questions (12 multiple choices and 3 opened-ended) and all respondents belong to 19-24 group. The majority was female's 65% and

35% males; the participants come from various schools and departments such as: 22.14% belonged to Business Administration (BA), 7.14% to Computer Science Engineering (CSE), 7.14% to Electrical Engineering (EE), 11.43% to Biotechnology (BT), 7.86% to Mathematics (MA), 7.86% to Physics (PH), 8.57% to English (EN), 8.57% to Industrial & Systems Engineering (ISE), 12.86% to Biomedical Engineering (BME) and 6.43% to Chemistry (CH) - a small Department which belongs to School of Biotechnology.

4 Data Analysis

We were considering 2016 - 2017 academic year for both semesters one and two (summer semester excluded).

The following tables show the comparison of overall mean scores between two semesters.

Table 1. Mean scores between Semester 1 & 2.

Order	School/Dept	Sem 1	Sem 2	Comment
1	Chemistry	4.5	4.13	Not improved
2	Physics	4.14	4.05	Not improved
3	English	4.12	4.04	Not improved
4	Computer Science Engineering	4.10	4.12	Improved
5	Biomedical Engineering	4.09	4.43	**Dramatically Improved**
6	Biotechnology	4.08	4.03	Not improved
7	Industrial & Systems Engineering	4.03	4.0	Not improved
8	Electrical Engineering	4.02	4.0	Not improved
9	Business Administration	3.99	3.94	Not improved
10	Mathematics	3.71	3.78	Improved
11	**University X**	**4.04**	**3.99**	Not improved

The second semester's average scores varied less than first semester did with a standard deviation of 0.158 compared to 0.183 for Semester 1. Considering compulsory subjects such as Mathematics, Physics, Chemistry and English, only Mathematics was evaluated with the least score due to its complexity and compulsoriness. Some students from other Schools or Departments which are non-mathematics oriented most likely dislike this subject as they could not see its practicality and realness to their future career. As discussed in literature review, one of the teaching responsibilities was that the lecturers shall explain to their students how to link the theory and the practice during the courses and also convince them how to apply it in specific cases.

The quality of teachers was also evaluated and the following table will show an overall picture for the academic year 2016-2017 :

Table 2. Mean scores for quality of teachers between Semester 1 & 2.

Semester 1			Semester 2	
X	Score	School/Dept	Score	X
4.03	3.97	BA	3.93	**3.98**
	4.07	BT	4.03	
	4.06	BME	4.42	
	4.42	CH	4.11	
	4.03	EE	3.98	
	4.09	EN	4.01	
	4.00	ISE	3.99	
	4.12	CSE	4.13	
	3.67	MA	3.77	
	4.17	PH	4.06	

In overall, the range of the mean scores is narrower in semester 2 which was [3.77, 4.42] compared to semester 1 was [3.67, 4.42], also almost Schools and Departments' scores tend to slightly decrease. Moreover, only a few average scores' improved such as of BME, CSE and MA's, and BME had remarkable changes. This worries the university a little as the quality of teaching in general and the quality of teachers in particular seems to become worse.

Teaching is a crucial tool in education as it transmits knowledge from one to others, thus the instructor must be knowledgeable and have strong teaching skills to help their students to gain the best understanding in no matter what subjects are.

Taking consideration on the fact that the instructor shall link the practice and the theory during teaching is also an important manner. Almost Schools and Departments did well with the average score above 80 per cent, yet only Mathematics Department achieved the lowest mark which is odd as this is a natural basic subject which is applied in many areas in real life and not all students were not told how to use the theory to solve the real life problem. Therefore, this Department only got 3.75 out of 5.0. This score is indeed very low as all students of this university had to study Mathematics in their degrees. This mark reflects on a whole population of X's students. In contrast, in second semester there were two more School and Department together with Mathematics Department achieved the lower average score than 80 per cent. In fact, it shows that the lecturers of those schools may have not clarified the necessity and applicability of the subjects they were delivering. The students need to be fully motivated by understanding the reasons why they should learn the subjects, especially basic subjects such as Mathematics, Physics and Chemistry.

Next, communicating in class is also an important factor in learning as the instructors can see whether their students are up to the level that they expected or not and if any gaps exist then they could fill in. Questions and answers are correlated with the aim of linking the lectures to the students' understandings and also to make the class's atmosphere more exciting, another words it is called 'interaction'. This has been discussed in the literature review. According to the results, only 4 out of 10 Schools/Departments achieved the average score for this factor which was higher than the X's mean score in the first semester and oppositely there was 6 out of 10 Schools/Departments got the higher mean score than the overall one in second semester. Encouraging students to communicate verbally is also a strategy in teaching because sometimes students may have clever questions which might make their lecturers think carefully before answering. Another issue is that the students are most likely shy to ask questions during lectures because they worry that their questions might be too simple. And the consequence for this issue is that they will fall behind the class time after time. As a result, the instructors are necessary to encourage their students to speak and interact as much as possible during the class.

In education, assessment is also needed. As it will reflect and evaluate what students have learnt and expect their applicable skills. Not only assess the theories but also test how they can deal with real world problem-solving. The overall mean score of a whole university for this question was under 80 per cent and most of Schools and Departments got higher scores than the overall mean score, except BA. In fact, there were in total of three Schools/Departments with the scores below the overall average and once again BA achieved the least score of nearly 75 per cent compared to 78 per cent in overall. Core subjects as Chemistry and Mathematics both got the average score higher than the overall one in both semesters. This fulfilled the students' needs. Setting assignments, projects or even homeworks are totally up to the instructors so they have full right to design and set their students in a few or many assessment pieces also depending on the subjects. The more assessment pieces involves the more work the lectures will do in checking, marking and providing feedback for them. Students will learn from their mistakes given that full comments and suggestions shall be given by their teachers. Moreover, students need to be marked equally and fairly with clear marking schemes given and no discrimination shall exist in any case. In overall, in both semester every School and Department got quite good score, however only Mathematics Department obtained a little lower average score than others for both semester.

Outside class assistance can be of importance to students for some reasons. First of all is to help the students to clarify misunderstanding or unsure problems occurred in class that the students did not dare to ask. Second is to guide them in doing projects or even some difficult homework. Third is to talk to the students and sometimes motivate them. Relying on the set class times is not always possible to answers all questions raised. The overall score for this matter was 3.96 in semester one and downed to 3.89 in semester two, in addition almost Schools and Departments achieved lower scores in second semester, except BME and CSE which have been improved a little. In general, in education there should have much volunteer support offered by the teachers to help learners to gain deep understanding together with sharing valuable experiences throughout teaching times.

The last issue needs to be taken consideration is the helpfulness of the courses. In fact, students will not be aware of its usefulness without clarifying and connecting what they were studying with the practice. In both semesters, MA and PH had the mean scores below the X's mean score which might be not acceptable as these subjects are basic nature courses that will be surely somehow applied in real life. For some courses such as BA's, BT's, EE's, etc, students may not fully understand how to apply the knowledge as the materials may be quite challenging and difficult to understand. Also, some subjects are pre-requisites for others so the students have to complete all of them according to the set curriculum. Hence, if they can be clarified with the reasons why they were taking the courses and how to apply the knowledge to the future careers then they will surely be motivated to learn the courses more effectively. All these tasks are definitely teachers' responsibilities.

Based on the above results, we may not be able to judge the quality of teaching as well as of lecturers and instructors. However, we use these facts to revise for a new version of "Course Evaluation Form" with some changes in order to suit students' needs and expectations. Please refer to Appendix.

First of all, the name of the form is recommended change from "Course Evaluation Form" to "Teaching Evaluation Form" as we should not combine "teaching" & "course" evaluations in the same form as before. The courses' contents are periodically reviewed each year and are done by the Academic Board so the students may not have enough experience to suggest how to change.

Secondly, most of questions have been revised, rephrased and re-ordered in a different way. These questions have been considered and suggested based on the discussed teaching perspectives in the literature review such as: transmission, apprenticeship, developmental, nurturing and social reform with the objective of defining the correlation between teaching perspectives and student evaluation. The students may not evaluate correctly without appropriate questions. These new questions mainly focus on teaching style and the interaction between lecturers and students as we are aware of the fact that it is not necessarily right to say a high level educated lecturer will be a good lecturer, it depends on how he or she interacts with students and also how to transmit their knowledge in the best way. Teaching is not an easy task as the lecturers have to be patient to deal with different level of students in the same class. Lecturers shall practice their skills every time they come to class as to improve their teaching skills better. In addition, they should achieve the scores that they have been evaluated by their students in order to revise what they have done good and what not too good. Sometimes they might get some comments and they should not ignore those.

However, there is a fact that we have to be aware of when doing this research is the use of feedback as it is sometimes dangerous to rely on and the reason for that is sometimes a strict lecturer will be badly evaluated which does not mean that that lecturer is not good and vice versus. But if the result badly repeated over and over again so it may prove the quality of that lecturer to be true.

Revising for a better version of the teaching evaluation form is important as the students' expectations vary from time to time. Feedback is just used as a tool to identify the gap between "teaching" and students not for evaluating the quality of teachers. In fact the outcome for this research is a revised teaching evaluation form to see whether it will be effective or not.

The results show that over 90 per cent of participants agreed that their lectures should have a mastery of the subjects they teach as well as being not too easy nor too strict in teaching as it may cause panic in learning. Also, more than 85 per cent people wanted to be clearly defined the objectives of any subjects they study at the beginning and how to apply in real life. About a third commented that their workload is currently quite heavy and they suggested that the weithing of assessments shall be reduced and the group assignments should be worth more because in that case the students will have more time to maximize their research's abilities as well as team work skills in order to produce the good projects. They mentionned "We do and we can learn from it, we can not learn from what we are assessed". Therefore, the lecturers may take this consideration to adjust the syllabi to suit their students' needs.

In addition, 100 per cent of participants agreed that the strictness of lecturers merely affected their learning. As a result, this issue should be carefully considered to improve the learning environment in the university. A super easy-going lecturer not necessarily means that he or she will teach well, the advantage for this person is easier to get closer to students so he or she may be able to assist the students better.

Interaction between the lecturers and the students are unfortunately not highly evaluated as students said that during lectures not many lecturers offered them a chance to speak so they did not dare to ask when needed. Interacting in class not only to clarify the confusion but also share knowledge in many dimensions. Moreover, about half of participants had the same feeling that in class they will only be told what to do and not to do in a single way not in multiple situations.

112 out of 140 students were satisfied with the assistance offered by their lecturers besides the official class times and tutorial classes. They found those extra helps very beneficial and valuable. There will be no better way to learn effectively if they do not fully understand the concepts or definitions. In education, "interaction" between the instructor and the learner is always crucial.

Table 3. Reliability Analysis

Variables	Cronbach's Alpha
Satisfaction (SA)	0.742
Mastery of subject (MA)	0.649
Strictness of lecturers (ST)	0.823
Interaction (IN)	0.775
Outside class assistance (OC)	0.861

According to standards, the reliability coefficient of 0.60 or higher is considered "reasonable acceptable" and in this study statistics shows above four variables are independent.

Pearson correlation allows us to determine the relationship between variables as it checks the "interdependency" of the variables considered in the model.

Table 4. Correlations

Correlation: Satisfaction, Mastery of subject, Strictness, Interaction, Outside class assistance

Correlations

	Satisfaction	Mastery of subje	Strictness	Interaction
Mastery of subje	0.114			
	0.180			
Strictness	0.029	0.064		
	0.738	0.453		
Interaction	-0.052	0.001	0.238	
	0.541	0.993	0.005	
Outside class as	-0.186	-0.015	-0.033	0.041
	0.028	0.864	0.701	0.633

Cell Contents
Pearson correlation
P-Value

The correlations table indicated that there exists a positive association between satisfaction and outside class assistance (r = -0.186, n = 140, p = 0.028) and there exists no statistically significant linkage between satisfaction and mastery of subject (r = -0.114, n = 140, p = 0.180), strictness (r = 0.029, n = 140, p = 0.738) and interaction (r = -0.052, n = 140, p = 0.541). Outside class assistance is statistically significantly associated with none of the independent variables. Interaction is statistically significantly associated with strictness (r = 0.238, n = 140, p = 0.005).

Table 5. Model summary

Model	R Square	Adjusted R Square
1	0.0519[a]	0.0238
2	0.0502[b]	0.0292
3	0.0482[c]	0.0343
4	0.0346[d]	0.0276
a. Predictors: (Constant), MA, ST, IN, OC		
b. Predictors: (Constant), MA, IN, OC		
c. Predictors: (Constant), MA, OC		
d. Predictors: (Constant), OC		
e. Predictors: (Constant), OC		

Model summary table shows suitability of the regression model to the specified data. Backward regression indicated fourth model to be the most significant one with R square value is 0.0346 this means outside class assistance can bring variation in satisfaction up to 3.46 per cent.

Table 6. Anova statistics

Model		Adjusted Sum of Squares	Degree of fredom	Adjusted Mean Square	F	p-value
1	**Regression**	4.9231	4	1.2308	1.85	0.123[a]
	Residual	89.8983	135	0.6659		
	Total	94.8214	139			
2	**Regression**	4.7574	3	1.5858	2.39	0.071[b]
	Residual	90.0640	136	0.6622		
	Total	94.8214	139			
3	**Regression**	4.570	2	2.2852	3.47	0.034[c]
	Residual	90.251	137	0.6588		
	Total	94.821	139			
4	**Regression**	3.2790	1	3.2790	4.94	0.028[d]
	Residual	91.5425	138	0.66335		
	Total	94.8214	139			
a. Predictors: (Constant), MA, ST, IN, OC						
b. Predictors: (Constant), MA, IN, OC						
c. Predictors: (Constant), MA, OC						
d. Predictors: (Constant), OC						
e. Dependent variable: SA						

Focusing on the fourth model, it can be found that outside class assistance can have a strong impact on the satisfaction $F(1, 138) = 4.94$, $p<0.05$.

Table 7. Coefficients statistics

Model		Coefficient	SE Coef	T-value	p-value
1	**(Constant)**	5.040	0.502	10.05	0.000
	MA	-0.1053	0.0741	-1.42	0.158
	ST	0.0351	0.0704	0.50	0.619
	IN	-0.0446	0.0703	-0.63	0.527
	OC	-0.1487	0.0678	-2.19	0.030
2	**(Constant)**	5.135	0.463	11.10	0.000
	MA	-0.1029	0.0737	-1.40	0.165
	IN	-0.0362	0.0681	-0.53	0.596
	OC	-0.1502	0.0676	-2.22	0.028
3	**(Constant)**	5.003	0.389	12.86	0.000
	MA	-0.1030	0.0735	-1.40	0.164
	OC	-0.1516	0.0674	-2.25	0.026
4	**(Constant)**	4.601	0.264	17.46	0.000
	OC	-0.1503	0.0676	-2.22	0.028

Considering conclusions of hypotheses made for this study as follows:

Hypothesis 1:

Independent variables were employed as mastery of subject, strictness, interaction and outside class assistance and satisfaction as an dependent variable. It turned out three variables in the model that are having insignificant values such as mastery of subject ($t = -1.42$, $p = 0.158$), interaction ($t = -0.63$, $p = 0.527$), and the most insignificant variable that is strictness ($t = 0.50$, $p = 0.619$) will be removed to improve the model. This concluded that there is no linkage between strictness and satisfaction.

Hypothesis 2:

In the second model, we still included three variables but two are again found to be not significant as mastery of subject ($t = -1.40$, $p = 0.165$) and the most insignificant variable interaction ($t = -0.53$, $p = 0.596$) will be excluded. This concluded that there is no linkage between interaction and satisfaction.

Hypothesis 3:

Although two variables have been eliminated, the model still needs to be reconsidered as another independent variable which is mastery of subject is found to be insignificant ($t = -1.40$, $p = 0.164$). This would make us accept our third hypothesis that is; there is no linkage between mastery of subject and satisfaction.

Hypothesis 4:

The last model only has outside class assistance exists to be found as the most significant one ($t = -2.22$, $p = 0.028$). It indicates that there is a statistical relationship between outside class assistance and satisfaction.

Hence, regression equation is as follows:

Satisfaction = 4.601 + 0.1503*Outside class assistance

5 Conclusion

Teaching that is not only a single task, but also a long process. A good teacher shall have a mastery of knowledge and know how motivate his/her students in learning process by giving clear explanations and linking theories to practice appropriately. More importantly, the teachers should define clearly the learning outcomes for their students at the beginning so they can see its importance and it will motivate them to complete the courses more efficiently. However, students' expectations may vary year after year, hence a revision of teaching evaluation form should be needed to match with the students' needs and expectations in learning. The outcome of this research is just simply a suggestion of a newer version of teaching evaluation form.

References

1. Dunn, R. & Griggs, S.A. *Multiculturalism and Learning Style*. Wesport, Connectticut, London (1998).
2. Ingvarson, L., & Rowe, K. *Conceptualising and Evaluating Teacher Quality: Substantive and Methodological Issues. The Australian Journal of Education, Vol. 52, No.1, Apr 2008: 5-35. Available at:* http://search.informit.com.au/documentSummary;dn=058594415630070;res=IELHSS *ISSN: 0004-9441* (2007).
3. Mitchell, K.J., Robinson, D.Z., Plake, B.S., & Knowles, K.T. *Testing Teacher Candidates – The role of licensure tests in improving teacher quality*. National Academy Press, Washington, DC: National Research Council (2001).
4. Qualters, D. *What's your perspective on teaching*. Issue No.4. St Thomas University (2002).

Hypothesis Testing on the Location Parameter of a Skew-Normal Distribution (SND) with Application

Phontita Thiuthad[1] and *Nabendu* Pal[2,*]

[1]Department of Mathematics and Statistics, Faculty of Science, Prince of Songkla University, Hatyai, Songkhla 90110, Thailand
[2]Department of Mathematics, University of Louisiana at Lafayette, Lafayette, Louisiana 70504, USA

Abstract. This work deals with testing a hypothesis on the location parameter (μ) of a skew-normal distribution (SND) based on a random sample of size n. The details of this work can be summarized in four major components: (a) First we review some useful results on SND, including the approximate probability distribution of the sample average. (b) Next, we develop several tests to test a hypothesis on μ based on the sample mean when the scale (σ) and shape (λ) parameters are known. (c) The tests for the known scale and shape are then extended for unknown scale and shape. (d) Finally, the test methods have been used for a real-life data set.

Keywords: Fisher information, Likelihood ratio test, Neyman-Pearson Lemma, Power function, Size of a test.
MSC 2010 Subject Classifications: 62F03, 62F05, 62E17.

1 Introduction

1.1 Preliminaries

A random variable W is said to have a skew-normal distribution with location parameter μ, scale parameter σ, and shape parameter λ, henceforth denoted by '$W \sim \text{SND}(\mu, \sigma, \lambda)$', provided its probability density function (pdf) is given as

$$f(w|\mu, \sigma, \lambda) = (2/\sigma)\phi((w-\mu)/\sigma)\Phi(\lambda(w-\mu)/\sigma), \quad -\infty < w < \infty, \tag{1.1}$$

where ϕ and Φ are the standard normal *pdf* and *cdf* respectively, for any $-\infty < \mu < \infty, -\infty < \lambda < \infty$, and $\sigma > 0$. The objective of this work is to develop test procedures for testing $H_0 : \mu = \mu_0$ vs. $H_A : \mu \neq \mu_0$. The two-sided alternative has been used only for convenience since an one-sided test can be developed easily with a suitable one-sided critical value.

*e-mail: nxp3695@louisiana.edu

Note that the above SND family includes the normal family as a special case (i.e., $N(\mu, \sigma^2)$) when the skew parameter $\lambda = 0$. Also, (1.1) includes positively skewed shape (when $\lambda > 0$) and negatively skewed shape (when $\lambda < 0$) which makes SND more appealing than the normal distribution.

The structure of SND was introduced in literature by several researchers independently (see Roberts (1966), O'Hagan and Leonard (1976), Aigner and Lovell (1977)), but it was given the present name by Azzalini (1985) who studied many characterization properties and applications of this distribution. For a nice review of many interesting properties of SND see Arnold and Lin (2004), and Gupta, Nguyen and Sanqui (2004) and other references therein. For applications of SND in reliability studies see Gupta and Brown (2001) which also used the SND to model IQ data taken from Roberts (1988). Figueiredo and Gomes (2013) used SND to model and monitor in statistical process control. Ngunkeng and Ning (2014) considered an information approach to detect changes in all three parameters in an SND, and then applied the proposed method to study Brazilian and Chilean stock market data (originally analyzed by Arellano-Valle et al. (2013)). For a multivariate generalization of the SND see Azzalini and Dalla Valle (1996), and Azzalini and Capitanio (1999).

1.2 Some Useful Properties of SND

The SND$(0, 1, \lambda)$ distribution is called the standard skew-normal distribution, and the following are some useful properties of the SND(μ, σ, λ):

(i) $W \sim \text{SND}(\mu, \sigma, \lambda)$ if and only if $Z = (W - \mu)/\sigma \sim \text{SND}(0, 1, \lambda)$;

(ii) $\lambda = 0$ corresponds to the normal distribution $N(\mu, \sigma^2)$;

(iii) If $W \sim \text{SND}(\mu, \sigma, \lambda)$, then $(-W) \sim \text{SND}(-\mu, \sigma, -\lambda)$;

(iv) As $\lambda \to \pm\infty$, SND$(0, 1, \lambda)$ tends to the half-normal distribution, i.e., the distribution of $\pm|Z_*|$ when $Z_* \sim N(0, 1)$;

(v) If $Z \sim \text{SND}(0, 1, \lambda)$, then $Z^2 \sim \chi_1^2$ (Chi-square with 1 degree freedom);

(vi) If $U, V \overset{iid}{\sim} N(0, 1)$, then $Z = (\lambda/\sqrt{1 + \lambda^2})|U| + (1/\sqrt{1 + \lambda^2})V \sim \text{SND}(0, 1, \lambda)$.

(vii) The MGF of the standard SND is $M(t) = 2exp(t^2/2)\Phi(\delta t) = 2\psi(t)\Phi(\delta t)$, where $\delta = \lambda/\sqrt{1 + \lambda^2}$ and $\psi(t)$ is the MGF of $Z_* \sim N(0, 1)$.

(viii) If $Z_* \sim N(0, 1)$ is independent of $Z \sim \text{SND}(0, 1, \lambda)$, then for any $a, b \in \mathbb{R}$

$$(aZ_* + bZ)/\sqrt{a^2 + b^2} \sim \text{SND}(0, 1, (b\lambda)/\sqrt{a^2(1 + \lambda^2) + b^2}).$$

(ix) The first three central moments of $W \sim \text{SND}(\mu, \sigma, \lambda)$ are $E(W) = \mu + \sigma\lambda\sqrt{2/\pi}/\sqrt{1 + \lambda^2}$, $\text{Var}(W) = \sigma^2\{1 - (2/\pi)(\lambda^2/(1 + \lambda^2))\}$, and $E(W - E(W))^3 = \sigma^3\sqrt{(2/\pi)}((4/\pi) - 1)\{\lambda/\sqrt{1 + \lambda^2}\}^3$, respectively.

The above property (vi) is important as it is often used to generate SND observations in simulation studies. Property (ix) will be useful in method of moments estimation (MME). For more properties and further generalizations, see Azzalini (2005), Henze (1986), and the other references therein.

1.3 Approximate Distribution of an SND Sample Average

In a recent study Thiuthad and Pal (2018) have shown that the exact distribution of the average of a random sample from SND, though extremely complicated, can be approximated fairly well by a suitable SND as described below.

Assume $W_1, W_2, ..., W_n$ are *iid* SND(μ, σ, λ). Let $\overline{W} = \sum W_i/n$. It is proposed that $\overline{W} \stackrel{\cdot}{\sim}$ SND$(\bar{\mu}, \bar{\sigma}, \bar{\lambda})$ where $\bar{\mu}, \bar{\sigma}, \bar{\lambda}$ are functions of (μ, σ, λ) and n, which can be found (with the details omitted here) as follows.

(A) First, obtain $\bar{\lambda}$ as

$$\bar{\lambda} = \text{sign}\,(\lambda)\left\{n^{1/3}[(1+\lambda^2)/\lambda^2 - (2/\pi)] - [1-(2/\pi)]\right\}^{-1/2} \tag{1.2}$$

where sign (λ) is > 0 if $\lambda > 0$; < 0 if $\lambda < 0$; and $= 0$, if $\lambda = 0$.

(B) Next, obtain $\bar{\sigma}$ as

$$\bar{\sigma} = \sqrt{\frac{\sigma^2\{1-(2/\pi)(\lambda^2/(1+\lambda^2))\}}{n\{1-(2/\pi)(\bar{\lambda}^2/(1+\bar{\lambda}^2))\}}}. \tag{1.3}$$

(C) Finally, $\bar{\mu}$ is obtained as

$$\bar{\mu} = \mu + (\sigma\lambda)\sqrt{2/\pi}/\sqrt{1+\lambda^2} - (\bar{\sigma}\bar{\lambda})\sqrt{2/\pi}/\sqrt{1+\bar{\lambda}^2}. \tag{1.4}$$

Remark 1.1 Note that $\bar{\mu}, \bar{\sigma}, \bar{\lambda}$ are functions of $(\mu, \sigma, \lambda, n)$. Further, $(\bar{\sigma}, \bar{\lambda})$ is dependent only on (σ, λ, n). Interestingly, it is easy to see that when $\lambda = 0$, then $\bar{\lambda} = 0$ and $\bar{\sigma} = \sigma/\sqrt{n}$ as well as $\bar{\mu} = \mu$, which perfectly matches with the result of a normal distribution. It has been shown through extensive simulation that the proposed approximation (of $\overline{W}$ by SND$(\bar{\mu}, \bar{\sigma}, \bar{\lambda})$) is very close even for small n.

The rest of the paper is organized as follows. Section 2 gives a brief summary of two types of point estimators as they will be essential in constructing the test statistics. Section 3 deals with various tests on μ with known (σ, λ), whereas Section 4 provides details in the most general case, that is with unknown (σ, λ). Finally, a real-life data set has been used in Section 5 as a demonstration.

2 A Brief Review of Point Estimation of μ

Thiuthad and Pal (2018) have provided the details of point estimation of μ under two separate cases as follows.

2.1 σ and λ are known

Here, the method of moments estimator (MME) of μ, say $\hat{\mu}_{MM}$, is given as

$$\hat{\mu}_{MM} = \overline{W} - \sigma\sqrt{2/\pi}(\lambda/\sqrt{1+\lambda^2}); \tag{2.1}$$

and the maximum likelihood estimator (MLE) of μ, say $\hat{\mu}_{ML}$, is found by solving

$$(\overline{W} - \mu)/\sigma = (\lambda/n)\sum_{i=1}^{n}\phi(\lambda(W_i - \mu)/\sigma)/\Phi(\lambda(W_i - \mu)/\sigma). \tag{2.2}$$

In this case, $\hat{\mu}_{MM}$ is unbiased for μ and $Var(\hat{\mu}_{MM}) = (\sigma^2/n)\{1 - (2/\pi)(\lambda^2/(1+\lambda^2))\}$. By CLT, $\hat{\mu}_{MM} \xrightarrow{d} \mathrm{N}(\mu, Var(\hat{\mu}_{MM}))$ as $n \to \infty$. By MLE's asymptotic property, $\hat{\mu}_{ML} \dot{\sim} \mathrm{N}(\mu, 1/(nI_0))$, where I_0 = Fisher information per observation = $(1/\sigma^2)\Big[1 + \lambda^2\mathrm{E}\Big\{\big(\phi(\lambda Z)/\Phi(\lambda Z)\big)^2\Big\}\Big]$.

2.2 σ and λ are unknown

The MME of μ depends on those of λ and σ. Therefore, the following steps are followed in a sequential order.

Let $m_1 = \sum_{i=1}^{n} W_i/n$ and $m_j = \sum_{i=1}^{n}(W_i - m_1)^j/n$, $j = 2, 3$, be the first three sample moments. First obtain $\hat{\lambda}$ by solving (provided a solution exists)

$$\{((1+\hat{\lambda}^2)/\hat{\lambda}^2) - (2/\pi)\}^3 - (2/\pi)((4/\pi) - 1)^2(m_2^3/m_3^2) = 0. \tag{2.3}$$

A solution of (2.3) exists provided $m_2^3/m_3^2 > (\pi - 2)^3/(2(4-\pi)^2) = c_0$ (say). If a solution of (2.3) in terms of $\hat{\lambda}^2$ does not exist, then $\hat{\lambda}^2$ is deemed to have taken the value ∞. Thus, the MME of λ, i.e., $\hat{\lambda}_{MM}$ is defined as

$$\hat{\lambda}_{MM} = \begin{cases} \mathrm{sign}(m_3) \times \hat{\lambda}, & \text{if (2.3) has a solution.} \\ \mathrm{sign}(m_3) \times \infty, & \text{otherwise.} \end{cases} \tag{2.4}$$

For practical purposes, in (2.4) replace '∞' by a sufficiently large value, say M which makes the estimated distribution almost half normal. After estimating λ as above, obtain $\hat{\sigma}_{MM}$ as

$$\hat{\sigma}_{MM} = \begin{cases} \sqrt{m_2/\{1 - (2/\pi)(\hat{\lambda}^2_{MM}/(1+\hat{\lambda}^2_{MM}))\}}, & \text{if (2.3) has a solution.} \\ \sqrt{m_2/\{1-(2/\pi)\}}, & \text{otherwise.} \end{cases} \tag{2.5}$$

Finally,

$$\hat{\mu}_{MM} = \begin{cases} m_1 - \hat{\sigma}_{MM}\sqrt{2/\pi}(\hat{\lambda}_{MM}/\sqrt{1+\hat{\lambda}^2_{MM}}), & \text{if (2.3) has a solution.} \\ m_1 - \hat{\sigma}_{MM}\sqrt{2/\pi}, & \text{otherwise.} \end{cases} \tag{2.6}$$

Note that even if $\hat{\lambda}_{MM}$ takes the infinite value with a positive probability, it does not prevent the MME of σ and λ from taking finite values with probability 1.

When all parameters are unknown, the MLE of μ may not be a viable option as it may result into infinite value with a positive probability. Hence, a maximum penalized likelihood estimate (MPLE) looks more reasonable which maximizes

$$L_{**} = n\,ln(2/\sigma) + \sum_{i=1}^{n} ln\phi((W_i - \mu)/\sigma) + \sum_{i=1}^{n} ln\Phi(\lambda(W_i-\mu)/\sigma) - c_1 ln(1 + c_2\lambda^2), \tag{2.7}$$

with $c_1 = 3\pi^2/32$ and $c_2 = 8/\pi^2$. The last term in (2.7) is the penalty function as suggested by Azzalini and Arellano-Valle (2012).

Thiuthad and Pal (2018) provided results of their extensive simulation study comparing $\hat{\mu}_{MM}$ and $\hat{\mu}_{MPL}$. Further, as noted by these authors, the Fisher information matrix $I = nI_0$ where I_0 is the 3×3 information matrix per observation) has at least three singularity points rendering I^{-1} useless as a benchmark dispersion matrix for the estimators of the parameter vector (μ, σ, λ).

3 Tests on the location of an SND with known (σ, λ)

3.1 The Case of a Single Observation

To develop the idea of testing on μ with known (σ, λ) we first consider the case of a single observation $W \sim \text{SND}(\mu, \sigma, \lambda)$. (The method developed here can easily be transported to a test based on $\overline{W} \stackrel{\cdot}{\sim} \text{SND}(\bar{\mu}, \bar{\sigma}, \bar{\lambda})$.) As mentioned earlier, we test $H_0 : \mu = \mu_0$ against a suitable alternative which, for convenience, has been taken as the two sided one, i.e., $H_A : \mu \neq \mu_0$. To develop the idea further, we consider a simple alternative as $H_A : \mu = \mu_A$, where $\mu_A \neq \mu_0$.

The likelihood ratio test (LRT) statistic for the above simple null against the simple alternative becomes

$$\begin{aligned} \Lambda &= \frac{f(W|\mu_0, \sigma, \lambda)}{f(W|\mu_A, \sigma, \lambda)} \\ &= \frac{\phi((W-\mu_0)/\sigma)\Phi(\lambda(W-\mu_0)/\sigma)}{\phi((W-\mu_A)/\sigma)\Phi(\lambda(W-\mu_A)/\sigma)}. \end{aligned} \tag{3.1}$$

It is more convenient to work with the natural logarithm of the LRT statistic, i.e.,

$$\Lambda_* \text{ (say)} = ln\Lambda = \frac{-(W-\mu_0)^2}{2\sigma^2} + \frac{(W-\mu_A)^2}{2\sigma^2} + ln\Phi(\lambda(W-\mu_0)/\sigma) - ln\Phi(\lambda(W-\mu_A)/\sigma). \tag{3.2}$$

Write $g_0(W) = ln\Phi(\lambda(W-\mu_0)/\sigma)$, which has the following Taylor series expression

$$\begin{aligned} g_0(W) &= g(\mu_0) + (W-\mu_0)g'(\mu_0) + \frac{(W-\mu_0)^2}{2}g''(\mu_0) + \ldots \\ &\approx -ln2 + \frac{\lambda}{\sigma}\sqrt{\frac{2}{\pi}}(W-\mu_0) - \frac{\lambda^2}{\pi\sigma^2}(W-\mu_0)^2, \end{aligned} \tag{3.3}$$

by retaining up to the second order terms. Similarly,

$$g_A(W) = ln\Phi(\lambda(W-\mu_A)/\sigma) \approx -ln2 + \frac{\lambda}{\sigma}\sqrt{\frac{2}{\pi}}(W-\mu_A) - \frac{\lambda^2}{\pi\sigma^2}(W-\mu_A)^2. \tag{3.4}$$

Using (3.3) and (3.4), we get a second order approximation of Λ_*, say $\Lambda_*^{(2)}(W)$, as

$$\Lambda_*(W) \approx \Lambda_*^{(2)}(W) := (\mu_0 - \mu_A)\Big\{(2W - \mu_0 - \mu_A)\Big(\frac{1}{2\sigma^2} + \frac{\lambda^2}{\pi\sigma^2}\Big) - \frac{\lambda}{\sigma}\sqrt{\frac{2}{\pi}}\Big\}. \tag{3.5}$$

Let $\delta = \mu_0 - \mu_A$ and $W' = W - \mu_0$. Then, $W' \sim \text{SND}(0, \sigma, \lambda))$ under H_0 and,

$$\Lambda_*^{(2)}(W) = \Lambda_*^{(2)}(W') = \delta\Big\{(2W' + \delta)\Big(\frac{1}{2\sigma^2} + \frac{\lambda^2}{\pi\sigma^2}\Big) - \frac{\lambda}{\sigma}\sqrt{\frac{2}{\pi}}\Big\}. \tag{3.6}$$

Using the Neyman-Pearson Lemma, retain H_0 if $\Lambda_* \geq k$, and reject H_0 otherwise. Therefore, using the second order approximation, retain H_0 if $\Lambda_*^{(2)} \geq k$, and reject H_0 if $\Lambda_*^{(2)} < k$, where k is found by the level condition $P(\text{Type-I error}) = \alpha$. This is considered in two cases:

Case (a) If $\mu_A > \mu_0$, then $\delta = \mu_0 - \mu_A < 0$. Hence, $\Lambda_*^{(2)} < k$ if and only if $W' > k_1'$.

Case (b) If $\mu_A < \mu_0$, then $\delta = \mu_0 - \mu_A > 0$. Hence, $\Lambda_*^{(2)} < k$ if and only if $W' < k_2'$.

Note that the cut-off points k_1' and k_2' depends only on the level condition, not the exact value of μ_A on either side of μ_0. Therefore, for a two sided alternative $H_A : \mu \neq \mu_0$, we reject H_0 based on the second order LRT statistic if $W' < k_2'$ or $W' > k_1'$, and retain H_0 if $k_2' \leq W' \leq k_1'$, where k_1' and k_2' can be chosen such that

$$G(k_2') = \int_{-\infty}^{k_2'} (2/\sigma)\phi(u/\sigma)\Phi(\lambda(u/\sigma))du = (\alpha/2), \tag{3.7}$$

where, G represents the cdf of $\text{SND}(0, \sigma, \lambda)$; i.e., $k_2' = G^{-1}(\alpha/2)$. Similarly,

$$1 - G(k_1') = \int_{k_1'}^{\infty} (2/\sigma)\phi(u/\sigma)\Phi(\lambda(u/\sigma))du = (\alpha/2), \tag{3.8}$$

i.e., $k_1' = G^{-1}(1 - (\alpha/2))$.

Further, note that the cut-off points k_1' and k_2' can be expressed as $k_i' = \sigma k_{i*}'$, $i = 1, 2$, where k_{i*}' $(i = 1, 2)$ are found by the following equations.

$$\int_{-\infty}^{k_{2*}'} 2\phi(u_*)\Phi(\lambda u_*)du_* = (\alpha/2) \text{ and } \int_{-\infty}^{k_{1*}'} 2\phi(u_*)\Phi(\lambda u_*)du_* = 1 - (\alpha/2). \tag{3.9}$$

The following Table 3.1 tabulates the values of k_{i*}' $(i = 1, 2)$ for various values of α. Note that when $\lambda = 0$, i.e., $\text{SND}(0, \sigma, 0)$, the cut-off points are exactly those of $N(0, \sigma^2)$. Also note that for a one sided alternative one should use only one of k_{1*}' and k_{2*}' with $(\alpha/2)$ replaced by α in one of the equations in (3.9).

Remark 3.1 The above Table 3.1 lists the two sided cut-off points only for $\lambda > 0$. Note that $(-W') = -(W - \mu_0) \sim \text{SND}(0, \sigma, -\lambda)$. Therefore, for $\lambda < 0$, one can get the cut-off points easily by interchanging k_{1*}' by k_{2*}', and k_{2*}' by k_{1*}' with signs reversed.

3.2 The Case of a Sample of Size n (≥ 1)

We first look at the two point estimators, i.e., MME and MLE, given earlier as $\hat{\mu}_{MM}$ and $\hat{\mu}_{ML}$ respectively. The exact variance of $\hat{\mu}_{MM}$ is

$$V(\hat{\mu}_{MM}) = V(\overline{W}) = (\sigma^2/n)\{1 - (2/\pi)(\lambda^2/(1 + \lambda^2))\}. \tag{3.10}$$

Table 3.1: The two-sided cut-off points (k'_{2*}, k'_{1*}) satisfying (3.9) (i.e., each tail probability = $\alpha/2$) for various values of $\lambda > 0$.

λ	$\alpha = 0.01$		$\alpha = 0.02$		$\alpha = 0.05$		$\alpha = 0.10$		$\alpha = 0.20$	
	k'_{2*}	k'_{1*}	k'_{2*}	k'_{1*}	k'_{2*}	k'_{1*}	k'_{2*}	k'_{1*}	k'_{2*}	k'_{1*}
0.0	-2.576	2.576	-2.326	2.326	-1.960	1.960	-1.645	1.645	-1.282	1.282
0.5	-2.030	2.786	-1.801	2.548	-1.464	2.199	-1.173	1.900	-0.838	1.556
1.0	-1.471	2.807	-1.282	2.575	-1.002	2.239	-0.760	1.955	-0.478	1.632
1.5	-1.079	2.807	-0.925	2.576	-0.696	2.241	-0.496	1.960	-0.261	1.644
2.0	-0.823	2.807	-0.695	2.576	-0.503	2.241	-0.334	1.960	-0.134	1.645
2.5	-0.652	2.807	-0.543	2.576	-0.378	2.241	-0.232	1.960	-0.056	1.645
3.0	-0.532	2.807	-0.437	2.576	-0.292	2.241	-0.163	1.960	-0.006	1.645
3.5	-0.445	2.807	-0.360	2.576	-0.232	2.241	-0.115	1.960	0.028	1.645
4.0	-0.379	2.807	-0.303	2.576	-0.186	2.241	-0.080	1.960	0.051	1.645
4.5	-0.328	2.807	-0.259	2.576	-0.152	2.241	-0.054	1.960	0.068	1.645
5.0	-0.288	2.807	-0.224	2.576	-0.125	2.241	-0.035	1.960	0.081	1.645
5.5	-0.255	2.807	-0.196	2.576	-0.104	2.241	-0.019	1.960	0.090	1.645
6.0	-0.228	2.807	-0.173	2.576	-0.087	2.241	-0.007	1.960	0.097	1.645
6.5	-0.205	2.807	-0.154	2.576	-0.073	2.241	0.004	1.960	0.103	1.645
7.0	-0.186	2.807	-0.137	2.576	-0.061	2.241	0.012	1.960	0.107	1.645
7.5	-0.170	2.807	-0.123	2.576	-0.051	2.241	0.019	1.960	0.111	1.645
8.0	-0.155	2.807	-0.112	2.576	-0.042	2.241	0.024	1.960	0.113	1.645
8.5	-0.143	2.807	-0.101	2.576	-0.035	2.241	0.029	1.960	0.116	1.645
9.0	-0.132	2.807	-0.092	2.576	-0.029	2.241	0.033	1.960	0.117	1.645
9.5	-0.122	2.807	-0.084	2.576	-0.023	2.241	0.037	1.960	0.119	1.645
10.0	-0.114	2.807	-0.077	2.576	-0.018	2.241	0.040	1.960	0.120	1.645

But the exact variance expression of $\hat{\mu}_{ML}$ is not known. Hence we can use the asymptotic variance (AV) which is quite close to the exact one (see Thiuthad and Pal (2018)).

$$AV(\hat{\mu}_{ML}) = 1/(nI_0). \tag{3.11}$$

Based on the above expressions (3.10) and (3.11), and the asymptotice distributions presented in Subsection 2.1, we propose two tests, say T_1 and T_2 respectively, defined as follows:

$$T_1 : \text{ Rej } H_0 \text{ if } |\hat{\mu}_{MM} - \mu_0|/\sqrt{V(\hat{\mu}_{MM})} > z_{\alpha/2} \tag{3.12}$$

$$T_2 : \text{ Rej } H_0 \text{ if } |\hat{\mu}_{ML} - \mu_0|/\sqrt{AV(\hat{\mu}_{ML})} > z_{\alpha/2} \tag{3.13}$$

A third test, say T_3, can be constructed as follows. Based on a random sample of size n from SND(μ, σ, λ), we use $\overline{W}$ to test $H_0 : \mu = \mu_0$ versus $H_A : \mu \neq \mu_0$. From Subsection 1.3 it is easily seen that $\overline{W} \stackrel{\cdot}{\sim}$ SND($\bar{\mu}, \bar{\sigma}, \bar{\lambda}$) where $\bar{\mu}$ is given in (1.4), and $\bar{\mu} = \mu+$ (an expression dependent solely on (σ, λ)). Therefore, testing the above H_0 vs. H_A is equivalent to testing $\overline{H}_0 : \bar{\mu} = \bar{\mu}_0$ where $\bar{\mu}_0$ is the same expression in (1.4) with μ replaced by μ_0.

The rest of the methodology is same as the above case (Subsection 3.1) of a single observation, i.e., retain $\overline{H}_0$ if $\overline{W}' = (\overline{W} - \bar{\mu}_0)$ is between $\bar{k}'_2$ and $\bar{k}'_1$, and reject $\overline{H}_0$ otherwise, where $\bar{k}'_i = \bar{\sigma}\bar{k}'_{i*}$, $i = 1, 2$, with $\bar{k}'_{i*}$ satisfying (3.9) with λ replaced by $\bar{\lambda}$, i.e.,

$$T_3 = \begin{cases} \text{retain } \overline{H}_0, & \text{if } \bar{k}'_2 < \overline{W}' < \bar{k}'_1, \\ 0, & \text{otherwise.} \end{cases} \tag{3.14}$$

The following Tables 3.2 - 3.7 provide the simulated power (or size) of the above three tests, based on 2.5×10^4 replications as μ deviates from $\mu_0 = 0$ (w.l.g.) for various values of λ. All the tables have been constructed using $\alpha = 0.05$. Similar tables for other values of α have been provided in the Appendix A.1.

Table 3.2: Power (or size, when $\mu = \mu_0 = 0$) of three tests for various μ with fixed $\lambda = 0.0$

μ	$n = 5$			$n = 10$			$n = 25$		
	T_1	T_2	T_3	T_1	T_2	T_3	T_1	T_2	T_3
0.0	0.0512	0.0512	0.0512	0.0492	0.0492	0.0492	0.0485	0.0485	0.0485
0.2	0.0747	0.0747	0.0747	0.0966	0.0966	0.0966	0.1715	0.1715	0.1715
0.4	0.1490	0.1490	0.1490	0.2452	0.2452	0.2452	0.5120	0.5120	0.5120
0.6	0.2685	0.2685	0.2685	0.4756	0.4756	0.4756	0.8536	0.8536	0.8536
0.8	0.4314	0.4314	0.4314	0.7144	0.7144	0.7144	0.9780	0.9780	0.9780
1.0	0.6102	0.6102	0.6102	0.8829	0.8829	0.8829	0.9988	0.9988	0.9988
2.0	0.9940	0.9940	0.9940	1.0000	1.0000	1.0000	1.0000	1.0000	1.0000
3.0	1.0000	1.0000	1.0000	1.0000	1.0000	1.0000	1.0000	1.0000	1.0000
4.0	1.0000	1.0000	1.0000	1.0000	1.0000	1.0000	1.0000	1.0000	1.0000
6.0	1.0000	1.0000	1.0000	1.0000	1.0000	1.0000	1.0000	1.0000	1.0000
8.0	1.0000	1.0000	1.0000	1.0000	1.0000	1.0000	1.0000	1.0000	1.0000

Table 3.3: Power (or size, when $\mu = \mu_0 = 0$) of three tests for various μ with fixed $\lambda = 1.0$

μ	$n = 5$			$n = 10$			$n = 25$		
	T_1	T_2	T_3	T_1	T_2	T_3	T_1	T_2	T_3
0.0	0.0487	0.0481	0.0484	0.0502	0.0502	0.0504	0.0494	0.0491	0.0495
0.2	0.0732	0.0758	0.0701	0.0923	0.0926	0.0887	0.1553	0.1575	0.1528
0.4	0.1468	0.1507	0.1414	0.2458	0.2489	0.2395	0.5110	0.5123	0.5050
0.6	0.2983	0.3042	0.2890	0.5117	0.5136	0.5034	0.8788	0.8778	0.8762
0.8	0.5051	0.5115	0.4944	0.7876	0.7887	0.7822	0.9918	0.9911	0.9916
1.0	0.7248	0.7288	0.7159	0.9455	0.9437	0.9427	0.9997	0.9997	0.9997
2.0	0.9999	0.9999	0.9999	1.0000	1.0000	1.0000	1.0000	1.0000	1.0000
3.0	1.0000	1.0000	1.0000	1.0000	1.0000	1.0000	1.0000	1.0000	1.0000
4.0	1.0000	1.0000	1.0000	1.0000	1.0000	1.0000	1.0000	1.0000	1.0000
6.0	1.0000	1.0000	1.0000	1.0000	1.0000	1.0000	1.0000	1.0000	1.0000
8.0	1.0000	1.0000	1.0000	1.0000	1.0000	1.0000	1.0000	1.0000	1.0000

Remark 3.2 The first three tables (Tables 3.2 - 3.4) provide the power/size as a function of μ for fixed λ, whereas the last three tables (Tables 3.5 - 3.7) provide the same as a function of λ for fixed μ. Therefore, the entire Table 3.5 provides size and the Tables 3.6 - 3.7 provide pure power of the three proposed tests. On the other hand, the first row of Tables 3.2 - 3.4 provide the size, and the remaining rows of these tables show power. All the three tests adhere to the level condition very closely, even for a sample size as small as 5. The patterns emerging from these tables are as follows.

Table 3.4: Power (or size, when $\mu = \mu_0 = 0$) of three tests for various μ with fixed $\lambda = 2.0$

μ	$n = 5$			$n = 10$			$n = 25$		
	T_1	T_2	T_3	T_1	T_2	T_3	T_1	T_2	T_3
0.0	0.0474	0.0506	0.0481	0.0497	0.0507	0.0496	0.0498	0.0495	0.0496
0.2	0.0650	0.0782	0.0584	0.0808	0.0889	0.0743	0.1202	0.1268	0.1122
0.4	0.1288	0.1476	0.1133	0.1909	0.2070	0.1747	0.3956	0.4020	0.3781
0.6	0.2548	0.2753	0.2280	0.4253	0.4349	0.4006	0.7831	0.7797	0.7700
0.8	0.4382	0.4549	0.4049	0.7014	0.7046	0.6785	0.9721	0.9685	0.9683
1.0	0.6560	0.6654	0.6217	0.9014	0.8967	0.8904	0.9989	0.9987	0.9987
2.0	0.9998	0.9996	0.9996	1.0000	1.0000	1.0000	1.0000	1.0000	1.0000
3.0	1.0000	1.0000	1.0000	1.0000	1.0000	1.0000	1.0000	1.0000	1.0000
4.0	1.0000	1.0000	1.0000	1.0000	1.0000	1.0000	1.0000	1.0000	1.0000
6.0	1.0000	1.0000	1.0000	1.0000	1.0000	1.0000	1.0000	1.0000	1.0000
8.0	1.0000	1.0000	1.0000	1.0000	1.0000	1.0000	1.0000	1.0000	1.0000

Table 3.5: Size of three tests for various λ with fixed $\mu = 0.0$

λ	$n = 5$			$n = 10$			$n = 25$		
	T_1	T_2	T_3	T_1	T_2	T_3	T_1	T_2	T_3
0.0	0.0466	0.0466	0.0466	0.0492	0.0492	0.0492	0.0485	0.0485	0.0485
0.2	0.0504	0.0503	0.0503	0.0495	0.0494	0.0495	0.0491	0.0491	0.0491
0.4	0.0486	0.0488	0.0487	0.0508	0.0510	0.0509	0.0491	0.0488	0.0491
0.6	0.0480	0.0478	0.0480	0.0477	0.0477	0.0477	0.0483	0.0483	0.0483
0.8	0.0490	0.0492	0.0491	0.0508	0.0498	0.0504	0.0510	0.0509	0.0507
1.0	0.0504	0.0505	0.0498	0.0515	0.0513	0.0520	0.0517	0.0521	0.0521
2.0	0.0475	0.0519	0.0476	0.0496	0.0503	0.0488	0.0491	0.0494	0.0495
3.0	0.0478	0.0578	0.0486	0.0475	0.0528	0.0473	0.0505	0.0502	0.0494
4.0	0.0483	0.0697	0.0485	0.0489	0.0580	0.0496	0.0502	0.0532	0.0490
6.0	0.0453	0.0934	0.0484	0.0487	0.0678	0.0480	0.0489	0.0574	0.0479
8.0	0.0458	0.1173	0.0483	0.0479	0.0826	0.0490	0.0493	0.0622	0.0497

Table 3.6: Power of three tests for various λ with fixed $\mu = 1.0$

λ	$n = 5$			$n = 10$			$n = 25$		
	T_1	T_2	T_3	T_1	T_2	T_3	T_1	T_2	T_3
0.0	0.6070	0.6070	0.6070	0.8857	0.8857	0.8857	0.9986	0.9986	0.9986
0.2	0.6668	0.6667	0.6667	0.9176	0.9176	0.9176	0.9996	0.9996	0.9996
0.4	0.6998	0.7010	0.6988	0.9356	0.9358	0.9353	0.9999	0.9999	0.9999
0.6	0.7175	0.7195	0.7145	0.9448	0.9454	0.9436	1.0000	1.0000	1.0000
0.8	0.7216	0.7245	0.7163	0.9452	0.9459	0.9438	0.9998	0.9998	0.9998
1.0	0.7257	0.7290	0.7152	0.9455	0.9437	0.9427	0.9997	0.9997	0.9997
2.0	0.6575	0.6648	0.6222	0.9043	0.9016	0.8938	0.9991	0.9990	0.9990
3.0	0.5841	0.5978	0.5370	0.8397	0.8251	0.8166	0.9954	0.9923	0.9946
4.0	0.5284	0.5366	0.4751	0.7922	0.7595	0.7642	0.9890	0.9781	0.9864
6.0	0.4661	0.4890	0.4082	0.7061	0.6434	0.6684	0.9661	0.9207	0.9601
8.0	0.4273	0.4780	0.3684	0.6548	0.5727	0.6118	0.9447	0.8556	0.9357

(a) When $\lambda = 0$, i.e., the distribution is $N(\mu, \sigma^2)$, then all the three tests become the identical normal test (based on the sample mean), and hence they provide identical

Table 3.7: Power of three tests for various λ with fixed $\mu = 2.0$

λ	$n=5$			$n=10$			$n=25$		
	T_1	T_2	T_3	T_1	T_2	T_3	T_1	T_2	T_3
0.0	0.9936	0.9936	0.9936	1.0000	1.0000	1.0000	1.0000	1.0000	1.0000
0.2	0.9989	0.9989	0.9989	1.0000	1.0000	1.0000	1.0000	1.0000	1.0000
0.4	0.9998	0.9998	0.9998	1.0000	1.0000	1.0000	1.0000	1.0000	1.0000
0.6	0.9998	0.9998	0.9998	1.0000	1.0000	1.0000	1.0000	1.0000	1.0000
0.8	0.9997	0.9997	0.9997	1.0000	1.0000	1.0000	1.0000	1.0000	1.0000
1.0	0.9999	0.9999	0.9998	1.0000	1.0000	1.0000	1.0000	1.0000	1.0000
2.0	0.9994	0.9995	0.9992	1.0000	1.0000	1.0000	1.0000	1.0000	1.0000
3.0	0.9982	0.9974	0.9977	1.0000	1.0000	1.0000	1.0000	1.0000	1.0000
4.0	0.9961	0.9926	0.9948	1.0000	1.0000	1.0000	1.0000	1.0000	1.0000
6.0	0.9905	0.9710	0.9860	1.0000	0.9989	1.0000	1.0000	1.0000	1.0000
8.0	0.9862	0.9488	0.9810	0.9999	0.9962	0.9999	1.0000	1.0000	1.0000

performance.

(b) For λ close to 0, the test T_2 seems to have the best power performance. But note that T_2 becomes very liberal (with size exceeding α substantially) as λ gets large (≥ 3).

(c) The test T_3 has good local power for small λ (≤ 2), but its performance deteriorates as λ gets larger. This is not surprising because $\overline{W}$ is not a sufficient statistic for μ, and as shown in Thiuthad and Pal (2018), the amount of information about μ carried by $\overline{W}$ diminishes rapidly as λ deviates away from 0.

(d) The mechanisms of the subsection 3.2 can be followed easily even for unknown (σ, λ) provided one has a large sample which can give reliable estimates of σ and λ. In such a case follow the methods of this section by replacing (σ, λ) by $(\hat{\sigma}, \hat{\lambda})$, for a large n, and then pretend that the nuisance parameters were known.

4 Hypothesis Testing on μ with unknown (σ, λ)

In order to develop tests on μ for unknown σ and λ, we can think of replacing the nuisance parameters by their suitable estimates (MME and/or MPLE) and call the resultant tests (i.e., variants of T_1, T_2, T_3 as developed in Subsection 3.2) as T_1', T_2', T_3' respectively. But the main difficulty we face here is that the asymptotic distributions of $\hat{\mu}_{MM}$ and $\hat{\mu}_{MPL}$ are far from clear, and the normal cut-off point as used in Section 4 produces bad results.

We can use a parametric bootstrap version of T_1 and T_2, i.e., based on $\hat{\mu}_{MM}$ and $\hat{\mu}_{MPL}$, to create two tests, say T_1^* and T_2^*, respectively. The basic idea of a parametric bootstrap (PB) test based on a point estimate is described in the following.

Let $\hat{\mu}$ be an estimator based on the given data $(W_1, W_2, ..., W_n)$ ($\hat{\mu}$ can be either $\hat{\mu}_{MM}$ (to create T_1^*) or $\hat{\mu}_{MPL}$ (to create T_2^*)). Our objective is to test $H_0 : \mu = \mu_0$ versus $H_A : \mu \neq \mu_0$. The general steps of the PB method are as follows:

Step-1: For a given set of n *iid* observations, obtain $\hat{\mu}$ and compute a deviant metric $D = (\hat{\mu} - \mu_0)$, which measures the deviation of the point estimator $\hat{\mu}$ from the null value of μ.

Step-2: Assume that H_0 is true, i.e., $\mu = \mu_0$. Here, we obtain the estimates of σ and λ under H_0 which are called the restricted estimates (REs). For this step, we first derive the restricted method of moments (RMM) and the restricted maximum penalized likelihood (RMPL) estimators of $(\hat{\sigma}, \hat{\lambda})$ (for the details see in Appendix A.2), whichever is needed for the estimation method of μ. [If we use $D_{MM} = (\hat{\mu}_{MM} - \mu_0)$ as our deviant measure D, then we need to use corresponding $\hat{\sigma}_{RMM}$ and $\hat{\lambda}_{RMM}$ under H_0 for the PB test to work. Similarly, if we use $D_{MPL} = (\hat{\mu}_{MPL} - \mu_0)$ as our metric D, then we should use $\hat{\sigma}_{RMPL}$ and $\hat{\lambda}_{RMPL}$ accordingly.]

Step-3: Generate *iid* observations under H_0, i.e., $W_1^*, W_2^*, ..., W_n^* \sim \text{SND}(\mu_0, \hat{\sigma}_{RE}, \hat{\lambda}_{RE})$. Obtain $\hat{\mu}$ based on $(W_1^*, W_2^*, ..., W_n^*)$, and call it $\hat{\mu}^*$. Recalculate D as $D^* = (\hat{\mu}^* - \mu_0)$.

Step-4: Repeat the above Step-3 a large number of times, say N times. This gives us N copies of D^*, i.e., $D^{*1}, D^{*2}, ..., D^{*N}$. Order them as $D^{*(1)} \leq ... \leq D^{*(N)}$. Then the cut off points for D are to be found by the (approximate) probability distribution of D under H_0, i.e., the left cut-off point $D_L = D^{*((\alpha/2)N)}$ and the right cut-off point $D_U = D^{*((1-\alpha/2)N)}$ such that $\alpha \simeq \text{P(Rej } H_0 | H_0 \text{ is true)}$. If $D \in (D_L, D_U)$, then we retain H_0. If $D < D_L$ or $D > D_U$, then we reject H_0. Alternatively, we can compute the p-value as $p_{PB} = 2\min\{p_L, p_U\}$ where $p_L = \{\text{number of } D^{*j} < D\}/N$ and $p_U = \{\text{number of } D^{*j} > D\}/N$.

Apart from T_1^* and T_2^* (based on D_{MM} and D_{MPL}, respectively), we propose a few other tests as follows. Though not all of them may not show good performance, the idea behind them seems reasonable, and hence worth trying.

It is tempting to modify the third test T_3 in (3.14) and propose T_3' as follows:

$$T_3' = \begin{cases} \text{retain } \overline{H}_0, & \text{if } \bar{k}_2'' < \overline{W}' < \bar{k}_1'', \\ 0, & \text{otherwise,} \end{cases} \tag{4.1}$$

where $\overline{W}' = (\overline{W} - \mu_0)$, $\bar{k}_i'' = \hat{\hat{\sigma}}_{RE}\bar{k}_{i*}''$, $i = 1, 2$, with $\bar{k}_{i*}''$ satisfying (3.9) with λ replaced by $\hat{\hat{\lambda}}_{RE}$ where $\hat{\hat{\sigma}}_{RE}$ and $\hat{\hat{\lambda}}_{RE}$ satisfy (1.2) and (1.3) with λ replaced by $\hat{\lambda}_{RE}$ and σ replaced by $\hat{\sigma}_{RE}$, respectively. (The subscript 'RE' indicates the restricted estimate of the nuisance parameters under the null hypothesis.) Here we can create two versions of T_3' by using either RMM or RMPL estimators. The two versions of T_3' are henceforth referred to as $T_{3(RMM)}'$ and $T_{3(RMPL)}'$, respectively. However, both the versions of this test show a very bad and strange Type-I probability performance (either 0 or 1). Therefore we opt for a parametric bootstrap version of T_3, call it T_3^*, as follows.

Similar to T_1^* and T_2^*, we propose T_3^* with the deviant metric $D = (\overline{W} - \mu_0) = \overline{W}'$ whose cut-off points are to be found through a parametric bootstrap method. The steps to implement T_3^* are identical to those of T_1^* and T_2^* as mentioned above. However, in step 3, the *iid* bootstrap observations W_i^* are to be generated from SND$(\mu_0, \hat{\sigma}_{RE}, \hat{\lambda}_{RE})$, where $\hat{\sigma}_{RE}$ and $\hat{\lambda}_{RE}$ are the restricted estimates (RE) of σ and λ under H_0. This gives rise to two subversions of T_3^*, called here as $T^*_{3(RMM)}$ and $T^*_{3(RMPL)}$, depending on the type of RE used. If we use the method of moments estimates of σ and λ under the restriction of H_0 then it is called $T^*_{3(RMM)}$ and similarly $T^*_{3(RMPL)}$.

We propose two other tests based on the likelihood ratio (LR) statistic as detailed below. In order to test $H_0 : \mu = \mu_0$ vs. $H_A : \mu \neq \mu_0$, the likelihood ratio statistic

$$\Lambda = \frac{\sup_{H_0} L(\mu, \sigma, \lambda | \underset{\sim}{W})}{\sup L(\mu, \sigma, \lambda | \underset{\sim}{W})} = \frac{L(\mu_0, \hat{\sigma}_{RMPL}, \hat{\lambda}_{RMPL})}{L(\hat{\mu}_{MPL}, \hat{\sigma}_{MPL}, \hat{\lambda}_{MPL})} \tag{4.2}$$

Note that the above Λ is supposed to be a likelihood based deviant metric which measures the compatibility of H_0 with the data.

Asymptotically, i.e., as $n \to \infty$, under H_0,

$$\Lambda_* = (-2ln\Lambda) \dot{\sim} \chi_1^2.$$

So we use $\Lambda_* = 2\sum_{i=1}^n [lnf(W_i|\hat{\mu}_{MPL}, \hat{\sigma}_{MPL}, \hat{\lambda}_{MPL}) - lnf(W_i|\mu_0, \hat{\sigma}_{RMPL}, \hat{\lambda}_{RMPL})]$ as our deviant measure. This leads us to the fourth test T_4' as follows:

$$T_4' := \begin{cases} \text{retain } H_0, & \text{if } \Lambda_* \leq \chi^2_{1,\alpha}, \\ \text{reject } H_0, & \text{if } \Lambda_* > \chi^2_{1,\alpha}, \end{cases} \tag{4.3}$$

which is the asymptotic LRT. This T_4' is expected to work well for a large sample size.

For small samples, we propose a parametric bootstrap version for our fourth test to create the PB LRT, say T_4^*, where the cut-off point is found by simulation. The method comes with the following steps:

Step-1: Given the original data $W_1, W_2, ..., W_n$, compute Λ_*.

Step-2: Under the null hypothesis H_0, we get $\mu = \mu_0$, $\sigma = \hat{\sigma}_{RMPL}$, and $\lambda = \hat{\lambda}_{RMPL}$.

Step-3: Generate *iid* observations under H_0, i.e., $W_1^*, W_2^*, ..., W_n^* \sim$ SND$(\mu_0, \hat{\sigma}_{RMPL}, \hat{\lambda}_{RMPL})$. Based on $\underset{\sim}{W}^* = (W_1^*, ..., W_n^*)$, obtain all the unrestricted and restricted estimates of the parameters marked with the superscript '*' for easy identification, i.e., $\hat{\mu}^*_{MPL}, \hat{\sigma}^*_{MPL}, \hat{\lambda}^*_{MPL}, \hat{\sigma}^*_{RMPL}$, and $\hat{\lambda}^*_{RMPL}$, and get $\Lambda_*^* = 2\sum_{i=1}^n [lnf(W_i^*|\hat{\mu}^*_{MPL}, \hat{\sigma}^*_{MPL}, \hat{\lambda}^*_{MPL}) - lnf(W_i^*|\mu_0, \hat{\sigma}^*_{RMPL}, \hat{\lambda}^*_{RMPL})]$.

Step-4: Repeat the above Step-3 a large number of times, (N times). This gives us N copies of Λ_*^*, i.e., $\Lambda_*^{*1}, \Lambda_*^{*2}, ..., \Lambda_*^{*N}$. Order them as $\Lambda_*^{*(1)} \leq ... \leq \Lambda_*^{*(N)}$. Then the cut-off points for Λ_* are to be found by the (approximate) probability distribution of Λ_* under H_0, i.e., $\Lambda_{*U} = \Lambda_*^{*((1-\alpha)N)}$ such that $\alpha \simeq$ P(Rej $H_0|H_0$ is true). If $\Lambda_* > \Lambda_{*U}$, then we will reject H_0. Alternatively, the p-value = $(1/N)$(Number of $\Lambda_*^{*j} > \Lambda_*$).

Now, to study the power/size of a test, we have to generate the original data $(W_1, W_2, ..., W_n)$ multiple times, say Q times, and then the power/size = proportion (out of Q times) of times H_0 is rejected. This simulation is quite time consuming since all the PB tests are subjected to double loops, and hence we had to restrict the number of replications to $N = Q = 1000$. The results of our simulation study for $n = 10$ have been provided in the following Tables 4.1 - 4.2.

Table 4.1: Size of six tests using $n = 10$ and $\alpha = 0.05$

λ	$\mu = 0$ (size)					
	T_1^*	T_2^*	$T_{3(RMM)}^*$	$T_{3(RMPL)}^*$	T_4'	T_4^*
0	0	0.0220	0	0	0.0010	0.0210
1	0.0140	0.0250	0	0	0.0230	0.0380
2	0.0240	0.0200	0	0	0.0660	0.0700

Table 4.2: Power of six tests using $n = 10$ and $\alpha = 0.05$

λ	$\mu = 1.0$ (power)					
	T_1^*	T_2^*	$T_{3(RMM)}^*$	$T_{3(RMPL)}^*$	T_4'	T_4^*
0	0.0260	0.0140	0	0	0.0920	0.1050
1	0.0310	0.0090	0	0	0.6240	0.6020
2	0.0050	0.0050	0	0	0.9280	0.9060

Remark 4.1 Strangely, the above simulation (Tables 4.1 - 4.2) shows the strange behavior of $T_{3(RMM)}^*$ and $T_{3(RMPL)}^*$, and hence they should not be considered as reliable test methods. In terms of size T_1^* appears to be very conservative. The tests T_4' and T_4^* appear to be a bit liberal when λ gets large (i.e., SND becomes very skewed). On the other hand, the test T_2^* seems to be more reliable, and appears to maintain the level consistently, though a bit conservative. Therefore, we recommend using T_2^* for hypothesis testing on μ. However, a more comprehensive simulation with higher values of N and Q will be needed along with other values of n.

5 An application with a Real-life Data Set

In this section our objective is the usage of the tests provided in Section 4 for a realistic data set. For this purpose we use Krung Thai Bank (KTB)'s quarterly stock price data set taken from Thiuthad and Pal (2018) where the quarterly price change (QPC) in % has been observed over 16 quarters. Note that μ indicates the location

Table 5.1: Quarterly stock price change of KTB from Thai stock market (Thiuthad and Pal (2018))

Period	1	2	3	4	5	6	7	8
QPC(%)	20.65	11.17	9.72	-5.42	3.17	-25.33	2.40	-0.60
Period	9	10	11	12	13	14	15	16
QPC(%)	16.88	-11.89	4.76	-0.56	13.33	-8.29	0.00	1.05

shift of the distribution, but not the mean of the distribution. Also, SND was found to have a good fit for this KTB data.

The objective is to test $H_0 : \mu = 0$ vs. $H_A : \mu \neq 0$. The following are the p-value of all the six tests as discussed in the previous section.

Test	T_1^*	T_2^*	$T_{3(RMM)}^*$	$T_{3(RMPL)}^*$	T_4'	T_4^*
p-value	0.2480	0.7732	0.9908	0.6831	0.9081	0.3984

As mentioned earlier, T_2^* seems to be the most reliable test, and for the given data its p-value = 0.7732 which suggest that we retain H_0.

References

[1] Aigner, D.J. and Lovell, C.A.K.. Formulation and estimation of stochastic frontier production function model. *Econometrics* **12**: 21-37 (1977).

[2] Arellano-Valle, R. B., Castro, L. M. and Loschi, R. H.. Change Point Detection in the Skew-Normal Model Parameters. *Communications in Statistics-Theory and Methods* **42**: 603-618 (2013).

[3] Arnold, B.C. and Lin, G.D.. Characterizations of the Skew-normal and Generalized Chi Distributions. *Sankhyā: The Indian Journal of Stat.* **66**: 593-606 (2004).

[4] Azzalini, A.. A class of distribution which includes the normal ones. *Scan. J. Statist.* **12**: 171-178 (1985).

[5] Azzalini, A.. The Skew-normal Distribution and Related Multivariate Families. *Scan. J. Statist.* **32**: 159-188 (2005).

[6] Azzalini, A. and Capitanio, A.. Statistical applications of the multivariate skew-normal distribution. *J. Roy. Statist. Soc. B* **61**: 579-602 (1999).

[7] Azzalini, A. and Dalla Valle, A.. The multivariate skew-normal distribution. *Biometrika* **83**: 715-726 (1996).

[8] Figueiredo, F. and Gomes, M.I.. The Skew-Normal Distribution in SPC. *REVSTAT – Statistical Journal* **11**: 83-104 (2013).

[9] Gupta, A.K., Nguyen, T.T. and Sanqui, J.T.. Characterization of the Skew-Normal Distribution. *Ann. Inst. Statist. Math.* **56**: 351-360 (2004).

[10] Gupta, R.C. and Brown, N.. Reliability Studies of the Skew Normal Distribution and Its Application to a Strength-Stress Model. *Commun. Statist.-Theory Meth.* **11**: 2427-2445 (2001).

[11] Henze, N.. A probabilistic representation of the skew-normal distribution. *Scand. J. Statist.* **13**: 271-275 (1986).

[12] Ngunkeng, G. and Ning, W.. Information Approach for the Change-Point Detection in the Skew Normal Distribution and Its Applications. *Sequential Analysis*, **33**: 475-490 (2014).

[13] O'Hagan, A. and Leonard, T.. Bayes estimation subject to uncertainly about parameter constraints, *Biometrika*, **63**: 201-202 (1976).

[14] Roberts, C.. A correlation model useful in the study of twins. *Journal of the American Statistical Association*. **61**: 1184-1190 (1966).

[15] Roberts, H.V.. *Data Analysis for Managers with Minitab*. (Scientific Press: Redwood City, CA, 1988).

[16] Thiuthad, P. and Pal, N.. On Point Estimation of the Location Parameter of a Skew-Normal Distribution. *Submitted* (2018).

A Appendix

A.1 Power and size tables for Section 3 with other α values

The following Tables A.1 - A.6 provide the simulated power (or size) of the above three tests, based on 2.5×10^4 replications as μ deviates from $\mu_0 = 0$ (w.l.g.) for various values of λ. All the tables have been constructed using $\alpha = 0.01$.

Table A.1: Power (or size, when $\mu = \mu_0 = 0$) of three tests for various μ with fixed $\lambda = 0.0$

μ	$n = 5$			$n = 10$			$n = 25$		
	T_1	T_2	T_3	T_1	T_2	T_3	T_1	T_2	T_3
0.0	0.0093	0.0093	0.0093	0.0092	0.0092	0.0092	0.0102	0.0102	0.0102
0.2	0.0168	0.0168	0.0168	0.0274	0.0274	0.0274	0.0577	0.0577	0.0577
0.4	0.0478	0.0478	0.0478	0.0978	0.0978	0.0978	0.2806	0.2806	0.2806
0.6	0.1120	0.1120	0.1120	0.2481	0.2481	0.2481	0.6631	0.6631	0.6631
0.8	0.2146	0.2146	0.2146	0.4845	0.4845	0.4845	0.9233	0.9233	0.9233
1.0	0.3634	0.3634	0.3634	0.7174	0.7174	0.7174	0.9919	0.9919	0.9919
2.0	0.9714	0.9714	0.9714	0.9999	0.9999	0.9999	1.0000	1.0000	1.0000
3.0	1.0000	1.0000	1.0000	1.0000	1.0000	1.0000	1.0000	1.0000	1.0000
4.0	1.0000	1.0000	1.0000	1.0000	1.0000	1.0000	1.0000	1.0000	1.0000
6.0	1.0000	1.0000	1.0000	1.0000	1.0000	1.0000	1.0000	1.0000	1.0000
8.0	1.0000	1.0000	1.0000	1.0000	1.0000	1.0000	1.0000	1.0000	1.0000

Table A.2: Power (or size, when $\mu = \mu_0 = 0$) of three tests for various μ with fixed $\lambda = 1.0$

μ	$n = 5$			$n = 10$			$n = 25$		
	T_1	T_2	T_3	T_1	T_2	T_3	T_1	T_2	T_3
0.0	0.0104	0.0103	0.0104	0.0091	0.0089	0.0087	0.0103	0.0107	0.0106
0.2	0.0191	0.0200	0.0169	0.0265	0.0276	0.0239	0.0516	0.0523	0.0493
0.4	0.0518	0.0539	0.0464	0.0990	0.1001	0.0910	0.2812	0.2829	0.2711
0.6	0.1372	0.1395	0.1239	0.2902	0.2913	0.2749	0.7168	0.7137	0.7069
0.8	0.2955	0.2986	0.2760	0.5963	0.5951	0.5784	0.9629	0.9622	0.9610
1.0	0.5181	0.5201	0.4944	0.8464	0.8463	0.8370	0.9986	0.9984	0.9986
2.0	0.9988	0.9986	0.9986	1.0000	1.0000	1.0000	1.0000	1.0000	1.0000
3.0	1.0000	1.0000	1.0000	1.0000	1.0000	1.0000	1.0000	1.0000	1.0000
4.0	1.0000	1.0000	1.0000	1.0000	1.0000	1.0000	1.0000	1.0000	1.0000
6.0	1.0000	1.0000	1.0000	1.0000	1.0000	1.0000	1.0000	1.0000	1.0000
8.0	1.0000	1.0000	1.0000	1.0000	1.0000	1.0000	1.0000	1.0000	1.0000

Table A.3: Power (or size, when $\mu = \mu_0 = 0$) of three tests for various μ with fixed $\lambda = 2.0$

μ	$n = 5$			$n = 10$			$n = 25$		
	T_1	T_2	T_3	T_1	T_2	T_3	T_1	T_2	T_3
0.0	0.0100	0.0119	0.0085	0.0109	0.0112	0.0100	0.0091	0.0106	0.0090
0.2	0.0191	0.0234	0.0134	0.0238	0.0278	0.0182	0.0378	0.0412	0.0311
0.4	0.0466	0.0554	0.0330	0.0770	0.0846	0.0589	0.1927	0.1973	0.1708
0.6	0.1167	0.1294	0.0884	0.2226	0.2336	0.1868	0.5737	0.5680	0.5394
0.8	0.2469	0.2623	0.1987	0.4918	0.4889	0.4400	0.9119	0.9012	0.8971
1.0	0.4553	0.4612	0.3914	0.7782	0.7650	0.7384	0.9944	0.9922	0.9927
2.0	0.9976	0.9967	0.9958	1.0000	1.0000	1.0000	1.0000	1.0000	1.0000
3.0	1.0000	1.0000	1.0000	1.0000	1.0000	1.0000	1.0000	1.0000	1.0000
4.0	1.0000	1.0000	1.0000	1.0000	1.0000	1.0000	1.0000	1.0000	1.0000
6.0	1.0000	1.0000	1.0000	1.0000	1.0000	1.0000	1.0000	1.0000	1.0000
8.0	1.0000	1.0000	1.0000	1.0000	1.0000	1.0000	1.0000	1.0000	1.0000

Table A.4: Size of three tests for various λ with fixed $\mu = 0.0$

λ	$n = 5$			$n = 10$			$n = 25$		
	T_1	T_2	T_3	T_1	T_2	T_3	T_1	T_2	T_3
0.0	0.0095	0.0095	0.0095	0.0094	0.0094	0.0094	0.0101	0.0101	0.0101
0.2	0.0099	0.0099	0.0099	0.0100	0.0100	0.0100	0.0098	0.0098	0.0097
0.4	0.0111	0.0112	0.0111	0.0102	0.0101	0.0101	0.0094	0.0094	0.0094
0.6	0.0104	0.0104	0.0108	0.0095	0.0096	0.0096	0.0112	0.0110	0.0112
0.8	0.0100	0.0103	0.0099	0.0100	0.0098	0.0100	0.0099	0.0092	0.0097
1.0	0.0111	0.0106	0.0106	0.0113	0.0110	0.0110	0.0104	0.0101	0.0104
2.0	0.0100	0.0121	0.0093	0.0112	0.0118	0.0107	0.0107	0.0113	0.0107
3.0	0.0112	0.0164	0.0095	0.0102	0.0128	0.0101	0.0103	0.0112	0.0098
4.0	0.0106	0.0241	0.0083	0.0105	0.0177	0.0088	0.0091	0.0113	0.0087
6.0	0.0103	0.0426	0.0070	0.0096	0.0227	0.0079	0.0106	0.0171	0.0093
8.0	0.0122	0.0631	0.0086	0.0111	0.0352	0.0093	0.0099	0.0192	0.0093

Table A.5: Power of three tests for various λ with fixed $\mu = 1.0$

λ	$n = 5$			$n = 10$			$n = 25$		
	T_1	T_2	T_3	T_1	T_2	T_3	T_1	T_2	T_3
0.0	0.3654	0.3654	0.3654	0.7201	0.7201	0.7201	0.9926	0.9926	0.9926
0.2	0.4276	0.4278	0.4274	0.7846	0.7846	0.7844	0.9969	0.9969	0.9969
0.4	0.4656	0.4667	0.4642	0.8277	0.8275	0.8262	0.9982	0.9982	0.9982
0.6	0.4992	0.5022	0.4922	0.8434	0.8432	0.8406	0.9991	0.9990	0.9991
0.8	0.5110	0.5130	0.4970	0.8518	0.8511	0.8456	0.9991	0.9992	0.9991
1.0	0.5168	0.5189	0.4924	0.8464	0.8464	0.8370	0.9986	0.9983	0.9986
2.0	0.4616	0.4673	0.3957	0.7798	0.7694	0.7438	0.9945	0.9928	0.9932
3.0	0.3914	0.4020	0.3092	0.6929	0.6625	0.6298	0.9802	0.9693	0.9737
4.0	0.3493	0.3648	0.2654	0.6204	0.5741	0.5455	0.9572	0.9276	0.9429
6.0	0.2956	0.3358	0.2124	0.5243	0.4621	0.4425	0.9048	0.8085	0.8782
8.0	0.2637	0.3430	0.1852	0.4720	0.4063	0.3878	0.8662	0.7015	0.8313

Table A.6: Power of three tests for various λ with fixed $\mu = 2.0$

λ	$n = 5$			$n = 10$			$n = 25$		
	T_1	T_2	T_3	T_1	T_2	T_3	T_1	T_2	T_3
0.0	0.9697	0.9697	0.9697	1.0000	1.0000	1.0000	1.0000	1.0000	1.0000
0.2	0.9909	0.9909	0.9909	1.0000	1.0000	1.0000	1.0000	1.0000	1.0000
0.4	0.9966	0.9966	0.9965	1.0000	1.0000	1.0000	1.0000	1.0000	1.0000
0.6	0.9985	0.9986	0.9985	1.0000	1.0000	1.0000	1.0000	1.0000	1.0000
0.8	0.9989	0.9989	0.9989	1.0000	1.0000	1.0000	1.0000	1.0000	1.0000
1.0	0.9987	0.9989	0.9984	1.0000	1.0000	1.0000	1.0000	1.0000	1.0000
2.0	0.9977	0.9970	0.9962	1.0000	1.0000	1.0000	1.0000	1.0000	1.0000
3.0	0.9933	0.9889	0.9886	1.0000	1.0000	1.0000	1.0000	1.0000	1.0000
4.0	0.9867	0.9729	0.9768	1.0000	0.9997	0.9998	1.0000	1.0000	1.0000
6.0	0.9752	0.9322	0.9552	0.9998	0.9955	0.9996	1.0000	1.0000	1.0000
8.0	0.9611	0.8964	0.9341	0.9995	0.9845	0.9989	1.0000	1.0000	1.0000

The following Tables A.7 - A.12 provide the simulated power (or size) of the above three tests, based on 2.5×10^4 replications as μ deviates from $\mu_0 = 0$ (w.l.g.) for various values of λ. All the tables have been constructed using $\alpha = 0.10$.

Table A.7: Power (or size, when $\mu = \mu_0 = 0$) of three tests for various μ with fixed $\lambda = 0.0$

μ	$n = 5$			$n = 10$			$n = 25$		
	T_1	T_2	T_3	T_1	T_2	T_3	T_1	T_2	T_3
0.0	0.1008	0.1008	0.1008	0.0993	0.0993	0.0993	0.0972	0.0972	0.0972
0.2	0.1306	0.1306	0.1306	0.1662	0.1662	0.1662	0.2642	0.2642	0.2642
0.4	0.2324	0.2324	0.2324	0.3570	0.3570	0.3570	0.6383	0.6383	0.6383
0.6	0.3856	0.3856	0.3856	0.5987	0.5987	0.5987	0.9103	0.9103	0.9103
0.8	0.5613	0.5613	0.5613	0.8107	0.8107	0.8107	0.9906	0.9906	0.9906
1.0	0.7219	0.7219	0.7219	0.9327	0.9327	0.9327	0.9997	0.9997	0.9997
2.0	0.9979	0.9979	0.9979	1.0000	1.0000	1.0000	1.0000	1.0000	1.0000
3.0	1.0000	1.0000	1.0000	1.0000	1.0000	1.0000	1.0000	1.0000	1.0000
4.0	1.0000	1.0000	1.0000	1.0000	1.0000	1.0000	1.0000	1.0000	1.0000
6.0	1.0000	1.0000	1.0000	1.0000	1.0000	1.0000	1.0000	1.0000	1.0000
8.0	1.0000	1.0000	1.0000	1.0000	1.0000	1.0000	1.0000	1.0000	1.0000

Table A.8: Power (or size, when $\mu = \mu_0 = 0$) of three tests for various μ with fixed $\lambda = 1.0$

μ	$n = 5$			$n = 10$			$n = 25$		
	T_1	T_2	T_3	T_1	T_2	T_3	T_1	T_2	T_3
0.0	0.0987	0.0984	0.0991	0.0996	0.1001	0.0999	0.1024	0.1021	0.1028
0.2	0.1262	0.1309	0.1242	0.1570	0.1606	0.1547	0.2413	0.2445	0.2390
0.4	0.2318	0.2394	0.2267	0.3523	0.3572	0.3473	0.6226	0.6257	0.6201
0.6	0.4123	0.4210	0.4059	0.6336	0.6359	0.6289	0.9285	0.9285	0.9278
0.8	0.6276	0.6324	0.6222	0.8660	0.8662	0.8633	0.9969	0.9963	0.9969
1.0	0.8176	0.8206	0.8140	0.9713	0.9716	0.9704	0.9999	0.9999	0.9999
2.0	1.0000	1.0000	1.0000	1.0000	1.0000	1.0000	1.0000	1.0000	1.0000
3.0	1.0000	1.0000	1.0000	1.0000	1.0000	1.0000	1.0000	1.0000	1.0000
4.0	1.0000	1.0000	1.0000	1.0000	1.0000	1.0000	1.0000	1.0000	1.0000
6.0	1.0000	1.0000	1.0000	1.0000	1.0000	1.0000	1.0000	1.0000	1.0000
8.0	1.0000	1.0000	1.0000	1.0000	1.0000	1.0000	1.0000	1.0000	1.0000

Table A.9: Power (or size, when $\mu = \mu_0 = 0$) of three tests for various μ with fixed $\lambda = 2.0$

μ	$n = 5$			$n = 10$			$n = 25$		
	T_1	T_2	T_3	T_1	T_2	T_3	T_1	T_2	T_3
0.0	0.0992	0.0994	0.0997	0.1002	0.1023	0.1015	0.1018	0.1026	0.1024
0.2	0.1204	0.1336	0.1144	0.1411	0.1522	0.1360	0.1882	0.2028	0.1827
0.4	0.1992	0.2212	0.1862	0.2826	0.3009	0.2703	0.5070	0.5199	0.4972
0.6	0.3487	0.3746	0.3297	0.5390	0.5510	0.5231	0.8524	0.8529	0.8466
0.8	0.5444	0.5662	0.5240	0.7883	0.7904	0.7775	0.9852	0.9847	0.9844
1.0	0.7488	0.7597	0.7322	0.9430	0.9398	0.9382	0.9996	0.9994	0.9995
2.0	0.9998	0.9999	0.9998	1.0000	1.0000	1.0000	1.0000	1.0000	1.0000
3.0	1.0000	1.0000	1.0000	1.0000	1.0000	1.0000	1.0000	1.0000	1.0000
4.0	1.0000	1.0000	1.0000	1.0000	1.0000	1.0000	1.0000	1.0000	1.0000
6.0	1.0000	1.0000	1.0000	1.0000	1.0000	1.0000	1.0000	1.0000	1.0000
8.0	1.0000	1.0000	1.0000	1.0000	1.0000	1.0000	1.0000	1.0000	1.0000

Table A.10: Size of three tests for various λ with fixed $\mu = 0.0$

λ	$n = 5$			$n = 10$			$n = 25$		
	T_1	T_2	T_3	T_1	T_2	T_3	T_1	T_2	T_3
0.0	0.1008	0.1008	0.1008	0.0993	0.0993	0.0993	0.0972	0.0972	0.0972
0.2	0.0953	0.0953	0.0954	0.1002	0.1002	0.1002	0.0983	0.0984	0.0983
0.4	0.1040	0.1042	0.1040	0.1002	0.0996	0.1003	0.0995	0.0996	0.0993
0.6	0.1018	0.1011	0.1016	0.1049	0.1042	0.1051	0.1036	0.1028	0.1039
0.8	0.0996	0.0985	0.0992	0.0998	0.0984	0.0994	0.0989	0.0988	0.0988
1.0	0.1009	0.1024	0.1014	0.1030	0.1022	0.1035	0.1028	0.1016	0.1024
2.0	0.1002	0.1011	0.1002	0.0999	0.1024	0.1016	0.1017	0.1009	0.1022
3.0	0.0985	0.1086	0.1008	0.0979	0.1004	0.0986	0.1008	0.1017	0.1006
4.0	0.0959	0.1176	0.1014	0.0971	0.1095	0.0992	0.0971	0.0990	0.0980
6.0	0.0932	0.1426	0.0989	0.0950	0.1176	0.0968	0.0996	0.1043	0.1004
8.0	0.0944	0.1702	0.1013	0.0992	0.1302	0.1016	0.0999	0.1104	0.1006

Table A.11: Power of three tests for various λ with fixed $\mu = 1.0$

λ	$n = 5$			$n = 10$			$n = 25$		
	T_1	T_2	T_3	T_1	T_2	T_3	T_1	T_2	T_3
0.0	0.7226	0.7226	0.7226	0.9339	0.9339	0.9339	0.9996	0.9996	0.9996
0.2	0.7706	0.7709	0.7705	0.9567	0.9568	0.9567	0.9998	0.9998	0.9998
0.4	0.7953	0.7962	0.7949	0.9682	0.9685	0.9681	1.0000	1.0000	1.0000
0.6	0.8121	0.8137	0.8112	0.9728	0.9729	0.9727	1.0000	1.0000	1.0000
0.8	0.8164	0.8189	0.8136	0.9736	0.9735	0.9731	1.0000	1.0000	1.0000
1.0	0.8176	0.8206	0.8140	0.9713	0.9716	0.9704	0.9999	0.9999	0.9999
2.0	0.7527	0.7610	0.7363	0.9453	0.9423	0.9411	0.9997	0.9996	0.9996
3.0	0.6825	0.6921	0.6559	0.8994	0.8913	0.8915	0.9981	0.9975	0.9978
4.0	0.6252	0.6369	0.5958	0.8544	0.8352	0.8417	0.9949	0.9902	0.9944
6.0	0.5603	0.5764	0.5271	0.7856	0.7378	0.7672	0.9831	0.9566	0.9808
8.0	0.5184	0.5538	0.4809	0.7447	0.6764	0.7222	0.9683	0.9106	0.9650

Table A.12: Power of three tests for various λ with fixed $\mu = 2.0$

λ	$n = 5$			$n = 10$			$n = 25$		
	T_1	T_2	T_3	T_1	T_2	T_3	T_1	T_2	T_3
0.0	0.9980	0.9980	0.9980	1.0000	1.0000	1.0000	1.0000	1.0000	1.0000
0.2	0.9996	0.9996	0.9996	1.0000	1.0000	1.0000	1.0000	1.0000	1.0000
0.4	0.9998	0.9998	0.9998	1.0000	1.0000	1.0000	1.0000	1.0000	1.0000
0.6	1.0000	1.0000	1.0000	1.0000	1.0000	1.0000	1.0000	1.0000	1.0000
0.8	0.9999	1.0000	0.9999	1.0000	1.0000	1.0000	1.0000	1.0000	1.0000
1.0	1.0000	1.0000	1.0000	1.0000	1.0000	1.0000	1.0000	1.0000	1.0000
2.0	0.9998	0.9999	0.9998	1.0000	1.0000	1.0000	1.0000	1.0000	1.0000
3.0	0.9993	0.9990	0.9992	1.0000	1.0000	1.0000	1.0000	1.0000	1.0000
4.0	0.9981	0.9954	0.9977	1.0000	1.0000	1.0000	1.0000	1.0000	1.0000
6.0	0.9960	0.9859	0.9947	1.0000	0.9995	1.0000	1.0000	1.0000	1.0000
8.0	0.9922	0.9700	0.9906	1.0000	0.9985	1.0000	1.0000	1.0000	1.0000

A.2 The restricted estimates (REs)

A.2.1 The restricted method of moments estimators (RMME) of σ and λ under $H_0 : \mu = \mu_0$:

The estimators of σ and λ are found by solving

$$\mu_0 + \sigma\lambda\sqrt{2/\pi}/\sqrt{1+\lambda^2} = \overline{W} = m_1 \text{ (say)}, \tag{A.1}$$

$$\sigma^2\{1 - (2/\pi)(\lambda^2/(1+\lambda^2))\} = (1/n)\sum_{i=1}^{n}(W_i - \overline{W})^2 = m_2 \text{ (say)}, \tag{A.2}$$

Note that, in equation (A.1), the estimated parameter $\hat{\lambda}_{RMM}$ carries the sign of $(m_1 - \mu_0)$. Simplification of (A.1) - (A.2) yields

$$\hat{\sigma}_{RMM} = \sqrt{m_2 + (m_1 - \mu_0)^2} \tag{A.3}$$

and solve

$$(1 + \hat{\lambda}^2)/\hat{\lambda}^2 = (2/\pi)\{1 + m_2/(m_1 - \mu_0)^2\} \tag{A.4}$$

to obtain $\hat{\lambda}$ which is the value of $\hat{\lambda}_{RMM}$. Since the LHS in (A.4) is > 1, a finite solution of (A.4) exists only if

$$\begin{aligned} &(2/\pi)\{1 + m_2/(m_1 - \mu_0)^2\} > 1, \\ \text{i.e.,}\quad & m_2/(m_1 - \mu_0)^2 > (\pi/2) - 1 = 0.570796.... \end{aligned} \tag{A.5}$$

Here, we suggest an estimate of λ based on the RMME approach as

$$\hat{\lambda}_{RMM} = \begin{cases} \text{sign}(m_1 - \mu_0) \times \hat{\lambda}, & \text{if (A.5) holds.} \\ \text{sign}(m_1 - \mu_0) \times \infty, & \text{if (A.5) does not hold.} \end{cases} \tag{A.6}$$

The inequality (A.5) can be treated as an acceptance region for testing $H_0 : |\lambda| < \infty$ against $H_A : |\lambda| = \infty$. If H_0 is retained, then λ gets estimated by a finite $\hat{\lambda}$ value. Otherwise, λ is estimated by $\pm\infty$. However, for practical considerations, one should replace ∞ by a large M in (A.6) so that the SND with $\hat{\lambda} = \pm M$ becomes very close to the intended half normal distribution.

A.2.2 The restricted maximum penalized likelihood estimator (RMPLE) of σ and λ under $H_0 : \mu = \mu_0$:

The penalized log-likelihood function is

$$L_{**}(\mu_0) = L_*(\mu_0) - c_1 ln(1 + c_2\lambda^2), \tag{A.7}$$

where $L_*(\mu_0)$ is the log-likelihood function with $\mu = \mu_0$ and c_1, c_2 are constants mentioned in (2.7). So, the RMPLE should be found by solving

$$0 = \frac{\partial L_{**}(\mu_0)}{\partial \sigma} = -((\lambda/\sigma^2)\sum_{i=1}^{n}(W_i - \mu_0)\phi(\lambda(W_i - \mu_0)/\sigma)/\Phi(\lambda(W_i - \mu_0)/\sigma)) - (n/\sigma) + \sum_{i=1}^{n}(W_i - \mu_0)^2/\sigma^3, \quad \text{(A.8)}$$

$$0 = \frac{\partial L_{**}(\mu_0)}{\partial \lambda} = \sum_{i=1}^{n}((W_i - \mu_0)/\sigma)\phi(\lambda(W_i - \mu_0)/\sigma)/\Phi(\lambda(W_i - \mu_0)/\sigma) - 2c_1c_2\lambda/(1 + c_2\lambda^2). \quad \text{(A.9)}$$

The range of a nonatomic vector measure in $\mathbb{R}^3$

Maciej Wilczyński[1,*]

[1]Faculty of Pure and Applied Mathematics,
Wrocław University of Science and Technology,
Wybrzeże Wyspiańskiego 27, 50-370 Wrocław, Poland

Abstract. Using the Neyman-Pearson lemma we give a method for finding the range of three nonatomic probability measures. Next we use this result to construct an explicit form of this range in the case, where the densities corresponding to these measures satisfy an additional assumption.

1 Introduction

Let $(X, \mathcal{B})$ be a measurable space and let $\mu_1, \ldots, \mu_n$ be nonatomic probability measures on $(X, \mathcal{B})$ possessing densities $f_1, \ldots, f_n$ with respect to some measure ν. The subset $\mathscr{R}(\mu_1, \ldots, \mu_n)$ of $\mathbb{R}^n$, defined by

$$\mathscr{R}(\mu_1, \ldots, \mu_n) = \{ (\mu_1(A), \ldots \mu_n(A)) \ : \ A \in \mathcal{B} \},$$

is called the range of the vector measure (or a zonoid). The Lyapounov Convexity Theorem (see Lyapounov [1]) states that the set $\mathscr{R}(\mu_1, \ldots, \mu_n)$ is convex and compact in $\mathbb{R}^n$. The problem is how to find this set for given nonatomic probability measures $\mu_1, \ldots, \mu_n$. In the case of $n = 2$ this question was answered by Legut and Wilczyński [2]. For the sake of completeness we recall here this result. Then we give some partial results concerning the structure of the range of the vector measure in $\mathbb{R}^3$.

Without loss of generality, we suppose throughout that $(X, \mathcal{B}) = (\mathbb{R}, \mathcal{B}_{\mathbb{R}})$, where $\mathcal{B}_{\mathbb{R}}$ is the Borel σ-field on the real line $\mathbb{R}$. Moreover, we assume that the measures μ_1, μ_2 and μ_3 are absolutely continuous with respect to the Lebesgue measure λ and we put $f_1 = d\mu_1/d\lambda$, $f_2 = d\mu_2/d\lambda$ and $f_3 = d\mu_3/d\lambda$ for the corresponding Lebesgue probability density functions (p.d.f.'s). We denote by X_1, X_2 and X_3 three random variables, defined on some probability space $(\Omega, \mathscr{F}, P)$, which are distributed according to f_1, f_2 and f_3, respectively. The corresponding cumulative distribution functions (c.d.f.'s) we denote by F_1, F_2 and F_3. We define the quantile function F_1^{-1} (the quasi inverse of F_1) by

$$F_1^{-1}(x) = \inf\{y \ : \ F_1(y) \geq x\} \quad \text{for all} \quad 0 < x < 1. \tag{1}$$

Finally, we use the symbol $\mathscr{R}(f_1, f_2, f_3)$ instead of $\mathscr{R}(\mu_1, \mu_2, \mu_3)$ for the range of a vector measure (μ_1, μ_2, μ_3), i.e. we put

$$\mathscr{R}(f_1, f_2, f_3) = \left\{ \left(\int_A f_1 \, dt, \int_A f_2 \, dt, \int_A f_3 \, dt \right) \ : \ A \in \mathcal{B} \right\}.$$

*e-mail: maciej.wilczynski@pwr.edu.pl

2 Lyapounov set in $\mathbb{R}^2$

In this section we recall the results of Legut and Wilczyński [2] concerning the set $\mathscr{R}(f_1, f_2)$. We will use these facts in inference concerning Lyapounov set in $\mathbb{R}^3$.

The compactness of the set $\mathscr{R}(f_1, f_2)$ implies that for any $x \in [0,1]$ there exists a set $D_x \in \mathcal{B}$ which maximizes $\int_A f_2(t)\,dt$ over all Borel subsets A of $\mathbb{R}$ satisfying $\int_A f_1(t)\,dt = x$. This means that

$$\int_{D_x} f_2(t)\,dt = \sup_{A\in\mathcal{B}} \left\{ \int_A f_2(t)\,dt \ : \ \int_A f_1(t)\,dt = x \right\}, \tag{2}$$

Let the function $G : [0,1] \to [0,1]$ be defined by

$$G(x) = \int_{D_x} f_2(t)\,dt, \quad x \in [0,1].$$

It is clear that for each $x \in [0,1]$, the point $(x, G(x))$ lies on the boundary of the set $\mathscr{R}(f_1, f_2)$. Moreover, since $\mathscr{R}(f_1, f_2)$ is centrally symmetric about the point $(\frac{1}{2}, \frac{1}{2})$ it follows that $(x,y) \in \mathscr{R}(f_1, f_2) \iff (1-x, 1-y) \in \mathscr{R}(f_1, f_2)$. Hence, the convexity of $\mathscr{R}(f_1, f_2)$ implies the following result.

Theorem 1. *The Lyapounov set $\mathscr{R}(f_1, f_2)$ has the form*

$$\mathscr{R}(f_1, f_2) = \left\{(x,y) \in \mathbb{R}^2 \ : \ 0 \le x \le 1, \ 1 - G(1-x) \le y \le G(x)\right\}.$$

□

To obtain the Lyapounov set $\mathscr{R}(f_1, f_2)$ explicitly, it is necessary to determine each of the sets D_x, $x \in [0,1]$. To do this we use the Neyman-Pearson lemma (see Lehmann and Romano [3] p. 60). This lemma implies the following corollary.

Corollary 1. *For each $x \in [0,1]$, a set D_x satisfies (2) if and only if there exists a number $k \ge 0$, which depends on x, such that*

$$\{t \in \mathbb{R} : f_2(t) > k f_1(t)\} \subset D_x \subset \{t \in \mathbb{R} : f_2(t) \ge k f_1(t)\}. \tag{3}$$

Obviously, D_x is uniquely determined (up to sets of Lebesgue measure zero) except on the set on which $f_2(t) = k f_1(t)$. On this set, D_x can be defined arbitrarily provided that it has μ_1 measure x.

In some cases the function G, and hence the shape of the set $\mathscr{R}(f_1, f_2)$, can easily be found. This is illustrated in the proposition below, which follows immediately from (3) (cf. Legut and Wilczyński [2]).

Proposition 1. *Assume that $\{x : f_2(x) > 0\} \subset \{x : f_1(x) > 0\} = (0,1)$. Let F_1^{-1} be the inverse function of F_1 and let r be the likelihood ratio given by*

$$r(x) := \frac{f_2(x)}{f_1(x)} I_{(0,\infty)}(f_1(x)), \quad x \in \mathbb{R}. \tag{4}$$

Then the following statements hold:

1. *If the ratio $r(x)$ is decreasing in x on $(0,1)$, then $D_x = \left(0, F_1^{-1}(x)\right)$ and hence $G(x) = F_2(F_1^{-1}(x))$.*

2. *If the ratio $r(x)$ is increasing in x on $(0,1)$, then $D_x = \left(F_1^{-1}(1-x), 1\right)$ and hence $G(x) = 1 - F_2(F_1^{-1}(1-x))$.*

3. *If the ratio $r(x)$ is symmetric about $x_0 = 1/2$ and decreasing in x on $(0,1/2)$, then $D_x = \left(0, F_1^{-1}(x/2)\right) \cup \left(F_1^{-1}(1-x/2), 1\right)$ and hence $G(x) = F_2\left(F_1^{-1}(x/2)\right) + 1 - F_2\left(F_1^{-1}(1-x/2)\right)$.*

4. *If the ratio $r(x)$ is symmetric about $x_0 = 1/2$ and increasing in x on $(0,1/2)$, then $D_x = \left(F_1^{-1}\left(\frac{1-x}{2}\right), F_1^{-1}\left(\frac{1+x}{2}\right)\right)$ and hence $G(x) = F_2\left(F_1^{-1}(\frac{1+x}{2})\right) - F_2\left(F_1^{-1}(\frac{1-x}{2})\right)$.*

When the likelihood ratio r, defined by (4), satisfies none of the above properties, finding the shape of the set $\mathscr{R}(f_1, f_2)$ is much more complicated. However, Legut and Wilczyński [2] proved that for any Lebesgue p.d.f.s f_1, f_2 on $(\mathbb{R}, \mathcal{B}_{\mathbb{R}})$ there exist Lebesgue p.d.f.s f_1^*, f_2^* on $(\mathbb{R}, \mathcal{B}_{\mathbb{R}})$ such that:

1. f_1^* is the uniform density on $(0,1)$, i.e. $f_1^*(x) = I_{(0,1)}(x)$, $x \in \mathbb{R}$,
2. f_2^* is nonincreasing in x on $(0,1)$,
3. $\mathscr{R}(f_1, f_2) = \mathscr{R}(f_1^*, f_2^*)$.

To describe the relationship between f_2^* and the densities f_1, f_2 we denote by $\overline{H}$ the survival function of the random variable $r(X_1)$, i.e. we put

$$\overline{H}(y) = P(\, r(X_1) > y\,), \quad y \in \mathbb{R}. \tag{5}$$

Let $\overline{H}^{-1} : (0,1) \to [0, \infty)$ be the quasi inverse of $\overline{H}$, defined by

$$\overline{H}^{-1}(x) := \inf\{y \geq 0 \,:\, \overline{H}(y) \leq x\} \quad \text{for all} \quad 0 < x < 1. \tag{6}$$

Clearly, $\overline{H}^{-1}$ is nonnegative, nonincreasing and right continuous function on the interval $(0,1)$. Moreover, if U is uniformly distributed on $(0,1)$, then the random variables $\overline{H}^{-1}(U)$ and $r(X_1)$ have the same distribution (see Shorack [4], p. 111–117). Hence,

$$\int_0^1 \overline{H}^{-1}(u)\,du = \mathbb{E}\left[\overline{H}^{-1}(U)\right] = \mathbb{E}[\,r(X_1)\,] = \int_{\{x: f_1(x)>0\}} r(x) f_1(x)\,dx = \int_{\{x: f_1(x)>0\}} f_2(x)\,dx \leq 1.$$

Now, let f_2^* be any Lebesgue p.d.f. on $(\mathbb{R}, \mathcal{B}_{\mathbb{R}})$, which satisfies

$$f_2^*(x) = \overline{H}^{-1}(x) \quad \text{for all} \quad x \in (0,1).$$

These useful properties of $\overline{H}^{-1}$, and hence of f_2^*, were used in Legut and Wilczyński [2] to prove the following theorem.

Theorem 2. *Let f_1, f_2 be fixed probability Lebesgue densities on $(\mathbb{R}, \mathcal{B}_{\mathbb{R}})$ and let f_1^* and f_2^* be the corresponding Lebesgue densities defined above. Then $\mathscr{R}(f_1, f_2) = \mathscr{R}(f_1^*, f_2^*)$. Moreover,*

$$\mathscr{R}(f_1^*, f_2^*) = \left\{(x,y) \in \mathbb{R}^2 \,:\, 0 \leq x \leq 1,\ 1 - G(1-x) \leq y \leq G(x)\right\},$$

where the function $G : [0,1] \to [0,1]$ has the form

$$G(x) = \int_{\{t: f_1(t)=0\}} f_2(t)\,dt + \int_0^x f_2^*(t)\,dt \quad \text{for all} \quad x \in [0,1]. \tag{7}$$

Proof. (The rough sketch) The proof is based on the following two facts. First, $r(X_1) \overset{D}{=} f_2^*(U)$ and hence, for each $y > 0$,

$$\mathbb{E}\left[\, I\{r(X_1) > y\}\,\right] = \mathbb{E}\left[\, I\{f_2^*(U) > y\}\,\right],$$

$$\mathbb{E}\left[\, r(X_1) I\{r(X_1) > y\}\,\right] = \mathbb{E}\left[\, f_2^*(U) I\{f_2^*(U) > y\}\,\right].$$

where $I(A)$ represents the indicator function of the event A. Second, the function f_2^* is nonincreasing and right continuouson on the interval $(0, 1)$, which implies that for each $y \geq 0$ there exists $u_y \in [0, 1]$ such that

$$\{u \in [0, 1] : f_2^*(u) > y\} = [0, u_y], \quad a.e.\ \mu_1.$$

□

Remark 1. The function G defined by (7) is the c.d.f. of a random variable which is continuously spread over the interval $(0, 1)$ with the density f_2^* and takes value 0 with the probability $\int\limits_{\{t: f_1(t)=0\}} f_2(t)\, dt$.

3 Lyapounov set in $\mathbb{R}^3$

We start this section by introducing some auxiliary results which are analogous to those in Theorem 1 and Corollary 1.

3.1 Auxiliary results

The compactness of the set $\mathscr{R}(f_1, f_2, f_3)$ implies that for any point (x, y) from $\mathscr{R}(f_1, f_2)$ there exists a set $D_{x,y} \in \mathcal{B}$, such that

$$\mu_3(D_{x,y}) = \sup_{A \in \mathcal{B}} \{\mu_3(A) \ : \ \mu_1(A) = x,\ \mu_2(A) = y\}, \tag{8}$$

Let the function $G_3 : \mathscr{R}(f_1, f_2) \to [0, 1]$ be defined by

$$G_3(x, y) = \mu_3(D_{x,y}), \quad (x, y) \in \mathscr{R}(f_1, f_2). \tag{9}$$

It is clear that for each $(x, y) \in \mathscr{R}(f_1, f_2)$ the point $(x, y, G_3(x, y))$ lies on the boundary of $\mathscr{R}(f_1, f_2, f_3)$.

Definition 1. *The set of the form* $\{(x, y, G_3(x, y)) : (x, y) \in \mathscr{R}(f_1, f_2)\}$ *is a surface in* $\mathbb{R}^3$, *which we will call the upper boundary of* $\mathscr{R}(f_1, f_2, f_3)$.

Since $\mathscr{R}(f_1, f_2, f_3)$ is centrally symmetric about the point $(\frac{1}{2}, \frac{1}{2}, \frac{1}{2})$ it follows that $(x, y, z) \in \mathscr{R}(f_1, f_2, f_3) \iff (1 - x, 1 - y, 1 - z) \in \mathscr{R}(f_1, f_2, f_3)$. This implies the following result, the proof of which is trivial.

Theorem 3. *The Lyapounov set* $\mathscr{R}(f_1, f_2, f_3)$ *has the form*

$$\begin{aligned} \mathscr{R}(f_1, f_2, f_3) \quad = \quad & \{(x, y, z) \in \mathbb{R}^3 \ : \ (x, y) \in \mathscr{R}(f_1, f_2), \\ & 1 - G_3(1 - x, 1 - y) \leq z \leq G_3(x, y)\}. \end{aligned} \tag{10}$$

The last theorem provides an explicit description of the set $\mathscr{R}(f_1, f_2)$ only in the case when the function G_3 can be given explicitly. The next result, which follows from the Neyman-Pearson lemma, is helpful in finding the structure each of the sets $D_{x,y}$, $(x, y) \in \mathscr{R}(f_1, f_2)$. Clearly, knowledge of this structures implies knowledge of the function G_3.

Corollary 2. *Suppose that (x, y) is an inner point of the Lyapounov set $\mathscr{R}(f_1, f_2, f_3)$. Then a set $D_{x,y}$ satisfies (8) if and only if there exist constants k_1, k_2, which depend on x, y, such that λ almost everywhere*

$$\left\{x \in \mathcal{X} : f_3(x) > \sum_{i=1}^{2} k_i f_i(x)\right\} \subset D_{x,y} \subset \left\{x \in \mathcal{X} : f_3(x) \geq \sum_{i=1}^{2} k_i f_i(x)\right\}.$$

□

It follows from Theorem 2 that in the problem of finding the shape of $\mathscr{R}(f_1, f_2)$, one can assume without loss of generality that f_1 is the uniform density on $(0, 1)$ (see Theorem 2). The next lemma shows that an analogous result for the case of $n = 3$.

Lemma 1. *Let f_1, f_2, f_3 be any Lebesgue p.d.f.'s on $(\mathbb{R}, \mathcal{B}_{\mathbb{R}})$ and let F_1 be the c.d.f. corresponding to f_1. If $\{x : f_k(x) > 0\} \subset \{x : f_1(x) > 0\}$ for $k = 2, 3$ then*

$$\mathscr{R}(f_1, f_2, f_3) = \mathscr{R}(\tilde{f}_1, \tilde{f}_2, \tilde{f}_3),$$

where $\tilde{f}_1, \tilde{f}_2, \tilde{f}_3$ are Lebesgue p.d.f.'s on $(0, 1)$ defined by

$$\tilde{f}_k(u) = \frac{f_k(F_1^{-1}(u))}{f_1(F_1^{-1}(u))}, \quad u \in (0, 1), \quad k = 1, 2, 3.$$

Proof. It follows from the properties of the quantile function that $F_1^{-1}(U) \overset{D}{=} X_1$, where X_1 is any random variable with c.d.f. F_1 (see Shorack [4], p. 111–117). This implies that

$$\begin{aligned}\int_0^1 \tilde{f}_k(u)\,du &= \mathbb{E}\left[\tilde{f}_k(U)\right] = \mathbb{E}\left[\frac{f_k(F_1^{-1}(U))}{f_1(F_1^{-1}(U))}\right] = \mathbb{E}\left[\frac{f_k(X_1)}{f_1(X_1)}\right] = \int_0^\infty \frac{f_k(x)}{f_1(x)} f_1(x)\,dx \\ &= \int_0^\infty f_k(x)\,dx = 1.\end{aligned}$$

Hence, $\tilde{f}_1, \tilde{f}_2, \tilde{f}_3$ are Lebesgue p.d.f.'s on $(0, 1)$ and $\tilde{f}_1$ is the uniform density on $(0, 1)$.

To prove that $\mathscr{R}(f_1, f_2, f_3) = \mathscr{R}(\tilde{f}_1, \tilde{f}_2, \tilde{f}_3)$ we fix two real numbers k_1, k_2 and define sets

$$A = \left\{x \in \mathbb{R} : \frac{f_3(x)}{f_1(x)} > k_2 \frac{f_2(x)}{f_1(x)} + k_1\right\}, \quad \widetilde{A} = \{u \in (0, 1) : \tilde{f}_3(u) > k_2 \tilde{f}_2(u) + k_1\}.$$

From Corrolary 2 we deduce that the points $\left(\int_A f_1(x)\,dx, \int_A f_2(x)\,dx, \int_A f_3(x)\,dx\right)$ and $\left(\int_{\widetilde{A}} \tilde{f}_1(u)\,du, \int_{\widetilde{A}} \tilde{f}_2(u)\,du, \int_{\widetilde{A}} \tilde{f}_3(u)\,du\right)$ lie on the upper boundaries of the sets $\mathscr{R}(f_1, f_2, f_3)$ and $\mathscr{R}(\tilde{f}_1, \tilde{f}_2, \tilde{f}_3)$, respectively. Moreover, for each $k = 1, 2, 3$, the following equality holds

$$\begin{aligned}\int_A f_k(x)\,dx &= \int_A \frac{f_k(x)}{f_1(x)} f_1(x)\,dx = \mathbb{E}\left[\frac{f_k(X_1)}{f_1(X_1)} I(X_1 \in A)\right] \\ &= \mathbb{E}\left[\frac{f_k(F_1^{-1}(U))}{f_1(F_1^{-1}(U))} I(F_1^{-1}(U) \in A)\right] = \int_{\widetilde{A}} \tilde{f}_k(u)\,du.\end{aligned}$$

The same reasoning applies when $>$ in A and $\widetilde{A}$ is replaced by $\geq$. This completes the proof.

□

3.2 Main result

The arguments given in Section 3.1 show that to find the shape of $\mathscr{R}(f_1, f_2, f_3)$ we can use throughout the following assumption:

Assumption A1. The functions f_1, f_2, f_3 are Lebesgue p.d.f.'s on $(0, 1)$ and f_1 is the uniform density on $(0, 1)$, i.e. $f_1(x) = I_{(0,1)}(x)$, $x \in \mathbb{R}$.

Unfortunately, in these general settings the problem of finding the shape of $\mathscr{R}(f_1, f_2, f_3)$ seems to be quite a tricky affair. Therefore, to simplify this problem we impose later two assumptions on the densities f_1, f_2, f_3.

Theorem 2 implies that to find the Lyapounov set in $\mathbb{R}^2$ we can assume that one of the densities corresponds to the uniform distribution on $(0, 1)$ while the second is nonincreasing function on $(0, 1)$. The following theorem shows that an analogous result holds for the Lyapounov set in $\mathbb{R}^3$. To state this result we adopt the notation used in Section 2. Let U be a random variable with the uniform distribution over the interval $(0, 1)$. We write $\overline{H}_2$ for the survival function of the random variable $f_2(U)$ and we denote by $\overline{H}_2^{-1}$ the quasi inverse of $\overline{H}_2$ (cf. (5) and (6)) . Moreover, we define the function $f_2^* : (0, 1) \to [0, \infty)$ by

$$f_2^*(u) = \overline{H}_2^{-1}(u), \quad \text{for all} \quad u \in (0, 1).$$

Using the same arguments as those used in inference concerning $\overline{H}^{-1}$ (see Section 2) we obtain that f_2^* is the Lebesgue p.d.f. on $(0, 1)$ and that $f_2^*(U) \stackrel{D}{=} f_2(U)$.

With this notation we have the following result.

Theorem 4. *Let f_1, f_2, f_3 be any Lebesgue p.d.f.'s on $(0, 1)$ such that f_1 is the uniform density on $(0, 1)$. If there exists a measurable function ϕ such that*

$$f_3(u) = \phi(f_2(u)) \quad \text{almost everywhere } \lambda, \tag{11}$$

then $\mathscr{R}(f_1, f_2, f_3) = \mathscr{R}(f_1^*, f_2^*, f_3^*)$, where $f_1^*(u) = 1$ and $f_3^*(u) = \phi(f_2^*(u))$, $u \in (0, 1)$.

Proof. The random vectors $(1, f_2(U), \phi(f_2(U)))$ and $(1, f_2^*(U), \phi(f_2^*(U)))$ have the same distribution, because $f_2^*(U) \stackrel{D}{=} f_2(U)$. Since $f_1 = f_1^* \equiv 1$, this implies that $\boldsymbol{Y} \stackrel{D}{=} \boldsymbol{Y}^*$, where

$$\boldsymbol{Y} = (f_1(U), f_2(U), f_3(U)), \quad \boldsymbol{Y}^* = (f_1^*(U), f_2^*(U), f_3^*(U)).$$

Let k_1, k_2 be any real numbers and let A and A^* be two subsets of $(0, 1)$, given by

$$A = \{u : f_3(u) > k_2 f_2(u) + k_1 f_1(u)\}, \quad A = \{u : f_3^*(u) > k_2 f_2^*(u) + k_1 f_1^*(u)\}.$$

Corollary 2 implies that to prove the theorem it suffices to show that for any $k_1, k_2 \in \mathbb{R}$

$$\left(\int_A f_1(u)\, du, \int_A f_2(u)\, dx, \int_A f_3(u)\, du \right) = \left(\int_{A^*} f_1^*(u)\, du, \int_{A^*} f_2^*(u)\, dx, \int_{A^*} f_3^*(u)\, du \right).$$

This equality holds, because $\boldsymbol{Y} \stackrel{D}{=} \boldsymbol{Y}^*$ and hence, for each $j = 1, 2, 3$,

$$\int_A f_j(u)\, du = \mathbb{E}\Big[Y_j\, I\{\boldsymbol{Y} \in B\} \Big] = \mathbb{E}\Big[Y_j^*\, I\{\boldsymbol{Y}^* \in B\} \Big] = \int_{A^*} f_j^*(u)\, du.$$

where $B = \{(y_1, y_2, y_3) \in \mathbb{R}^3 : y_3 > k_2 y_2 + k_1 y_1\}$. The same reasoning applies when $>$ in A, A^* and B is replaced by $\geq$. This completes the proof of the theorem. □

Remark 2. The assumption (11) is satisfied when the density f_2 is one-to-one on $(0, 1)$ or, more generally, when the random variable $f_3(U)$ is measurable with respect to the σ-field generated by the random variable $f_2(U)$.

We use the last theorem to prove the main result of the paper.

Theorem 5. *Under the assumptions of Theorem 4, let ϕ be a strictly convex function on $[0, \infty)$. Then the upper boundary of $\mathscr{R}(f_1, f_2, f_3)$ is the parametric surface of the form:*

$$\vec{r}(s,t) = \left(s + 1 - t, F_2^*(s) + 1 - F_2^*(t), F_3^*(s) + 1 - F_3^*(t)\right), \; 0 \le s \le t \le 1,$$

where F_2^* and F_3^* are the distribution functions corresponding to the densities f_2^* and f_3^*, respectively.

Proof. Since $\mathscr{R}(f_1, f_2, f_3) = \mathscr{R}(f_1^*, f_2^*, f_3^*)$, it suffices to show that the surface $\vec{r}$ is the upper boundary of $\mathscr{R}(f_1^*, f_2^*, f_3^*)$. For this purpose we first show that this boundary is generated by sets from the family $\mathcal{A}^* = \{(0, s) \cup (t, 1) : 0 \le s \le t \le 1\}$.

Let $\left(\int_{A^*} f_1^*(u)\,du, \int_{A^*} f_2^*(u)\,du, \int_{A^*} f_3^*(u)\,du\right)$ be any point which belongs to the upper boundary of the set $\mathscr{R}(f_1^*, f_2^*, f_3^*)$. Suppose first that $\left(\int_{A^*} f_1^*(u)\,du, \int_{A^*} f_2^*(u)\,du\right)$ lies on the upper boundary of $\mathscr{R}(f_1^*, f_2^*)$. Then, by Theorem 2, there exist a real number k such that

$$\{u \in (0, 1) : f_2^*(u) > k\} \subset A^* \subset \{u \in (0, 1) : f_2^*(u) \ge k\}.$$

Since f_2^* is nonincreasing, there exists $s \in (0, 1)$ such that $A^* = (0, s)$ almost everywhere λ. Using the same argument, one can show that if $\left(\int_{A^*} f_1^*(u)\,du, \int_{A^*} f_2^*(u)\,du\right)$ belongs to the lower boundary of the set $\mathscr{R}(f_1^*, f_2^*)$ then $A^* = (t, 1)$ for some $t \in (0, 1)$.

Suppose now, that $\left(\int_{A^*} f_1^*(u)\,du, \int_{A^*} f_2^*(u)\,du\right)$ is an inner point of $\mathscr{R}(f_1^*, f_2^*)$. Then, by Corollary 2, there exist real numbers k_1, k_2 such that

$$\{u : \phi(f_2^*(u)) > k_2 f_2^*(u) + k_1\} \subset A^* \subset \{u : \phi(f_2^*(u)) \ge k_2 f_2^*(u) + k_1\}. \tag{12}$$

Since the function $\psi(x) = \phi(x) - (k_1 + k_2 x)$ is strictly convex on $[0, \infty)$, the equation $\psi(x) = 0$ has at most two nonnegative roots. Let $B = \{x : \psi(x) > 0\}$. If $x_1 < x_2$ are two solutions of $\psi(x) = 0$, then from convexity of ψ it follows that $B = (-\infty, x_1) \cup (x_2, \infty)$. If x_1 is the unique solution of this equation, then $B = (-\infty, x_1)$ or $B = (x_1, \infty)$. Clearly, $B = (-\infty, \infty)$ or $B = \emptyset$, if $\psi(x) = 0$ has no solution. We consider only the first case, because the other two can be treated by analogy. So, let $B = (-\infty, x_1) \cup (x_2, \infty)$. Then, $A^* = \{u \in (0, 1) : f_2^*(u) < x_1 \vee f_2^*(u) > x_2\}$. Since $f_2^*(u)$ is nonincreasing in $u \in [0, 1]$, there exist $0 \le s < t \le 1$ such that

$$A^* = (0, s) \cup (t, 1) \quad \text{almost everywhere } \lambda.$$

To prove the converse, suppose that $A^* = (0, s) \cup (t, 1)$, where $0 < s < t < 1$ are any given points. Let $x_1 = f_2^*(s)$ and $x_2 = f_2^*(t)$. If $x_1 \ne x_2$ we define $y = k_1 + k_2 x$ to be the straight line going through the points $(x_1, \phi(x_1))$ and $(x_2, \phi(x_2))$. Otherwise, let $y = k_1 + k_2 x$ be the supporting line (hyperplane) to the boundary point $(x_1, \phi(x_1))$ of the convex epigraph of ϕ. In both cases, the set A^* satisfies (12). The same argument applies to the case, where $A^* = (0, s)$ or $A^* = (t, 1)$.

To complete the proof note that if $A^* = (0, s) \cup (t, 1)$, for some $0 < s < t < 1$, then

$$\int_{A^*} f_i^*(u)\,du = \int_0^s f_i^*(u)\,du + \int_t^1 f_i^*(u)\,du = F_i^*(s) + 1 - F_i^*(t).$$

□

Remark 3. The assumption that ϕ is stricly convex is quite restrictive. Clearly, it can be replaced by equally restrictive assumption that ϕ is strictly concave.

To illustrate more explicitly how Theorem 5 can be used in deriving the shape of $\mathscr{R}(f_1, f_2, f_3)$, we consider the following example.

Example 1. Let the Lebesgue p.d.f.'s on (0, 1) be given by

$$f_1(u) = 1,\ f_2(u) = 2u,\ f_3(u) = 3u^2,\ u \in (0, 1).$$

The conditions of Theorem 5 are clearly satisfied and $\phi(u) = 3u^2/4,\ u \in (0, 1)$. A straightforward calculation shows that the densities f_1^*, f_2^* and f_3^* have the form

$$f_1^*(u) = 1,\ f_2^*(u) = 2(1-u),\ f_3^*(u) = 3(1-u)^2,\ u \in (0, 1),$$

and hence, by Theorem 2,

$$\mathscr{R}(f_1, f_2) = \mathscr{R}(f_1^*, f_2^*) = \left\{(x, y) \in \mathbb{R}^2 \ :\ 0 \le x \le 1,\ x^2 \le y \le 1 - (1-x)^2\right\},$$

To find $\mathscr{R}(f_1, f_2, f_3)$ we note first that the c.d.f.'s, corresponding to f_1^*, f_2^*, f_3^*, have the form

$$F_1^*(u) = u, \quad F_2^*(u) = 1 - (1-u)^2, \quad F_3^*(u) = 1 - (1-u)^3,\ u \in (0, 1).$$

This implies that the upper boundary of $\mathscr{R}(f_1, f_2, f_3)$ is the parametric surface, given by:

$$\vec{r}(s, t) = \left(s + 1 - t, 1 - (1-s)^2 + (1-t)^2, 1 - (1-s)^3 + (1-t)^3\right),\ 0 \le s \le t \le 1,$$

To determine this boundary explicitly, we need to find the function G_3, defined by (9). Let (x, y) be any point from $\mathscr{R}(f_1, f_2)$. Since $(x, y, G_3(x, y))$ lies on this boundary, we must have

$$\begin{cases} x = s + 1 - t \\ y = 1 - (1-s)^2 + (1-t)^2 \\ G_3(x, y) = 1 - (1-s)^3 + (1-t)^3. \end{cases}$$

Solving this system of equations we get

$$G_3(x, y) = 1 - \left[\frac{(1-y) + (1-x)^2}{2(1-x)}\right]^3 + \left[\frac{(1-y) - (1-x)^2}{2(1-x)}\right]^3.$$

Using this solution in (10) we obtain the explicit form of $\mathscr{R}(f_1, f_2, f_3)$.

References

[1] A. Lyapounov, Bulletin of the Academy of Sciences of the USSR **4**, 465 (1940)

[2] J. Legut, M. Wilczyński, J. Math. Anal. Appl. **394**, 102 (2012)

[3] E. Lehmann, J. Romano, *Testing Statistical Hypotheses* (Springer Science+Business Media, Inc., New York, 2005)

[4] G. Shorack, *Probability for Statisticians*, Springer-Verlag (Springer, New York, Inc., 2000)

AUTHOR INDEX